Wearable AI in Psychotherapy

Asegul Hulus
Association for Computing Machinery, Cyprus

Hemachandran Kannan
Woxsen University, India

Raul Villamarin Rodriguez
Woxsen University, India

Manjeet Rege
University of St. Thomas, USA

Mir Aadil
Woxsen University, India

IGI Global
Scientific Publishing
Publishing Tomorrow's Research Today

Vice President of Editorial	Melissa Wagner
Director of Acquisitions	Mikaela Felty
Director of Book Development	Jocelynn Hessler
Production Manager	Mike Brehm
Cover Design	Jose Rosado

Published in the United States of America by
IGI Global Scientific Publishing
701 East Chocolate Avenue
Hershey, PA, 17033, USA
Tel: 717-533-8845
Fax: 717-533-7115
Website: https://www.igi-global.com E-mail: cust@igi-global.com

Library of Congress Cataloging-in-Publication Data

Names: Hulus, Asegul, 1995- editor | Kannan, Hemachandran, 1985- editor | Rodriguez, Raul Villamarin editor | Rege, Manjeet editor | Aadil, Mir, 1988- editor
Title: Wearable AI in psychotherapy / edited by Asegul Hulus, Hemachandran Kannan, Raul Villamarin Rodriguez, Manjeet Rege, Mir Aadil.
Other titles: Wearable artificial intelligence in psychotherapy
Description: Hershey, PA : Medical Information Science Reference, [2026] | Includes bibliographical references and index.
Identifiers: LCCN 2025005831 (print) | LCCN 2025005832 (ebook) | ISBN 9798337304670 hardcover | ISBN 9798337304687 paperback | ISBN 9798337304694 ebook
Subjects: LCSH: Psychotherapy--Data processing | Artificial intelligence--Medical applications | Biosensors | Psychotherapy--Moral and ethical aspects
Classification: LCC RC489.D35 W43 2025 (print) | LCC RC489.D35 (ebook)
LC record available at https://lccn.loc.gov/2025005831
LC ebook record available at https://lccn.loc.gov/2025005832

British Cataloguing in Publication Data
A Cataloguing in Publication record for this book is available from the British Library.

Table of Contents

Detailed Table of Contents

Virtual Reality Exposure Therapy (VRET) has revolutionized phobia treatment by combining traditional therapy with advanced technology. It immerses patients in virtual environments tailored to simulate specific fears, offering a safe, customizable, and accessible alternative to conventional exposure therapy. VRET is particularly beneficial for those with severe phobias or physical limitations. Its flexibility allows therapists to adjust exposure intensity, enhancing engagement and accelerating desensitization. Challenges include high equipment costs, cybersickness, and specialized training needs. As VR technology becomes more affordable, VRET has the potential to transform the treatment of anxiety disorders, expanding the accessibility and effectiveness of psychological care. This chapter explores VRET's principles, applications, challenges, and future directions.

Feature engineering plays a pivotal role in developing accurate and effective mental health applications, as it enables the extraction of meaningful insights from raw data for precise mental health monitoring. This study explores the processes of data collection and preprocessing, focusing on transforming diverse data sources—including behavioral, physiological, and contextual data—into usable formats. Key strategies for identifying relevant features for mental health assessment are discussed, including the use of machine learning techniques to pinpoint indicators strongly correlated with specific mental health conditions. The primary research question addressed is: What strategies can be employed to enhance feature selection for precise mental health monitoring? By examining advanced methods such as automated feature selection, dimensionality reduction, and domain-specific feature engineering, this study aims to optimize the accuracy and reliability of mental health assessments.

Preface

In recent years, the intersection of artificial intelligence (AI), wearable technology, and mental health care has opened new frontiers for innovation in psychotherapy. As the global burden of mental illness continues to rise, so too does the urgency to explore scalable, data-driven, and person-centered solutions. This edited volume, *Wearable AI in Psychotherapy*, brings together pioneering research and clinical perspectives to examine how AI-enhanced wearable technologies are reshaping therapeutic practices, from diagnosis and monitoring to treatment and beyond.

Our aim in curating this collection was to present a comprehensive yet nuanced view of the capabilities and challenges of wearable AI in mental health care. The chapters included range from foundational explorations of AI's role in psychotherapy to specialized applications like Cognitive Behavioral Therapy (CBT), mindfulness interventions, Virtual Reality Exposure Therapy (VRET), and trauma-informed approaches. The authors contributing to this volume represent a multidisciplinary and international community of scholars, clinicians, and technologists, reflecting the collaborative spirit necessary to address the complex dimensions of mental health innovation.

Importantly, this book does not shy away from the ethical, cultural, and technical challenges that accompany these advances. Whether it is algorithmic bias, patient privacy, or cultural compatibility, the contributors critically engage with the issues that must be addressed to ensure equitable and responsible implementation. The ultimate goal of this volume is to inform, provoke thoughtful dialogue, and guide future work in this rapidly evolving space.

Chapter 1: Introduction to AI in Psychotherapy

This chapter explores the evolution and growing impact of AI in psychotherapy, from early tech interventions to modern wearable tools. It highlights how machine learning and real-time biometric tracking are enhancing individualized care while raising important questions around privacy, engagement, and ethics.

Chapter 2: AI-Driven Cognitive Behavioural Therapy (CBT) using Wearables

Focusing on Cognitive Behavioural Therapy, this chapter examines how AI and wearables improve accessibility and adherence by enabling real-time feedback, symptom tracking, and digital cognitive restructuring, while also addressing challenges like data bias and the lack of human empathy.

Chapter 3: AI-Powered Wearables for Mental Health Monitoring

This chapter investigates how AI and IoT-based wearables support mental health by tracking conditions like depression, anxiety, and Parkinson's. It considers practical concerns of accessibility, diagnostic accuracy, and ethical implications in therapeutic and rehabilitation contexts.

Chapter 4: Efficacy of Wearable AI-Driven Mindfulness Practices in the Management of Stress

Through a systematic review, this chapter evaluates the effectiveness of AI-based wearables in mindfulness and stress reduction. Key themes include personalization, biofeedback, usability, and the need for further research to confirm long-term reliability.

Chapter 5: Virtual Reality Exposure Therapy (VRET) for Phobia Treatment

This chapter discusses how Virtual Reality Exposure Therapy (VRET) modernizes phobia treatment by immersing patients in customized virtual environments. It assesses the benefits, such as enhanced engagement, and challenges, including cost and training demands.

Chapter 6: Feature Engineering for Mental Health Applications

This chapter explains how feature engineering refines raw behavioral and physiological data for mental health monitoring. It outlines advanced strategies like dimensionality reduction and automated selection to improve the accuracy of AI-based assessments.

Chapter 7: The Moral Algorithm – Ethics in AI-Supported Mental Health

Exploring the ethical landscape, this chapter examines issues such as privacy, algorithmic fairness, and informed consent in AI-driven mental health services, while reviewing regulatory efforts aimed at safeguarding responsible care.

Chapter 8: Ethics in AI-Based Mental Health Care

This chapter highlights ethical risks in tech-based mental health tools, including algorithmic bias and discriminatory outcomes. It stresses the need for equity, transparency, and the protection of vulnerable populations in clinical applications.

Chapter 9: Responsible AI in Mental Health-Ethics, Advocacy, and Innovation

Echoing the previous chapter, this contribution expands on ethical imperatives, emphasizing the role of data justice and inclusive design in preventing harm and ensuring fair treatment across diverse populations using AI technologies.

Chapter 10: Adapting AI-Driven Psychotherapy – Cultural Challenges and Opportunities

Centering on the Indian context, this chapter examines cultural considerations in integrating AI into therapy. It discusses how language, social norms, and communication patterns may affect AI compliance and therapeutic success.

Chapter 11: Future Trends in AI and Wearable Technology for Mental Health

Looking ahead, this chapter explores emerging innovations like smart fabrics, AR/VR, and predictive algorithms. It outlines technical, ethical, and regulatory challenges to scaling AI-powered wearable mental health solutions worldwide.

As editors, we are grateful to the many scholars and practitioners who have contributed to this volume with such rigor and passion. Their work reflects not only the technical sophistication of this field but also a deep commitment to human well-being. It is our hope that this book will serve as a valuable resource for academics, clinicians, technologists, and policymakers who are invested in the future of mental health care.

We also recognize that the landscape of wearable AI is constantly evolving. New tools, techniques, and regulatory frameworks will continue to emerge, necessitating ongoing dialogue and adaptation. This book represents a snapshot—a point of convergence where current knowledge meets the momentum of ongoing inquiry. In offering this contribution to the field, we hope to foster more integrative, ethically sound, and culturally aware advancements.

Ultimately, *Wearable AI in Psychotherapy* is about more than machines or algorithms; it is about the people they aim to help. By blending technological innovation with therapeutic insight, we move closer to a vision of mental health care that is not only smarter but also more humane. We invite readers to engage with this volume critically and creatively, and to join us in shaping the future of psychotherapy in the age of intelligent technology.

Chapter 1
Introduction to AI in Psychotherapy

Basudha Bandyopadhyay
https://orcid.org/0000-0003-0330-6868
Ignitio Child Development Center, Hyderabad, India

Nistha Jain
https://orcid.org/0009-0009-7360-1097
Woxsen University, Hyderabad, India

ABSTRACT

AI is transforming mental health, particularly in psychotherapy, by enabling individualized, accessible, and scalable treatment solutions. This chapter explores AI's role in psychotherapy, alongside advancements in wearable technology, while tracing the history of technology in mental health care. Early innovations, such as telephone counseling and mobile mental health support, paved the way for modern tools that utilize machine learning, natural language processing, and real-time biometric data for therapeutic interventions. A key focus is the rise of wearable technology, which continuously monitors psychological and physical markers, providing objective measurements beyond traditional self-reported questionnaires. The chapter critically examines AI-driven therapy's effectiveness, patient engagement, privacy concerns, and ethical implications. By analyzing past and present applications and hinting at future directions, this chapter highlights AI's revolutionary potential in psychotherapy and mental health treatment.

DOI: 10.4018/979-8-3373-0467-0.ch001

1. INTRODUCTION: THE INTERSECTION OF AI AND PSYCHOTHERAPY

AI is no longer just a concept from any science friction; it is developing several fields, including mental health services. So, integrating artificial intelligence (AI) into psychotherapy is a revolution in mental health services, offering innovative approaches to diagnosis and intervention plans. Historically, Psychotherapy- the art and science of treating mental health conditions- has been a human endeavour. however, the way that the treatment is provided, accessible and viewed is changing as a result of the combination of AI and psychotherapy. A systematic review and meta-analysis published in NPJ Digital Medicines showed the effectiveness of AI-based technology in improving mental health services. The study explained that these AI-generated tools could be beneficial, especially for people with anxiety and depression by providing immediate support (Li et al., 2023).

Since the demand for mental health services increases, AI offers cost-effective support to practitioners and clients and even in certain cases as a substitute for traditional human-led interventions (D'Alfonso, 2020). Further research highlights the outcomes of AI to enhance psychotherapy interventions. To support this a review available on Research Gate shows how AI can enhance psychotherapy. Further research highlights the potential of AI to enhance therapeutic interventions through chatbots and online Cognitive Behavioural Therapy (CBT) offering and economic solutions for mental health care services (Beg et al., 2024).

Similar research shows how large language models (LLMs) such as Open AI, chat GPT, Google's Gemini and built-on artificial intelligence (AI) could change the future of behavioural healthcare. It's able to address the difficult symptoms of a few disorders which is undoubtedly helping the practitioner (Stade et al., 2024).

The proliferation of social media platforms has transformed communication, with over 4.9 billion users globally as of 2023 (Statista, 2023). While these platforms offer opportunities for connection, concerns have emerged about their impact on mental health. Similarly, integrating artificial intelligence (AI) in psychotherapy has introduced novel approaches to mental health care, raising questions about efficacy, accessibility, and ethics.

1.1 Historical Context

The integration of technology into psychotherapy has significantly transformed mental health care, enriching the accessibility, and effectiveness of diagnosis and treatment plans. This development can be traced through several key developments

1.1.1. Early Technological Interventions

The first integration of technology and psychotherapy appeared with the development of computer-assisted therapies. In 1966, Joseph Weizenbaum introduced ELIZA. ELIZA is a program designed to give prompts like a Rogerian Therapist or Person-Centered therapist by engaging users in text-based conversations. It was initially written in Michigan Algorithm Decoder (MAD-Slip) and many variations of the original scripts were made too. The most well-known variation was called DOCTOR. This one made to respond like a Rogerian therapist. Eliza was one of the 1st chatbots. It was a test to check the machine's ability to see the intelligence and behaviour equivalent to human behaviour. However, by today's standards, ELIZA fails to answer if one asks it a few complex questions (Integrating Technology Into Mental Healthcare, n.d.).

1.1.2. Emergence of Teletherapy

The 1990s played a very crucial role in the emergence of Tele therapy. It played a very major role in the field of psychotherapy using videoconferencing technology This innovative idea helped therapists to conduct sessions remotely, thereby increasing the accessibility of therapy, even in geographically isolated areas. By 2000 it was recognized as a successful and effective alternative method of in-person therapy. Which was proved cost-effective too.

But Tele therapy became most successful during the COVID-19 pandemic across the globe as 80% of therapists were offering teletherapy due to the unavailability of offline therapy and social isolation.

Though several studies have shown the effectiveness of teletherapy (Hamilton, 2023), there are certain limitations too, which cannot be overlooked:

1. Technical Difficulties: Unstable internet connections and technical glitches can interrupt the flow of the session and disturb sensitive conversation which can lead to extra frustration in both therapist and the client.
2. Limited physical Appearance: Building rapport is a bit difficult in the case of teletherapy as there is limited physical appearance.
3. Lack of Nonverbal Cues: In psychotherapy, nonverbal cues such as eye contact, body language, and tone of voice play a very major role. Virtual therapy may fail to interpret those cues sometimes.
4. Confidentiality Concerns: information transmission increases the potential risk in confidentiality and privacy.

1.1.3. Digital and Online Therapies

The rise of the Internet during the early 2000s led to the emergence of online therapy platforms which began to deliver multiple digital services including email counselling and chat-based interventions with video calling. Therapy platforms now provide flexible scheduling and easy access methods that cater to clients who find face-to-face treatment uncomfortable. The development of digital therapy keeps growing as different supportive tools become part of the treatment process (Pickens, 2021).

1.1.4. Mobile Applications and Virtual Reality

As smartphones became more popular people created mental health applications that included exercises for therapy alongside mood-tracking features and coping mechanisms. Virtual reality technology has become a tool for exposure therapy to address phobias and anxiety disorders. Clients face their fears within controlled settings through this method which boosts traditional therapy methods' effectiveness (Williams, 2020). Several mobile applications powered by artificial intelligence deliver automatic psychometric evaluations alongside continuous mental health service availability (The Use of Technology-Assisted Methods in Therapy and Psychological Treatments, 2023).

1.1.5. Artificial Intelligence in Psychotherapy

John McCarthy coined the term artificial intelligence in 1956. At the time, he held the first economic conference on the subject. He tried to understand if machines could do or think anything bigger than a human mind, which is how artificial intelligence came into the picture. It can be estimated that presently a huge population suffers from mental health disorders. According to WHO, one out of four families is likely to have at least one family member having some kind of behaviour or mental disorder during and after the COVID-19 pandemic, the whole world has experienced a huge surge in mental health issues. Unfortunately, not everyone was able to access the available mental health services due to lack of financial assistance, living in remote areas, fear of contamination, fear of getting stigmatized or lack of awareness. That is how the emergency of online mental health services could solve some of these problems as these are easily accessible to people from anywhere and it is also very much cost-effective. This also reduces the fear of being judged or levelled in front of society. That is how lots of efforts have been made today to integrate artificial intelligence with the traditional form of psychotherapy. AI-driven chatbots and virtual therapists offer instant support and intervention following cognitive behavioural principles.

AI-driven rules have made significant rights in analyzing Speech text facial expressions to identify early science of mental health disorders, such as depression, anxiety, et cetera. National language processing (NLP) techniques enable extracting valuable inside from written or spoken words. For example, sentiment analysis can identify new changes in individuals, and emotional conditions by examining social media, posts, chats, logs, written diaries, etc (Zhang et al., 2022).

This analysis can detect alterations in speech, compassion, and variation in tone rhythm, which might potentially serve as an identity of anxiety, depression, and other mental health conditions. Facial expression analysis often done with computer vision can provide insight into one individual's emotional state we can say that AI systems can detect expressions and subtle changes in facial features that may the indicators of underlying psychological conditions, this technology benefits, mental health techniques or services through video consultations or mobile app. The current trend in this domain involves advanced AI algorithms capable of detecting and comprehending these very minute indicators. As AI systems are improving, they hold the potential to provide early warning signs of mental health disorders and enable timely intervention and support (Jiang et al., 2024).

There is one AI-driven mental health app “Woebot” which helps to analyse user text input if a user constantly expresses sadness, hopelessness, and despair in their chat interaction. This app recognizes these patterns and offers guidance or recommends professional help (Woebot Health, 2025). There is another mental health app called “COGITO” which is an AI platform used in telehealth services that monitors patient speech changes during therapy sessions if there are shifts in peach tone or rhythm that may indicate depression or anxiety. This app alerts the therapist to address these emotional cues during the session (Cogito, 2025).

Facial expression analysis is also another variable tool in early detection. As we always say early detection is always beneficial. Another app called “Effectiva” which is a pioneer company in emotion. AI has developed failure, expression, and analysis tools. This tool has been employed in various mental health. Studios. For example, researchers use effective technology to examine the facial expressions of individuals with depression, shedding light on potential early diagnosis (Kuziemsky et al., 2019).

Artificial intelligence improves psychotherapy by enabling better data analysis efficiency and more effective visual representation of client progress while also helping to tailor individual intervention plans. Therapeutic practices have experienced recent advancements through the incorporation of artificial intelligence. AI-driven chatbots and virtual therapists offer instant support and intervention following cognitive behavioural principles. Therapists benefit from these tools which provide easy-to-access resources and enable client progress tracking via data analysis.

Research has investigated how AI integration into Cognitive Behavioural Therapy (CBT) can improve both accessibility and personalization for patients. A general review examines how artificial intelligence can improve CBT delivery through support and optimization while demonstrating its transformative ability to enhance personalized mental health interventions (Imel et al., 2017; Kearns, 2024).

Moreover, the integration of AI into Cognitive Behavioural Therapy (CBT) has been explored to enhance accessibility and personalization. A generic review discusses how AI can support, augment, and optimize CBT delivery, highlighting the transformative potential of AI in making mental health interventions more efficient and personalized (Jiang et al., 2024).

2. CURRENT LANDSCAPE OF AI AND PSYCHOTHERAPY

The roots of Artificial Intelligence (AI) have been expanding and engulfing almost every sector it touches, enhancing, automating, and providing intelligent decision-making simultaneously (Semmler & Rose, 2017). The fruits borne are relinquished by all sectors including healthcare where it has not stopped with prediction of the diseases, and analysis of symptoms but also ushered in groundbreaking tailor-made treatments and personalized medications. Amidst a worldwide problem marked by rising demand and limited resources, the mental health industry is seeing a dramatic paradigm shift made possible by AI, offering innovative strategies that pledge to transform conventional models of mental health care (Lee et al., 2021).

2.1. Artificial Intelligence (AI) and Machine Learning (ML)

The fascination with making computers intelligent that surpasses human beings has been prevalent for more than a decade now. Artificial Intelligence (AI) refers to the simulated human-like intelligence by machines using computers developing its capacity to engage in thought processes such as learning, reasoning, self-correction, etc (Kok et al., 2009; Samoili et al., 2021). Highly advanced yet entirely conceptualized machines are often envisioned to exhibit human-like cognitive abilities, encapsulating what is referred to as general AI (Sun et al., 2024). These systems are restricted to executing specific tasks, such as identifying patterns in speech to detect signs of depression or anxiety, without any capacity to replicate the depth or breadth of human reasoning (Müller, 2016). Both approaches, however, are rooted in the principles of machine learning.

According to Collins et al. (2021), machine learning (ML) is a subset of artificial intelligence (AI) that, like humans, processes large amounts of data using user-provided data and produces appropriate outputs with little to no guidance. To improve their performance over time, these systems use iteration, which is the act of running data through an algorithm several times. These algorithms progressively gain the capacity to make judgments on their own and improve their results by analyzing historical data from tasks that are similar but not identical (Janiesch et al., 2021).

AI and ML have rejigged the world and have seeped their way deep into almost every stratum letting the previously human-only chores being performed with unprecedented inventiveness and efficiency. For instance, the algorithms powered by MLs are used by AI-integrated chatbots on websites to ameliorate their participation and interactions over time. These algorithms familiarize themselves with past data and/or communication of the user and thus, offer more precise and tailor-made customer support (Howard, 2019). The interplay of AI and ML to analyze intricate medical imaging data in the healthcare sector enables earlier and more precise detection of illnesses like cancer by detecting subtle patterns that are unseen to the human eye (Rampton et al., 2020). In the ever-growing field of mental health, a convergence with technology, especially AI is pertinent to aid in its development. Online counselling services analyze user sentiment, behaviour, and even speech patterns in real-time using machine learning (ML) algorithms (Doorn et al., 2020). This makes it possible for AI systems to offer proactive and tailored answers.

2.2. AI Integrated Psychotherapy

The predominant goal of short-term psychotherapy is to facilitate individuals in understanding, identifying as well as reforming their behaviours while managing immediate emotional challenges. The approach stirs an impactful process via self-awareness, reflective listening, observation, and targeted interventions. Therapeutic approaches use various strategies and methods to fortify changes, with an emphasis on understanding both cognitive and emotional components (Lowenthal, 2024). Computational developments can help psychology as a field by supporting particular tasks. By providing insights into psychological concerns without departing from the fundamental ideas of psychological procedures, these tools function as extensions of traditional methods (Grodniewicz & Hohol, 2023). Instead of serving as treatments, these technologies help people become more self-aware, improve their judgment, and gather useful information to aid-in the therapists as well.

The use of cutting-edge technology in psychotherapy has revolutionized mental health treatment by increasing the effectiveness and accessibility of therapeutic interventions. Programs like ELIZA (Sharma et al., 2017), which imitated human-like speech and set the stage for later advancements, were among the first experiments in this area in the 1960s. Programs for computer-based cognitive behavioural therapy (CBT), which provide systematic, evidence-based therapies for common mental health conditions, were established during the late 20th century and are used widely in the present scenario (Jang et al., 2021). These early models laid the groundwork for the digital transformation of mental health services, despite the fact that they were primitive in comparison to modern technologies. These developments gradually grew to fill gaps in traditional therapy by including virtual therapists, individualized treatment programs, and instruments for early mental health problem detection.

2.2.1 AI-Powered Chatbots and Virtual Therapists

AI-powered chatbots and virtual therapists have become an essential kit in modern psychotherapy, that not only extends scalable, innovative support but also makes it accessible to a large number of populaces. Chatbots like Wysa, Woebot, Tess, and Rumi are widely used to provide real-time emotional validation and personalized coping mechanisms, making mental health care within the reach of individuals who may not have the convenience to get their hands on traditional therapy (Yeh et al., 2025). These tools leverage Natural Language Processing (NLP) and ML to engross users into sincere conversations that are worthwhile and, at the same time assist them in attending to their own cognitive biases and distortions, for example, overgeneralization, just world hypothesis, black-and-white thinking, etc. (Chen et al., 2023).

Inadequacy in true emotional perception is something that these AI-enhanced chatbots and therapists find consistently, even though they excel at pattern recognition and real-time quick response. Their predominant strength can be cited as their ability to provide adjunctive treatment, and immediate support and identify warning signs, especially at times when access to human and/or professional help is limited (Zipfel et al., 2024). Despite their effectiveness, ethical concerns such as data privacy, algorithmic bias, and the risk of over-reliance on technology remain significant challenges that must be addressed to ensure their responsible use in mental health care.

2.2.2 Socially Assistive Robots in Therapy

Amplification of communication and engagement is very important to establish and maintain a therapeutic alliance. For the same, socially assistive humanoid robots such as Pepper (Pandey et al., 2018) and Bandit (Khamassi et al., 2018) have emerged as beneficial tools. A study demonstrated in a higher education classroom setting that Pepper created a positive perception of learning and surged the learning gains by imparting the knowledge in a unique and interactive surrounding (Donnermann et al., 2020). Significant improvement in communication and increased participation in activities has been depicted in a research study with elderly schizophrenic patients and at the same time, the humanoid robot Pepper was reported to make the therapeutic process more enjoyable (Ujike et al., 2019; Team, 2023a).

Bandit is another humanoid robot that has been developed by the University of Southern California Interaction Lab and BlueSky Robotics (Team, 2023b). It is a multi-skilled, interactive, and assistive technology, designed to perform and excel at various social situations. Its effectiveness has been delivered in various social situations such a sat schools, offices, homes, and hospitals making it an effective tool for both adults and children (Schneider & Kummert, 2017). These robots not only enhance patient engagement but also provide a novel approach to therapy, particularly for individuals who may struggle with traditional methods.

2.2.3. Virtual Humans and Clinical Decision-Support Tools

A virtual human dialogue system, created in a virtual environment is another cutting-edge application of AI in psychotherapy. Ellie, a virtual human created with the goal of presenting a space and an interviewer who is a good listener, responds empathetically, and elicits tailor-made replies based on the specific questions asked (DeVault et al., 2014). Concomitantly it conducts semi-structured interviews to create interactive situations that facilitate the automatic assessment of distress indicators, such as verbal and non-verbal behaviours correlated with depression, anxiety, or post-traumatic stress disorder. By objectively measuring these behaviours, the virtual human interviewer provides valuable data for clinical decision-support tools, assisting specialists in diagnosing and tracking changes in symptoms over time.

Initiation of these virtual humans like Ellie has enhanced the clinical diagnostic process by offering objective measurements of psychological problems. Additionally, by storing past data, these tools enable practitioners to track the development of symptoms and more precisely customize interventions (Powell et al., 2019). Although AI-powered technologies and virtual persons have enormous potential to enhance mental health therapies, their application needs to be carefully weighed against ethical issues including data security, privacy, and maintaining the human aspect in therapy.

2.2.4. Convolution in Integration of AI

While AI is unlocking new potentials and reaching new heights every day making the lives of the users amicable it is pertinent to understand that it operates upon algorithms rather than emotional perception like humans. Their efficacy resides in their flexibility and ability to respond in real-time, even though they are excellent at assessing communication patterns and offering early therapeutic support. Because of these characteristics, AI tools are extremely useful in the treatment of mental health issues, especially when it comes to seeing early warning indicators and providing prompt assistance in circumstances when access to qualified assistance may be restricted (Zipfel et al., 2024).

Enhancing the patient's comprehension of their ideas and behaviours is the main objective of these tools, whether they are technological or traditional. They increase the precision and effectiveness of therapeutic activity in this way. But it's important to see these technologies as supplemental aids that work well with current therapy approaches rather than as stand-alone remedies. This method guarantees that AI improves therapy as a whole without taking precedence over the human component. AI's benefits in psychotherapy must be balanced with strong protections if its full potential is to be realized. To ensure that technology is used ethically in mental health care, issues including data security, privacy, and the danger of becoming overly dependent on it must be addressed. By doing so, AI can serve as a powerful tool to augment, rather than replace, the human connection that lies at the heart of effective psychotherapy.

3. CURRENT LANDSCAPE OF AI IN PSYCHOTHERAPY USING WEARABLES

The amalgamation of AI and wearable technology in healthcare more specifically psychotherapy can be regarded as an innovative and transformative shift in the sector of mental healthcare. The primary focus of the chapter is to explore the present landscape of AI-driven wearable systems in psychotherapy, focusing on their potential to enhance diagnosis, treatment, and monitoring of mental health conditions. These technologies provide creative answers to persistent problems in mental health treatment, like accessibility, cost, and subjective diagnostic procedures, by utilizing real-time data collecting, AI algorithms, and patient-centered design.

3.1 Personal Monitoring System

A wearable monitoring system, namely MONARCA, has been developed under the European project of the same name. This system has been constructed in such a way that succours the user to recognize early warning signs and make predictions of manic or depressive episodes in individuals with bipolar disorders while utilizing a smartphone-centered network of minimally invasive wearable sensors (Puiatti, 2011). This system showcases how wearable devices can offer continuous, longitudinal data without the requirement for more invasive or costly options. By tracking physiological and behavioral metrics like physical activity, sleep patterns, and social interactions, MONARCA empowers clinicians to identify subtle changes that might signal the start of an episode, making timely intervention possible.

In a research study, Piwek et al. (2016) highlight the potential of wearable devices to provide cost-effective simultaneously delivering a detailed report of mental health monitoring. It underscores the growing evidence that the data, particularly derived from wearable devices, including physical activity levels, sleep cycles (duration), and self-reported social interactions, can effectively gauge the gravity of the symptoms. This ability not only improves diagnostic precision but also aids in creating personalized treatment plans designed for specific needs.

3.2 Measurement-based care & wearable technology

The concept of measurement-based care has garnered attention as a framework for improving the quality and efficiency of mental health treatments. Sadeh-Sharvit & Hollon (2020) define this objective, measurement-based treatment as the practise of stabilizing clinical care in patient data collected through treatment. Wearable devices, by furnishing real-time data, sets in an advancement in the investigative process, in turn, pushing forward informed clinical decisions and treatments. For instance, raw data, minus the subjective influence of the clinician often associated with the traditional methods, leads to an impartial assessment and hence enables more accurate and personalized ones.

Moreover, the systematic barriers in mental health care such as sparse access to treatment, elevated costs, and the lack of unbiased methods for evaluating treatment progress can be easily addressed using wearable technology. By equipping both experts and patients with actionable insights, wearables facilitate more effective triage, decrease the burden on mental health services, and improve overall treatment outcomes.

3.3 Personalised Self-Tracking and Patient Empowerment

The One Button Tracker is a novel approach suited especially for psychotherapeutic treatment and research (Arendt et al., 2021). This wearable self-tracking instrument, unlike traditional mobile-based tools that are dependent upon preset rubrics, allows the patient to gain user-friendly experiences and focus upon their requisites, such as flashbacks, grounding techniques, and/or anger episodes (Riisager et al., 2024). Results from this patient-centric approach delineate the eminence of the upcoming rise in technology. Furthermore, it enhances engagement, and adherence and enables the therapists to customize the interventions that cater to the specific needs of each individual. The aforementioned study calls attention to OBT's success in supporting complex PTSD symptoms of refugees along with the potential of wearable devices to bridge gaps in healthcare, particularly for underserved populace.

3.4 Wearables in Anxiety and Panic Disorder Management

Management of anxiety and panic disorders has shown promising results using wearable devices (Chung et al., 2021). The research evaluated the feasibility of a biofeedback wearable device, namely the Lief Smart Patch, in conjunction with remote stress management coaching and demonstrated that the device effectively could reduce the symptoms of anxiety an depression. It underscores the potential of wearables to deliver scalable, non-invasive interventions.

Likewise, in another study, the use of a mobile health biofeedback (mHealth) system was used to explore its usability in the treatment of panic attacks. Although the sample size of the study was slightly small for the purpose of study, the preliminary results suggested the proficiency in using the system as it could assist in mitigating the severity during the act by recording the physiological data and thus can send alert to the user. It points to the possibility of wearable devices to serve as both diagnostic and supervision tool.

3.5 Challenges and Future Directions

Despite their deliverables, wearable devices are in their very preliminary stage, and it would not be very safe to uphold the thought of using it as a promising technology in furnishing psychotherapy independently. For instance, certain studies assessing the employment of electrocardiogram (EEG) data for panic disorder detection have reported conflicting results, and these variations are partly due to device specifications (Tsai et al., 2022). Additionally, the amalgamation of AI and ML models such as Natural Language Processing (NLPs) and the multi-layered ensemble deep learning approach used by Jacobson & Feng (2022), requires further validation to ensure their reliability and accuracy in a setting that involves various human emotions and not a mathematical equation which is consistent over time.

3.5.1 User Experience (UX) Design in AI-Based Psychotherapy Tools

UX design greatly impacts the acceptance and usage of AI-based psychotherapy tools. Good UX design makes sure that the interfaces are engaging, nuanced, and emotionally congruent with users' feelings. For example, Inkster et al. (2018) noted that users interacted more with AI chatbots when the conversational style used was empathetic and human-like. Interfaces that are impersonal or overly complicated, for instance featuring intricate systems of navigation, strip away trust and hurt the therapeutic alliance (Schueller et al., 2018). Some studies have shown that user experience is optimized through co-design with patients and professionals in mental health, which improves adherence and outcomes.

As it currently stands, more work needs to be done regarding accessibility aimed at a wider demographic, which includes age, ethnicity, and even technological proficiency. Older adults, for instance, may relate better to culturally sensitive language, while those of younger age may prefer large text and straightforward navigation (Czaja et al. 2019). Furthermore, the use of alternative text or voice commands, as well as haptic feedback, enables users with disabilities to partake in the interaction design. (Brewer et al., 2020).

Another critical aspect is the integration of personalization in UX design. AI tools that adapt to individual user preferences, such as adjusting the frequency of prompts or tailoring content based on user feedback, have been shown to foster greater engagement and therapeutic rapport (Cheung et al., 2019). For instance, machine learning algorithms can analyze user interactions to refine conversational flows, ensuring that responses feel relevant and emotionally attuned. However, designers must balance personalization with privacy concerns, as excessive data collection can erode trust (Lutz & Tamò-Larrieux, 2021).

Finally, the role of feedback loops in UX design cannot be overstated. Continuous user feedback, gathered through surveys, usage analytics, or direct input, allows developers to identify pain points and refine interfaces iteratively. Co-design approaches, where users and clinicians collaborate in the design process, have been particularly effective in aligning AI tools with real-world therapeutic needs (Thieme et al., 2020). By prioritizing empathy, accessibility, and iterative refinement, UX design can significantly enhance the adoption and effectiveness of AI-based psychotherapy tools

Future studies should focus on making a note of these limitations, addressing them by standardizing data collection methods, improving the explicability of AI algorithms, and conducting large-scale clinical trials. Ethical considerations, data privacy, and the prospect of over-reliance on technology must be carefully navigated to ensure that wearable devices complement, rather than replace the human elements in therapy.

Table 1. Technological Progression in Psychotherapy

Timeline	Technology Phase	Brief Description
Pre-1990s	Traditional talk therapy	In-person therapy with no digital tools; relied entirely on verbal interaction.
1990s	Digital record keeping	Transition to computer-based patient records; improved documentation and tracking.
Early 2000s	Teletherapy and online counselling	Emergence of remote therapy through phone and video calls, improving accessibility
2010-2018	Mental health Apps and chatbots	Rise of self-help apps, mood trackers, and basic AI chatbots like Woebot and Wysa.
2018-2022	AI-based Intervention	Integration of machine learning and basic NLP in therapy tools and clinical decision-making.
2022-2024	AI+ big data, NLP, Predictive analytics	Advanced tools leveraging data for personalised interventions and progress monitoring.
2025+ (Future)	Personalised AI, Ethical use, Hybrid therapy model	Predicted use of autonomous AI therapists, hybrid models, and a focus on ethical standards.

4. DISCUSSION

The amalgamation of AI in mental health care especially as a helping tool in psychotherapy is a catalyst offering novel solutions to direct the growing demand for easy-to-access and tailor-made treatment plans. Regardless of its propitious growth, AI's role in psychotherapy has raised high concerns about its effectiveness, ethical code, and humanistic side of the interventions. This section discusses the potential benefits, hurdles, and moral and ethical considerations that are associated with AI-powered therapies and how wearables can be a beneficial addition into that slot.

AI has shown a great deal of promise in expanding access to psychotherapy. AI-powered chatbots and virtual therapists make it possible for people who live far away or are reluctant to seek traditional treatment to get help. Research has demonstrated that AI-powered cognitive-behavioral therapy (CBT) technologies, especially chatbots and online platforms, are a helpful way to manage depression and anxiety (Beg et al., 2024). By evaluating user interactions and modifying therapy recommendations appropriately, AI also makes customized treatment regimens possible. Recent research have examined how AI might help therapy by real-time emotional assessments, showing increased therapeutic involvement (Gual-Montolio et al., 2022).

Furthermore, by giving affordable therapeutic alternatives, lowering the stigma attached to seeking mental health care, and providing round-the-clock support, AI has significantly improved accessibility. AI tools' scalability makes it possible for mental health services to reach more people, especially in underprivileged places (Dhiman, 2024). In order to ensure that treatment techniques are dynamically adjusted to patient development, machine learning algorithms have also been incorporated into therapy to customize interventions based on user feedback (Espejo et al., 2023). According to a comprehensive evaluation, chatbot-delivered psychotherapy effectively reduces symptoms of anxiety and depression, demonstrating the potential of artificial intelligence in digital mental health therapies (Lim et al., 2022).

AI is becoming more and more involved in clinical decision support and therapeutic oversight in addition to direct patient interaction. AI-powered tools help psychotherapists forecast treatment results, analyze session transcripts, and spot speech patterns in their patients (Sokolovskaya, 2024). For supervisors managing trainee therapists, AI models that evaluate patient interactions offer insightful information that guarantees adherence to optimal practices. Furthermore, practical research shows how AI can be used to lessen the workload of clinicians and increase the effectiveness of mental health care (Rollwage et al., 2023).

The application of predictive analytics is another exciting advancement in AI-driven treatment. AI can identify people who would need more extensive interventions, estimate patient relapse risk, and even suggest changes to therapy approaches by utilizing vast datasets (Richards, 2024). In addition to improving clinical decision-making, this predictive ability enables early interventions that can stop the progression of mental health issues (Creed et al., 2022). With AI-driven interventions increasing accessibility and patient involvement in trauma-related therapies, AI personalization in PTSD therapy has shown great promise (Malgaroli et al., 2021).

The use of AI in psychotherapy poses serious ethical issues notwithstanding its potential. Concerns of privacy, patient autonomy, and trust are brought up by the use of AI in mental health diagnosis and treatment (Richards, 2024). For AI models to work well, a lot of data is needed, which is frequently gathered through user interactions and could jeopardize anonymity. Research highlights the necessity of strict data security protocols to stop abuse and illegal access to private psychological information (Alfano et al., 2023). Furthermore, since AI lacks the innate capacity to offer profound emotional support, there is a continuous discussion regarding the function of human empathy in AI-driven treatment (Rubin et al., 2024).

Algorithmic bias raises further ethical concerns since AI systems may unintentionally perpetuate societal biases, resulting in disparate treatment outcomes for various groups. Thorough testing and ongoing algorithmic improvement are necessary to guarantee equity in AI-driven therapy and reduce inequities (Litwack, 2024). Maintaining patient trust also depends on AI decision-making being transparent, which means developers must make AI models accountable and interpretable (Nazareth et al., 2020). There are differing opinions among the public on AI in psychotherapy;

some are enthusiastic, while others are skeptical that AI can take the place of human therapists (Aktan et al., 2022).

The danger of being overly dependent on technology at the expense of interpersonal communication is a major worry in AI-driven psychotherapy. AI chatbots can be used to enhance mental health services, but they shouldn't take the place of conventional psychotherapy. Research warns against completely automating therapy, highlighting the fact that AI should supplement human interventions rather than replace them (Litwack, 2024). Furthermore, AI may find it difficult to decipher nonverbal clues and emotional subtleties that human therapists are accustomed to recognizing, which limits its efficacy in treating complicated mental health issues (Jurblum & Selzer, 2024).

The most successful model seems to be the combination of AI and human-led therapy. While human therapists offer sophisticated emotional understanding and therapeutic intervention, artificial intelligence (AI) can manage initial evaluations, symptom tracking, and behavioral pattern detection. These models have been effectively used in blended care strategies, where clinical judgments are not replaced by AI-driven recommendations, but rather informed by them (Rollwage et al., 2023). According to recent research, the combination of AI and human support can result in mental health outcomes that are on par with those attained by conventional human-led therapy (Palmer et al., 2024). Tess and Youper, two AI-driven therapies, have shown promise in lowering anxiety and depressive symptoms, indicating that AI-based mental health tools can be beneficial adjuncts to conventional therapy (Fulmer et al., 2018; Mehta et al., 2021).

5. DIVERSITY AND INCLUSION ISSUES

Diversity and inclusion are critical considerations in the development and deployment of AI psychotherapy tools. AI models trained on non-representative datasets may exhibit biases, leading to differential performance across age, language, or cultural groups. For example, Aguirre et al. (2021) highlighted that NLP models often underperform for non-English speakers or individuals from collectivist cultures due to training data skewed toward Western, English-speaking populations. Similarly, older adults may face barriers in using AI tools due to lower digital literacy or age-related cognitive declines (Czaja et al., 2019). Addressing these gaps requires inclusive dataset curation and culturally sensitive model design to ensure equitable access and efficacy.

6. ETHICAL AND REGULATORY FRAMEWORKS

The integration of AI in psychotherapy raises significant ethical and regulatory challenges. Compliance with frameworks such as the Health Insurance Portability and Accountability Act (HIPAA) in the United States and the General Data Protection Regulation (GDPR) in Europe is essential to protect user privacy and data security (Luxton, 2020). Additionally, the American Psychological Association's (APA) guidelines on digital ethics emphasize the need for transparency, informed consent, and accountability in AI-driven interventions (American Psychological Association, 2021). Ethical concerns also include the risk of over-reliance on AI tools, which may undermine the therapeutic relationship or lead to inappropriate self-diagnosis. Robust regulatory oversight and ethical guidelines are necessary to balance innovation with patient safety.

Beyond compliance with HIPAA and GDPR, developers of AI-based psychotherapy tools must navigate a complex landscape of ethical considerations to ensure responsible deployment. One pressing issue is the potential for algorithmic bias, which can arise from training data that inadequately represents diverse populations. Such biases may lead to inequitable treatment recommendations or misinterpretations of user needs, disproportionately affecting marginalized groups (Obermeyer et al., 2019). To mitigate this, developers should employ diverse datasets and conduct regular audits to identify and correct biases.

Another ethical concern is the transparency of AI decision-making processes. Users and clinicians must understand how AI tools generate recommendations or responses to foster trust and accountability. The APA's digital ethics guidelines underscore the importance of explainability, urging developers to provide clear, accessible explanations of AI functionalities (American Psychological Association, 2021). This is particularly critical in psychotherapy, where opaque systems could erode user confidence or lead to inappropriate interventions.

Regulatory frameworks also need to address the issue of liability. When AI tools provide therapeutic advice, determining responsibility for adverse outcomes—whether it lies with the developer, clinician, or platform—remains unclear. Emerging guidelines, such as those from the Food and Drug Administration (FDA) on software as a medical device, suggest a move toward stricter oversight of AI applications in healthcare (U.S. Food and Drug Administration, 2022). However, these regulations are still evolving, and global harmonization is needed to address cross-border use of AI tools.

Finally, ethical frameworks must prioritize user autonomy. AI psychotherapy tools should empower users to make informed decisions about their mental health rather than fostering dependency. By integrating robust ethical principles and regulatory compliance, AI psychotherapy tools can uphold patient safety and trust while advancing therapeutic innovation.

Similarly, AI-based psychotherapy tools show promise in improving access to care but face challenges related to user engagement and ethical concerns (Fitzpatrick et al., 2017). This review seeks to address Gaps in understanding platform-specific effects, AI tool inclusivity, and long-term outcomes.

7. FUTURE DIRECTIONS AND RESPONSIBLE AI IMPLEMENTATION

Future studies should concentrate on hybrid models that combine AI with human-led therapies in order to optimize AI's potential in psychotherapy. By automating routine assessments, AI-powered decision-support systems should enable therapists and free up professionals to work on more challenging cases. Regulations must guarantee that AI applications respect patient privacy, ethical norms, and the human element of treatment (Dhiman, 2024). In order to create algorithms that satisfy clinical requirements and ethical standards, AI developers should also work closely with mental health specialists (Espejo et al., 2023).

Future research in the crucial field of explainable AI (XAI) development is also necessary. XAI seeks to increase the transparency and comprehensibility of AI decision-making for patients and physicians alike. Enhancing interpretability can guarantee that treatment stays patient-centered while also boosting confidence in AI suggestions (Rubin et al., 2024). Additionally, real-time biofeedback can be obtained by combining wearable technologies, smartphone applications, and AI-driven treatments, enabling people to efficiently monitor and control their emotional states (Gual-Montolio et al., 2022). Furthermore, in order to ascertain how these technologies might effectively assist mental health practitioners, research must examine the possibilities of conversational AI in therapeutic contexts as AI chatbots advance in sophistication (Miner et al., 2019).

By improving accessibility, maximizing treatment customisation, and assisting with professional decision-making, artificial intelligence is revolutionizing psychotherapy. To guarantee that moral principles and human compassion continue to be at the center of mental health treatment, their integration must be done carefully. AI in psychotherapy has a bright future if it is used in a way that strikes a balance between utilizing technology breakthroughs and maintaining the core human connection that is necessary for effective therapy. To fully utilize AI while upholding moral and patient-centered care, interdisciplinary cooperation between AI developers, physicians, and legislators must continue.

8. CONCLUSION

The current landscape of AI in psychotherapy using wearables is marked by rapid innovation and growing evidence of their potential to revolutionize mental health care. The range lies between the detection of early signs and symptoms to furnishing specific interventions depending upon the individual's need making wearables augmented with AI extremely patient-centered and accessible at the remotest place to facilitate practical resources when first-aid takes time to reach. As it is at present an evolving field, the accuracy of psychotherapy is still questionable. Although with a more data-driven approach, equitable, and effective practice. The future of tech development explicitly lies in the ongoing collaboration among researchers, clinicians, and technologists and their expertise to challenge existing potholes and ensure the usability of these tools ethically and congruously, because with great power comes greater responsibility.

REFERENCES

Aguirre, C.. (2021). Addressing bias in AI-driven mental health interventions. *Journal of Medical Internet Research, 23*(9), e27894.

Aktan, M. E., Turhan, Z., & Dolu, İ. (2022). Attitudes and perspectives towards the preferences for artificial intelligence in psychotherapy. *Computers in Human Behavior, 133*, 107273. DOI: 10.1016/j.chb.2022.107273

Alfano, L., Malcotti, I., & Ciliberti, R. (2023). Psychotherapy, artificial intelligence and adolescents: Ethical aspects. *Journal of Preventive Medicine and Hygiene, 64*(4), E438–E442. DOI: 10.15167/2421-4248/jpmh2023.64.4.3135 PMID: 38379752

American Psychological Association. (2021). *APA guidelines for ethical conduct in the use of digital technologies in psychological practice*. https://www.apa.org/ethics/digital-guidelines

Arendt, I. T. P., Riisager, L. H. G., Larsen, J. E., Christiansen, T. B., & Moeller, S. B. (2021). Distinguishing between rumination and intrusive memories in PTSD using a wearable self-tracking instrument: A proof-of-concept case study. *Cognitive Behaviour Therapist, 14*, e15. Advance online publication. DOI: 10.1017/S1754470X2100012X

Beg, M. J., Verma, M., & Verma, M. K. (2024). Artificial intelligence for psychotherapy: A review of the current state and future directions. *Indian Journal of Psychological Medicine, 02537176241260819*. Advance online publication. DOI: 10.1177/02537176241260819

Brewer, R., Guberman, J., & Piper, A. M. (2020). Designing for disability: Evaluating the state of accessibility design in mental health apps. *Proceedings of the ACM on Human-Computer Interaction, 4*(CSCW2), 1–25. https://doi.org/DOI: 10.1145/3415184

Chen, Z., Lu, Y., & Wang, W. Y. (2023). Empowering psychotherapy with large language models: Cognitive distortion detection through diagnosis of thought prompting. *arXiv preprint arXiv:2310.07146*. https://doi.org//arXiv.2310.07146DOI: 10.18653/v1/2023.findings-emnlp.284

Cheung, F., Lucas, G. M., & Gratch, J. (2019). Personalizing mental health interventions through adaptive conversational agents. *Frontiers in Psychology, 10*, 2647. DOI: 10.3389/fpsyg.2019.02647

Chung, A. H., Gevirtz, R. N., Gharbo, R. S., Thiam, M. A., & Ginsberg, J. P. (2021). Pilot study on reducing symptoms of anxiety with a heart rate variability biofeedback wearable and remote stress management coach. *Applied Psychophysiology and Biofeedback, 46*(4), 347–358. DOI: 10.1007/s10484-021-09519-x PMID: 34308526

Cogito. (2025, February 14). *AI Training Data Company | Cogito Tech.* Cogitotech. https://www.cogitotech.com/

Collins, C., Dennehy, D., Conboy, K., & Mikalef, P. (2021). Artificial intelligence in information systems research: A systematic literature review and research agenda. *International Journal of Information Management, 60*, 102383. DOI: 10.1016/j.ijinfomgt.2021.102383

Creed, T. A., Salama, L., Slevin, R., Tanana, M., Imel, Z., Narayanan, S., & Atkins, D. C. (2022). Enhancing the quality of cognitive behavioral therapy in community mental health through artificial intelligence generated fidelity feedback (Project AFFECT): A study protocol. *BMC Health Services Research, 22*(1), 1177. DOI: 10.1186/s12913-022-08519-9 PMID: 36127689

Czaja, S. J., Boot, W. R., Charness, N., & Rogers, W. A. (2019). *Designing for older adults: Principles and creative human factors approaches* (3rd ed.). CRC Press., DOI: 10.1201/b22189

D'Alfonso, S. (2020). AI in mental health. *Current Opinion in Psychology, 36*, 112–117. DOI: 10.1016/j.copsyc.2020.04.005 PMID: 32604065

DeVault, D., Artstein, R., Benn, G., Dey, T., Fast, E., Gainer, A., & Morency, L. P. (2014, May). SimSensei Kiosk: A virtual human interviewer for healthcare decision support. In *Proceedings of the 2014 international conference on Autonomous agents and multi-agent systems* (pp. 1061-1068).

Dhiman, V. K. (2024). *The emergence of AI in mental health: A transformative journey.* World Journal of Advanced Research and Reviews., DOI: 10.30574/wjarr.2024.22.1.1298

Donnermann, M., Schaper, P., & Lugrin, B. (2020, August). Integrating a social robot in higher education–a field study. In *2020 29th IEEE international conference on robot and human interactive communication (RO-MAN)* (pp. 573-579). IEEE. DOI: 10.1109/RO-MAN47096.2020.9223602

Doorn, K. A., Kamsteeg, C., Bate, J., & Aafjes, M. (2020). A scoping review of machine learning in psychotherapy research. *Psychotherapy Research, 31*(1), 92–116. DOI: 10.1080/10503307.2020.1808729 PMID: 32862761

Espejo, G., Reiner, W., & Wenzinger, M. (2023). Exploring the role of artificial intelligence in mental healthcare: Progress, pitfalls, and promises. *Cureus, 15*. Advance online publication. DOI: 10.7759/cureus.44748 PMID: 37809254

Fitzpatrick, K. K.. (2017). Delivering cognitive behavior therapy to young adults with symptoms of depression and anxiety using a fully automated conversational agent (Woebot). *Journal of Medical Internet Research, 19*(6), e7785. PMID: 28588005

Fulmer, R., Joerin, A., Gentile, B., Lakerink, L., & Rauws, M. (2018). Using psychological artificial intelligence (Tess) to relieve symptoms of depression and anxiety: Randomized controlled trial. *JMIR Mental Health, 5*(4), e9782. DOI: 10.2196/mental.9782 PMID: 30545815

Grodniewicz, J. P., & Hohol, M. (2023). Waiting for a digital therapist: Three challenges on the path to psychotherapy delivered by artificial intelligence. *Frontiers in Psychiatry, 14*, 1190084. DOI: 10.3389/fpsyt.2023.1190084 PMID: 37324824

Gual-Montolio, P., Jaén, I., Martínez-Borba, V., Castilla, D., & Suso-Ribera, C. (2022). Using artificial intelligence to enhance ongoing psychological interventions for emotional problems in real- or close to real-time: A systematic review. *International Journal of Environmental Research and Public Health, 19*(13), 7737. Advance online publication. DOI: 10.3390/ijerph19137737 PMID: 35805395

Hamilton, A. (2023, May 5). *The Rise of Online Therapy: History and Challenges - masc*. Masc. https://mascmedical.com/healthcare-staffing-agencies-online-therapy-history-and-challenges

Howard, J. (2019). Artificial intelligence: Implications for the future of work. *American Journal of Industrial Medicine, 62*(11), 917–926. DOI: 10.1002/ajim.23037 PMID: 31436850

Imel, Z. E., Caperton, D. D., Tanana, M., & Atkins, D. C. (2017). Technology-enhanced human interaction in psychotherapy. *Journal of Counseling Psychology, 64*(4), 385–393. DOI: 10.1037/cou0000213 PMID: 28318277

Inkster, B., Sarda, S., & Subramanian, V. (2018). An empathy-driven, conversational artificial intelligence agent (Wysa) for digital mental well-being: Real-world data evaluation mixed-methods study. *JMIR mHealth and uHealth, 6*(11), e12106. DOI: 10.2196/12106 PMID: 30470676

Integrating Technology into Mental Healthcare. (n.d.). https://www.psychotherapy.net/article/integrating-technology

Jacobson, N. C., & Feng, B. (2022). Digital phenotyping of generalized anxiety disorder: Using artificial intelligence to accurately predict symptom severity using wearable sensors in daily life. *Translational Psychiatry*, *12*(1), 336. DOI: 10.1038/s41398-022-02038-1 PMID: 35977932

Jang, S., Kim, J. J., Kim, S. J., Hong, J., Kim, S., & Kim, E. (2021). Mobile app-based chatbot to deliver cognitive behavioral therapy and psychoeducation for adults with attention deficit: A development and feasibility/usability study. *International Journal of Medical Informatics*, *150*, 104440. DOI: 10.1016/j.ijmedinf.2021.104440 PMID: 33799055

Janiesch, C., Zschech, P., & Heinrich, K. (2021). Machine learning and deep learning. *Electronic Markets*, *31*(3), 685–695. DOI: 10.1007/s12525-021-00475-2

Jiang, M., Zhao, Q., Li, J., Wang, F., He, T., Cheng, X., Yang, B. X., Ho, G. W. K., & Fu, G. (2024). A Generic review of Integrating Artificial intelligence in cognitive Behavioral therapy. *arXiv (Cornell University)*. https://doi.org//arxiv.2407.19422DOI: 10.48550

Jurblum, M., & Selzer, R. (2024). Potential promises and perils of artificial intelligence in psychotherapy - The AI Psychotherapist (APT). *Australasian Psychiatry*. Advance online publication. DOI: 10.1177/10398562241286312 PMID: 39344214

Kearns, W. R., Bertram, J., Divina, M., Kemp, L., Wang, Y., Marin, A., Cohen, T., & Yuwen, W. (2024). Bridging the Skills Gap: Evaluating an AI-Assisted Provider Platform to Support Care Providers with Empathetic Delivery of Protocolized Therapy. *arXiv (Cornell University)*. https://doi.org//arxiv.2401.03631DOI: 10.48550

Khamassi, M., Velentzas, G., Tsitsimis, T., & Tzafestas, C. (2018). Robot fast adaptation to changes in human engagement during simulated dynamic social interaction with active exploration in parameterized reinforcement learning. *IEEE Transactions on Cognitive and Developmental Systems*, *10*(4), 881–893. DOI: 10.1109/TCDS.2018.2843122

Kok, J. N., Boers, E. J., Kosters, W. A., Van der Putten, P., & Poel, M. (2009). Artificial intelligence: definition, trends, techniques, and cases. *Artificial intelligence*, *1*(270-299), 51.

Kuziemsky, C., Maeder, A. J., John, O., Gogia, S. B., Basu, A., Meher, S., & Ito, M. (2019). Role of artificial intelligence within the telehealth domain. *Yearbook of medical informatics*, *28*(01), 035-040.

Lee, E. E., Torous, J., De Choudhury, M., Depp, C. A., Graham, S. A., Kim, H. C., Paulus, M. P., Krystal, J. H., & Jeste, D. V. (2021). Artificial intelligence for mental health care: Clinical applications, barriers, facilitators, and artificial wisdom. *Biological Psychiatry: Cognitive Neuroscience and Neuroimaging*, *6*(9), 856–864. DOI: 10.1016/j.bpsc.2021.02.001 PMID: 33571718

Li, H., Zhang, R., Lee, Y. C., Kraut, R. E., & Mohr, D. C. (2023). Systematic review and meta-analysis of AI-based conversational agents for promoting mental health and well-being. *NPJ Digital Medicine*, *6*(1), 236. DOI: 10.1038/s41746-023-00979-5 PMID: 38114588

Lim, S. M., Shiau, C. W. C., Cheng, L. J., & Lau, Y. (2022). Chatbot-delivered psychotherapy for adults with depressive and anxiety symptoms: A systematic review and meta-regression. *Behavior Therapy*, *53*(2), 334–347. DOI: 10.1016/j.beth.2021.09.007 PMID: 35227408

Litwack, E. B. (2024). *Chatbots, robots, and the ethics of automating psychotherapy*. Athens Journal of Philosophy., DOI: 10.30958/ajphil.3-3-1

Lutz, C., & Tamò-Larrieux, A. (2021). The privacy implications of conversational agents in mental health: A systematic review. *Journal of Medical Internet Research*, *23*(6), e27854. DOI: 10.2196/27854

Luxton, D. D. (2020). Ethical implications of conversational agents in mental health care. *Journal of Technology in Behavioral Science*, *5*(4), 346–354. DOI: 10.1007/s41347-020-00144-8

Malgaroli, M., Hull, T. D., & Schultebraucks, K. (2021). Digital Health and Artificial intelligence for PTSD: Improving treatment delivery through personalization. *Psychiatric Annals*, *51*(1), 21–26. DOI: 10.3928/00485713-20201203-01

Mehta, A., Niles, A. N., Vargas, J. H., Marafon, T., Couto, D. D., & Gross, J. J. (2021). Acceptability and Effectiveness of Artificial Intelligence Therapy for Anxiety and Depression (YOUPER): Longitudinal observational study. *Journal of Medical Internet Research*, *23*(6), e26771. DOI: 10.2196/26771 PMID: 34155984

Miner, A. S., Shah, N., Bullock, K. D., Arnow, B. A., Bailenson, J., & Hancock, J. (2019). Key considerations for Incorporating Conversational AI in Psychotherapy. *Frontiers in Psychiatry*, *10*, 746. Advance online publication. DOI: 10.3389/fpsyt.2019.00746 PMID: 31681047

Müller, V. C. (2016). Fundamental issues of artificial intelligence. In *Synthese Library/Synthese library*. https://doi.org/DOI: 10.1007/978-3-319-26485-1

Nazareth, S., Nussbaum, R. L., Siglen, E., & Wicklund, C. A. (2020). Chatbots & artificial intelligence to scale genetic information delivery. *Journal of Genetic Counseling*, *30*(1), 7–10. DOI: 10.1002/jgc4.1359 PMID: 33191601

Obermeyer, Z., Powers, B., Vogeli, C., & Mullainathan, S. (2019). Dissecting racial bias in an algorithm used to manage the health of populations. *Science*, *366*(6464), 447–453. DOI: 10.1126/science.aax2342 PMID: 31649194

Palmer, C. E., Marshall, E., Millgate, E., Warren, G., Ewbank, M. P., Cooper, E., Lawes, S., Bouazzaoui, M., Smith, A., Hutchins-Joss, C., Young, J., Margoum, M., Healey, S., Marshall, L., Mehew, S., Cummins, R., Tablan, V., Catarino, A., Welchman, A. E., & Blackwell, A. D. (2024). Combining AI and human support in mental health: a digital intervention with comparable effectiveness to human-delivered care. medRxiv *(Cold Spring Harbor Laboratory)*. DOI: 10.1101/2024.07.17.24310551

Pandey, A. K., Gelin, R., & Robot, A. M. P. S. H. (2018). Pepper: The first machine of its kind. *IEEE Robotics & Automation Magazine*, *25*(3), 40–48. DOI: 10.1109/MRA.2018.2833157

Pickens, J. N. PhD. (2021, October 13). Click "less worrying" or "improved sleep" to fine-tune your virtual therapy? *Psychology Today*. https://www.psychologytoday.com/us/blog/psychology-through-technology/202110/the-evolution-digital-therapy-technology

Piwek, L., Ellis, D. A., Andrews, S., & Joinson, A. (2016). The rise of consumer health wearables: Promises and barriers. *PLoS Medicine*, *13*(2), e1001953. DOI: 10.1371/journal.pmed.1001953 PMID: 26836780

Powell, W. A., Corbett, N., & Powell, V. (2019). The rise of the virtual human. In *Natural language processing*. (pp. 1459–1488). https://doi.org/DOI: 10.4018/978-1-7998-0951-7.ch070

Puiatti, A., Mudda, S., Giordano, S., & Mayora, O. (2011, August). *Smartphone-centred wearable sensors network for monitoring patients with bipolar disorder. In 2011 Annual International Conference of the IEEE Engineering in Medicine and Biology Society*. IEEE., DOI: 10.1109/IEMBS.2011.6090613

Rampton, V., Mittelman, M., & Goldhahn, J. (2020). Implications of artificial intelligence for medical education. *The Lancet. Digital Health*, *2*(3), e111–e112. DOI: 10.1016/S2589-7500(20)30023-6 PMID: 33328081

Richards, D. (2024). Artificial intelligence and psychotherapy: A counterpoint. *Counselling & Psychotherapy Research*. Advance online publication. DOI: 10.1002/capr.12758

Riisager, L. G., Christiansen, T. B., Moeller, S. B., Huniche, L., & Larsen, J. E. (2024, May). Snapping Out of It: How a Wearable for Self-Tracking Assisted Psychotherapy Bridges the Gap Between Thoughts and the World. In *Extended Abstracts of the CHI Conference on Human Factors in Computing Systems* (pp. 1-7). https://doi.org/DOI: 10.1145/3613905.3637112

Rollwage, M., Habicht, J., Juechems, K., Carrington, B., Viswanathan, S., Stylianou, M., Hauser, T. U., & Harper, R. (2023). Using conversational AI to facilitate mental health assessments and improve clinical efficiency within psychotherapy services: Real-world observational study. *JMIR AI, 2*, e44358. Advance online publication. DOI: 10.2196/44358 PMID: 38875569

Rubin, M., Arnon, H., Huppert, J., & Perry, A. (2024). Considering the role of human empathy in AI-driven therapy. *JMIR Mental Health, 11*, e56529. Advance online publication. DOI: 10.2196/56529 PMID: 38861302

Sadeh-Sharvit, S., & Hollon, S. D. (2020). Leveraging the power of nondisruptive technologies to optimize mental health treatment: Case study. *JMIR Mental Health, 7*(11), e20646. DOI: 10.2196/20646 PMID: 33242025

Samoili, S., Lopez, C. M., Delipetrev, B., Martinez-Plumed, F., Gomez, G. E., & De, P. G. (2021). *AI Watch. Defining Artificial Intelligence 2.0.* JRC Publications Repository., DOI: 10.2760/019901

Schneider, S., & Kummert, F. (2017, August). Exploring embodiment and dueling bandit learning for preference adaptation in human-robot interaction. In *2017 26th IEEE International Symposium on Robot and Human Interactive Communication (RO-MAN)* (pp. 1325-1331). https://doi.org/DOI: 10.1109/ROMAN.2017.8172476

Schueller, S. M.. (2018). Designing AI for mental health: The role of user experience. *Psychiatric Services (Washington, D.C.), 69*(12), 1201–1203.

Schueller, S. M., Neary, M., O'Loughlin, K., & Adkins, E. C. (2018). Discovery of and interest in health apps among those with mental health needs: Survey and focus group study. *Journal of Medical Internet Research, 20*(6), e10141. DOI: 10.2196/10141 PMID: 29891468

Semmler, S., & Rose, Z. (2017). Artificial intelligence: Application today and implications tomorrow. *Duke L. & Tech. Rev., 16*, 85.

Sharma, V., Goyal, M., & Malik, D. (2017). An intelligent behaviour shown by chatbot system. *International Journal of New Technology and Research, 3*(4), 263312.

Sokolovskaya, M. (2024). *Enhancing supervisory practice in Positive and Transcultural Psychotherapy through Artificial Intelligence*. The Global Psychotherapist., DOI: 10.52982/lkj253

Statista. (2023). Number of social media users worldwide. Retrieved from https://www.statista.com

Sun, K., Roy, A., & Tobin, J. M. (2024). Artificial intelligence and machine learning: Definition of terms and current concepts in critical care research. *Journal of Critical Care*, *82*, 154792. DOI: 10.1016/j.jcrc.2024.154792 PMID: 38554543

Team, R. (2023a, December 6). *Pepper*. ROBOTS: Your Guide to the World of Robotics. https://robotsguide.com/robots/pepper

Team, R. (2023b, December 19). *Bandit*. ROBOTS: Your Guide to the World of Robotics. https://robotsguide.com/robots/bandit/

The use of Technology-Assisted methods in therapy and psychological treatments. (2023, January 11). https://www.healthtechzone.com/topics/healthcare/articles/2023/01/11/454583-use-technology-assisted-methods-therapy-psychological-treatments.htm

Thieme, A., Wallace, J., Meyer, T. D., & Olivier, P. (2020). Designing for mental health: Understanding the interplay between technology and therapeutic practices. *Digital Health*, *6*, 1–15. DOI: 10.1177/2055207620936447\

Tsai, C. H., Chen, P. C., Liu, D. S., Kuo, Y. Y., Hsieh, T. T., Chiang, D. L., Lai, F., & Wu, C. T. (2022). Panic attack prediction using wearable devices and machine learning: Development and cohort study. *JMIR Medical Informatics*, *10*(2), e33063. DOI: 10.2196/33063 PMID: 35166679

Ujike, S., Yasuhara, Y., Osaka, K., Sato, M., Catangui, E., Edo, S., ... & Mifune, K. (2019). Encounter of Pepper-CPGE for the elderly and patients with schizophrenia: an innovative strategy to improve patient's recreation, rehabilitation, and communication. *The Journal of Medical Investigation*, *66*(1.2), 50-53. https://doi.org/DOI: 10.2152/jmi.66.50

U.S. Food and Drug Administration. (2022). *Software as a medical device (SaMD): Clinical evaluation*. https://www.fda.gov/regulatory-information/search-fda-guidance-documents/software-medical-device-samd-clinical-evaluation

Williams, M. D. (2020, August 27). *10 trends for technology in counseling and psychotherapy*. https://www.wolterskluwer.com/en/expert-insights/10-trends-for-technology-in-counseling-and-psychotherapy

Woebot Health. (2025, February 11). *Scalable enterprise solution for mental health | Woebot Health.* https://woebothealth.com/

Yeh, P. L., Kuo, W. C., Tseng, B. L., & Sung, Y. H. (2025). Does the AI-driven Chatbot Work? Effectiveness of the Woebot app in reducing anxiety and depression in group counseling courses and student acceptance of technological aids. *Current Psychology (New Brunswick, N.J.)*, *44*(9), 1–13. DOI: 10.1007/s12144-025-07359-0

Zhang, T., Schoene, A. M., Ji, S., & Ananiadou, S. (2022). Natural language processing applied to mental illness detection: A narrative review. *npj. Digital Medicine*, *5*(1), 46. DOI: 10.1038/s41746-022-00589-7 PMID: 35396451

Zipfel, S., Lutz, W., Schneider, S., Schramm, E., Delgadillo, J., & Giel, K. E. (2024). The future of enhanced psychotherapy: Towards precision psychotherapy. *Psychotherapy and Psychosomatics*, *93*(4), 230–236. DOI: 10.1159/000539022 PMID: 38934154

Chapter 2
AI-Driven Cognitive Behavioural Therapy (CBT) Using Wearables

Nistha Jain
https://orcid.org/0009-0009-7360-1097
Woxsen University, Hyderabad, India

Eshita Chatterjee
https://orcid.org/0009-0003-0088-3667
Woxsen University, Hyderabad, India

ABSTRACT

The integration of Artificial Intelligence (AI) into Cognitive Behavioural Therapy (CBT) enhances accessibility, affordability, and adherence in mental health care. Traditional CBT, while effective, faces limitations due to therapist shortages and geographic barriers. AI-driven CBT, supported by chatbots, machine learning, and wearable technology, enables real-time monitoring, cognitive restructuring, and behavioral activation. Wearables track physiological data, improving symptom management. However, challenges such as data privacy, algorithmic bias, and lack of human empathy remain. This chapter explores AI's role in CBT, advocating a hybrid model that blends AI efficiency with human expertise for optimal mental health care.

INTRODUCTION

With the rise of mental health issues, a range of therapeutic interventions have been formed and utilised to treat different types of disorders. According to a report

DOI: 10.4018/979-8-3373-0467-0.ch002

by WHO, in 2019 around 270 million people were suffering from a mental health problem, the most prevalent being anxiety and depression. Cognitive Behavioural Therapy (CBT) has set a benchmark for a number of conditions and has worked exceptionally well even in severe cases of anxiety, depression, dysthymia, OCD, PTSD and psychotic conditions such as schizophrenia (Hoffman et al., 2012).

Unfortunately, the traditional face-to-face treatment requires the provider to be present physically and thus, making it difficult for the patient and provider to reach. A restricted means of approach and a lack of resources frequently limit the way CBT is delivered. In a recent survey by the Health Resources and Services Administration (HRSA), around 160 million individuals in the USA alone live in areas with a lack of accessibility of practitioners. With the advancements in technology, it has become easier to reach out virtually and at a lower cost. Implementation of Internet-based Cognitive Behaviour Therapy (ICBT) has been proven to have similar functionality and effectiveness as traditional sessions (Andersson et al., 2019; Niles et al., 2021). Online platforms such as Bright Side (Cohelo & Hope, 2024) and other mobile platforms have come far in alleviating the obstacles of unavailability of professionals and geographical restrictions, yet the destination is still far away. These applications face many challenges, such as high dropout rates and low levels of interaction, despite their potential. Apart from that, their treatments frequently ignore each person's particular needs in favour of addressing general cognitive problems. To get over these limitations, new adaptable and flexible approaches to CBT are desperately needed, ones that can accommodate the various needs of various patient groups.

With the advent of Artificial Intelligence (AI), various sectors are exploring the technology and using its advantages to make things easier. The solutions that AI brings forth with its development of prediction, detection, analysis, pattern recognition, and data integration, offer valuable potentials to deliver the same in the sector of mental health care (D'Alfonso, 2020; Koutsouleris et al., 2022). It can be used as an aid-in tool for the practitioners as well as the patients by bridging the gap along with providing a tailor-made recommendation for personalised interventions and support (Assunção et al., 2022). Significant progress has been made since AI was incorporated into CBT, especially when it comes to large language models (LLMs). Not to add some already noteworthy developments in this field, including the creation of focused CBT prompts and their sharing for modeling, which enables the creation of CBT-LLM, a model designed especially for CBT (Na, 2024). These elements have played a significant role in the development of cognitive behavioural therapies, making them more precise and tailored to the various needs of patients.

This chapter has its focus on deliverable CBT integrated with AI. It also aims to identify the ways in which wearable technology when synthesized with AI, can be employed into CBT interventions and become a handy tool for both the mental health professionals and as a support system to the users. An important answer that the chapter seeks is to understand the comparable efficacy, adherence, and long-term outcomes of AI-driven CBT assisted through wearables to the traditional in-person CBT.

Artificial Intelligence (AI) is rapidly transforming mental health care, particularly in areas like Cognitive Behavioral Therapy (CBT). However, the integration of AI into psychotherapy faces several core challenges: achieving emotional nuance and empathy in AI responses, ensuring data privacy and ethical compliance, and maintaining the clinical validity of interventions. Understanding these challenges is essential for readers to grasp the complexity AI seeks to resolve in mental health settings.

The upcoming stage for AI-augmented CBT focuses on refining existing technologies to boost their accuracy, personalization, and user interaction. Advancements in natural language processing (NLP) and deep learning may result in more effective conversational AI agents that can respond with greater empathy and relevance to the concerns of users. For AI to be seamlessly incorporated into mental health care systems, it is essential to establish regulatory frameworks that address ethical and legal issues. To foster public trust and guarantee the responsible use of AI in mental health treatment, issues related to informed consent, data privacy, and accountability must be addressed. The use of wearable devices and real-time monitoring could enhance the efficacy and safety of AI-driven CBT, making it more customized, accessible, and data-driven. While AI improves operational efficiency, it cannot replace human compassion and clinical supervision. A synergistic approach that integrates AI capabilities with human expertise may provide the most effective means of leveraging the benefits of AI-based CBT while mitigating its limitations.

FROM THEORY TO TECHNOLOGY: A TIMELINE OF PSYCHOTHERAPY'S EVOLUTION

The trajectory of modern psychotherapy can be divided into six major developmental phases, each contributing to the foundation upon which AI-driven interventions are now being built.

- **Phase 1: Foundations of Psychodynamic Therapy (1890–1950)**

Sigmund Freud pioneered a structured theory of mental illness rooted in unconscious processes and early life experiences. His work not only laid the groundwork for modern psychotherapy but also introduced systematic techniques for therapeutic engagement. Parallel to these psychological advances, the emergence of pharmacology—spearheaded by Rudolf Buchheim and later Oswald

Schmiedeberg—initiated scientific inquiry into the biological dimensions of mental disorders (Yalom, 2019).

- **Phase 2: The Scientific Turn (1950–1970)**

The mid-20th century witnessed the rise of empirical research in psychotherapy. Randomized Controlled Trials (RCTs) became the gold standard for evaluating treatment efficacy, ushering in an era where psychological interventions were subjected to rigorous scientific validation (Beck, 1976).

- **Phase 3: Standardizing Therapy Techniques (1970–1980)**

To improve replicability, treatment manuals and protocols were developed, clearly specifying what therapists should and should not do. Supporting structures such as therapist certification, fidelity checks, and session-based reviews were introduced to monitor consistency and competence in therapeutic delivery (Eysenck, 1952; Lazarus, 1977).

- **Phase 4: Diagnostic and Procedural Standardization (1980–2005)**

The publication of the DSM-III by the American Psychiatric Association in 1980 revolutionized diagnostic practices by establishing uniform criteria for mental disorders. This movement toward standardization extended to clinical interviews and severity rating tools like the Hamilton Depression Rating Scale, which ensured consistent diagnosis and treatment planning (Kendell, 1980; Kabat-Zinn, 2003).

- **Phase 5: Digital Delivery (2005–2015)**

With advancements in internet infrastructure, smartphones, and cloud computing, psychotherapy transitioned into digital formats. Online platforms and mobile apps allowed real-time interventions, while security protocols like HIPAA ensured the protection of sensitive health data (Buyya et al., 2013; Proudfoot et al., 2003).

- **Phase 6: AI and Personalized Digital Therapeutics (2015–Present)**

Today, the field is characterized by the proliferation of digital mental health tools—an estimated range of 10,000 to 20,000 apps. New things like wearables, real-time data monitoring, and AI-powered conversational agents have come up that provide scalable, adaptive, and personalized therapy. Ecological momentary assessments and large-scale user data now enable continuous improvement in therapeutic accuracy and user engagement, motioning the frontier therapy forward (Agarwal et al., 2022; Rodgers et al., 2014).

NEW BOUNDARIES: WHAT'S THE PRESENT STAGE OF AI-AUGMENTED CBT?

AI-augmented CBT is the place where the previous therapeutic transformations are merged with the huge potential of machine learning. Natural Language Processing (NLP) and deep learning allow chatbots to be more conversational, empathetic and contextual. Real-time biofeedback received directly from wearables and mobile devices increases the level of personalization. However, introducing the aforementioned technologies creates new risks, such as the potential of algorithmic misinterpretation, the lack of certain human aspects, and regulatory gaps.

Over-relying on technical solutions is not advisable, as the relational and empathic part of psychotherapy should stay. Instead, it seems that a mixed model, i.e., AI as a way to reinforce rather than to substitute human therapists, is the most feasible one. Such collaboration can bring therapy to people who are not reached now and at the same time allow the necessary professional supervision and outcome evaluation.

Background

Understanding CBT

Cognitive Behavioural Therapy (CBT) is a structured, goal-oriented, time-limited, evidence-based form of psychotherapy. It focuses on the present in lieu of the past and uses cognitive techniques inculcated with behavioural elements. It is one of the most popular forms of therapy which was first introduced by Dr. Aron Beck (Beck, 1970). By altering maladaptive thought and behaviour patterns, CBT is a reliable method of addressing mental health issues. The core beliefs lie in the proposition that cognition, behaviour and emotions are interconnected. It is through the maladaptive interplay of the three, an individual's well-being is affected. The therapeutic approach of CBT aims to address these maladaptive patterns and extemporise that individual's overall way of living (Freeman, 1990).

It is believed that the sophisticated approach of CBT that is being used nowadays was initially inspired by Ellis' (1962) Rational Emotive Therapy (RET). The foundation of the therapy sessions lies upon experimental methods, so in the initial sessions, goals are formulated (initial hypothesis), psychoeducation is provided, and subsequently those formulations are worked out and modified if necessary. As stated earlier, CBT interventions focus on three main arenas: cognitive, conative and affective. The focal point in the cognitive domain is that of the thinking patterns, perceptions and faulty beliefs and automatic negative thoughts (such as blaming, overgeneralization etc). Using Cognitive Restructuring (CR) methods the therapist attempts to alter these patterns and aid the client with helpful thinking patterns and have a positive outlook. The conative part, like in behaviour therapy, deals with the maladaptive behaviours and habits. Employing behavioural methods such as role play, behavioural activation (BA), the ABC approach, the intervention strives to bring improvisation in the same (Freeman, 1990). CBT is also primarily concerned with helping clients recognize and manage negative emotions. It does this by helping them identify negative emotions, correcting their identification, and using appropriate cognitive and behavioural strategies to promote better emotional well-being (Baker et al., 2011).

Physiological effects on the three arenas and their impact upon an individual's well-being is also given importance in this approach. These factors can be problem specific for example increase of heartbeat, breathing issues during a panic attack or more generalised factors such as fatigue, increased levels of stress, which may affect a person's general arousal levels, which makes it imperative to take steps to help the patient in countering these as well. For this reason, methods for controlling physiological reactions such deep breathing, muscular relaxation, and motivating clients to engage in physical activity (Yang et al., 2020).

The efficacy of CBT has been studied and empirically proven in approximately more than 20 disorders. The list of the issues includes mood disorders: depression (Chand et al., 2018) & bipolar disorder (Ye, 2016), anxiety (Kodal et al., 2018), eating disorders: bulimia, binge eating & anorexia nervosa (Byrne, 2011), schizophrenia (Jauhar, 2014), substance use disorder (Zamboni et al., 2021), attention deficit and hyperactivity disorder (Young et al., 2020), obsessive-compulsive disorder(Olatunji, 2013), insomnia (Mitchell et al., 2012), marital distress (Aghdam, 2017), personality disorders (Matusiewicz et al., 2010) and many more. It has also been proven to aid-in chronic pain (Broderick et al., 2016) and fMRI studies have shown significant changes in the brain structures (König et al., 2024).

It is hence safe to say that CBT not only navigates an individual by engaging into the cognitive, affective and behavioural routes but also acknowledges the interaction between the mind and body. Typically, CBT customizes a variety of methods and approaches according to each person's unique requirements and circumstances, promoting resilience, enhancing mental health, and solving problems (Dobson & Dozois, 2021). In turn, it makes the approach a holistic one, making it widely adopted primary treatment for disorders.

Integration of Artificial Intelligence with CBT principles

Ambitious ideas about how new technologies can transform healthcare have been sparked by notable developments in artificial intelligence (AI) and machine learning (ML) (Johnson et al., 2021). Amalgamation of AI in mental healthcare has become a boon and an aid in bettering healthcare assistance world-wide making it more cost effective and overcome location constraints. Combining data gathered in electronic healthcare records (EHR) with ongoing trends in individual wellness tracking through wearables and mobile apps creates a wealth of behavioural and personal health data that can be used for monitoring, treatment, and assessment (Chikersal et al., 2020; McIntosh et al., 2016). It has infinite potentials, when applied in CBT principles and interventions which would not only help create an advanced therapeutic care but also enhance the effectiveness and utility of the therapy.

The seamless CBT framework makes it possible for technical feasibility and delivers remarkable results (Beck, 2017). The vade mecum of CBT advises the expert to commence the therapeutic relationship by establishing the correspondence through initial assessment, which entails assembling the case history, prevailing mental status, and future goals using standardized tools, structured and/or semi-structured interviews, and questionnaires (Dobson & Dozois, 2021). Further, CR and BA strategies are employed where active participation of the client is equally important and thus, creates opportunities for real-time platforms for implementing those strategies.

The technification of CBT goes way before AI came into picture. The entire spectrum of the course of treatment, including evaluation, symptom and homework tracking, cognitive, behavioural, and functional skills training, and relapse prevention, have seen dramatic expansion in the use of technology during the preceding thirty years (O'Donnell et al., 2022; Rizzo & Koenig, 2017).

1. **Diagnosis:** Agarwal (2024) using Large Language Models (LLMs) and prompt engineering and by integrating AI, developed an app named IlluminateAI that effectively diagnoses, explains and suggests therapeutic interventions but only in the case of depression. In another study, acoustics and a multi-layer perceptron (MLP) were used to analyse and detect disorders such as Major depressive disorder (MDD), bipolar disorder, schizophrenia and Generalized anxiety disorder (GAD) (Wanderley Espinola et al., 2022). Another study shows how the accuracy of symptom severity detection systems can be improved by combining paired data with sophisticated machine learning methods like cascading SVMs. It predicts improvisation in automated psychiatric evaluation techniques and provides useful advice for using machine learning in mental health evaluations (Goodwin et al., 2017). Even though the advancements in AI have tried to

make diagnosis easier, yet there lies a huge gap since most of them are centred around making the basic diagnosis and not taking into account the co-occurring conditions which are very quotidian.

2. **Cognitive Distortions:** The first step of CBT is the identification of cognitive distortions, which are unconscious automatic negative thought patterns that can affect one's well-being (Johnco, 2015). To train the Large Language models (LLMs) and/or any artificial intelligence models, the structured data set required is not available in the market, for which the textual classification techniques are primarily used. Typically, this type of text-data is collected by the use of personal blogs (Simms et al., 2017), social media texts (Tauscher et al., 2023) or by transcripts (Lalk et al., 2024). Singh et al. (2023) proposed a multi-tasking approach that inculcates all three text, audio and visual inputs to identify cognitive distortions with a promising outcome and improvements over the available state-of-the-art models. Another model namely the ICODLE model in order to detect cognitive distortions and improve the precision and effectiveness of CBT uses sophisticated deep learning techniques, including Word Embedding and Convolutional Neural Networks (CNNs), to examine patients' daily narratives and diaries (Xing et al., 2017). In an investigative analysis, the researchers trained a large language model to not only classify the thought distortions but also establish clusters plus classify the distortions together using keywords (Shickel et al., 2020). A prerequisite to train these models are the data inputs which are not easily available to public access because the actual narratives that take place between a client and therapist remain confidential. These publically available data might not always be accurate and thus a disadvantage.
3. **Cognitive Restructuring:** The cognitive restructuring (CR) in CBT refers to a collaborative approach by therapist and client which involves analysing, evaluating and modifying flawed thoughts that leads to distress in an individual's life (Dobson, 2016). In a study, researchers crowd-sourced self-reported daily thought records (DTRs) on platforms such as Prolific and getting responses in order to train the models based on their opinions (Furukawa et al., 2023). This model can be used to add a new cognitive restructuring module to an iCBT application, which would assist users in redefining their automatic thoughts to become more reflective of their feelings rather of their automatic ideas. Another model trained by crowd-sourced inputs using the T5 transformer approach created automated responses to aid in CR which are more positive and encouraging to cognitive distortions (de Toledo Rodriguez et al., 2021). Sharma et al. (2024) created a human language model interactive system which provides sustenance for CRs across different demographics at various stages.

Employing empathetic dialogues, HealMe, a chatbot, through adaptive language model addresses the deep-rooted negative thoughts at the same time fostering more rational and balanced perceptions (Xiao et al., 2024). Unlike any other textual models which provide immediate solutions, this bot utilizes a three step: segregation of facts from emotions, creating substitutive solutions by conjuring ideas along with the client, and giving an empathetic, non-judgemental, positive response to the client.

4. **Behavioural Activation (BA):** The therapeutic intervention of CBT is constituted of various smaller interventions and strategies and Behavioural Activation (BA) is one of them. The aim is to help the client get acquainted with the physical environment around them, decrease the ones that prelude positive behaviours and connect them with desirable/goal behaviours (Mazzucchelli et al., 2009). It is eminent to deal with the behaviour of the client along with the cognitive parts as well since mental health issues affect the patients with daily activities as well. To facilitate these activities and get the clients equipped with the latter, interventions have been developed.

By analysing speech and text, a study utilised a multi-modal artificial intelligence method for identifying behavioural activation activities for cognitive behavioural therapy (Madhu et al., 2022). Emotions were identified with an accuracy of 83%, sentiments were detected with an accuracy of 81%, and activities were identified with an accuracy of 82%. A smartphone app namely MUBS: Mobile-based Behavioural Activation System, has been developed for depression by Rohani et al. (2020). By providing activity directories and customized rewards, the system uses a novel content-based probabilistic recommendation algorithm to raise awareness of behavioural activation and motivate patients to undertake positive activities. Patients are more likely to engage regularly which will ultimately help them on their path to making significant behavioural changes. Juno, a prototype chatbot developed especially for pregnant women employs a 6-week intervention program intentioned to deploy a prevention behavioural activation program for them (Mancinelli et al., 2024). This study employed a multiple-case design to investigate pregnant women's experiences with a chatbot-based perinatal mental health intervention and found that while pregnant women expressed a common enthusiasm for the intervention's content and aesthetics, they also expressed concerns about technical difficulties and the need for more precise instructions. A few studies explored exclusive monitoring, emotional support and personalised behaviour activation for the users purely based on AI unlike the previous models that primarily focused only on prompts, assistance and/ or as a motivator for the patients (Rathnakya et al., 2022 and Agarwal et al., 2024).

5. **Emotional Analysis:** Although this approach has been defined as cognitive and behavioural, it incorporates emotional insights as well, strengthening the therapeutic alliance between the therapist and the client. However, despite the fact that many CBT practitioners understand the importance of addressing emotions within the therapeutic framework, emotional processing is often considered a secondary consequence rather than the primary objective (Baker et al., 2011). Yet, with the advancements in research and new insights, therapists have the realization that individuals are more likely to benefit from the treatment if they are able to manage their emotions (Provoost et al., 2019). There have been attempts at the inception of AI-driven approaches in emotional analysis.

The technique of emotion prediction presently is being used at various domains such a text analysis, speech analysis, image analysis (Hosseini & Yamaghani, 2024), facial emotional recognition (Lavanya et al., 2024), physiological signal (Vishwakarma & Sakalle, 2024) and eye tracking. CORTEX, a therapeutic chatbot was created in two languages: English and Polish and the results depicted satisfactory in Polish while a slightly better outcome in the former (Zygadło et al., 2021). According to a systematic review by Gual-Montolio et al. (2022), the most common techniques used are the conversational agents or the chatbots based on software that are easy to access from any smart-device. It also stated that most studies reported high satisfaction and engagement with improved emotional states especially when interaction occurred in real-time. Indication towards the assistive role of AI is crystal clear.

6. **Psycho-education:** Educating the client with respect to their problems, the fundamentals of the therapeutic treatment being provided to them and is a very crucial step in CBT before the commencement of the actual treatment process (Popovic, 2015). It increases the chances of the client's acceptability and credibility towards the treatment and in-turn the rapport with the therapist.

Numerous techniques, including printed materials, multimedia content like audio and video, and holding targeted seminars and workshops, can be utilized to deliver psychological education. Because of their accessibility and adaptability, mobile apps have created new avenues for providing psychoeducation. A self-guided psychoeducational chat-bot was run for a trial in comparison to the traditional method, and it appeared to be equally effective especially for ADHD patients (Selaskowski et al., 2023). In a review, researchers speculated that ChatGPT can be integrated within the treatment process and psychoeducation in CBT for its various applications involving its role in providing accessible, personalised and scalable support (Vahedifard et al., 2023). In a comparative study to understand the preferable mode of delivering psychoeducation between a textual or by a verbal virtual agent which concluded that the adherence is higher when delivered by the former (Tielman et al., 2017).

Numerous applications, chatbots, robots have been fabricated to empower the and integrate CBT principles into AI and leverage the latter to not only educate the clients but also in the promulgation of psychological education in public in turn creating self-awareness and better management of their mental health.

7. **Self-help tools:** AI technologies are being incorporated into mental health self-help platforms more and more to improve accessibility, engagement, and customization. Woebot and other AI-powered chatbots use conversational interfaces to deliver cognitive-behavioural therapy techniques, providing personalized interventions and real-time assistance (Jiang et al., 2024). Mobile apps, which frequently leverage AI for tailored recommendations or behavioural nudges, allow users to access psychoeducation, self-monitor symptoms, or practice mindfulness (Chan et al., 2017). By creating a safe space and non-judgmental environment, much evidence-guided, tech-based tools such as Wysa have been created which are integrated with various self-help tools designed to aid in various common mental health conditions such as depression, anxiety, low mood, chronic pain etc (Beatty et al., 2022). Another study depicted that mobile-based self-help chatbots achieved better therapeutic alliance than the traditional bibliotherapist simultaneously succeeding in providing relief in anxiety symptoms within 4 weeks (Liu et al., 2022). The participants were university students, besides the results set forth a tone that the new-generation are congenial to the new-age technology.
8. **Real-time Monitoring (Symptom and homework tracking):** The integration of digital tools and artificial intelligence (AI) into cognitive-behavioural therapy (CBT) has revolutionized symptom management and homework compliance. Homework assignments, essential for applying learned skills between sessions, now benefit from digital formats that improve accessibility and engagement. Simplification of therapists' analyses and consistent self-reflection can be easily fostered by personalised prompts by AI tools such as 'MindScape' that are trained upon behavioural data (Nepal et al., 2024). Perertz et al. (2023) created an app that can identify as well as monitor homework from conversations during the therapy session and thus can streamline assignment tracking. These developments address the traditional challenges, such as non-compliance to tasks assigned by therapists and time constraints, ensuring patients gain maximum therapeutic benefit.

AI-powered wearable devices and mobile applications further enhance CBT by enabling real-time monitoring of psychological states. Wearable sensors aligned with applications such as 'MoodTrainer' can detect loneliness and provide instant interventions, while the former (sensors) can assess the behavioural patterns, stress levels and provide solutions accordingly (Garcia-Ceja et al., 2015; Addepally & Purkayastha, 2017). In another study, Goodwin et al. (2019), demonstrated that biosensors can be used as a support for therapists and can forecast aggressive behaviours. By offering tailor-made insights, real-time monitoring, facilitating continuous engagement, these AI-powered tools can create improvements in delivering efficient CBT-based treatments.

9. **Relapse Prevention:** Relapse shortly after terminating CBT therapy sessions remains one of the hindrances resulting from the short duration of the intervention and limited durability of treatment effects. Remedies offered by AI-based applications put forward promising predictions and support sustained therapeutic outcomes. Early studies have bespoken the application of multivariate SVM-fMRI methods to speculate long-term outcomes of internet-based CBT (ICBT) for social anxiety disorder with an accuracy of 92% (Månsson et al., 2015). In a recent study, Lorimer et al. (2021) engaged a machine learning model such as XGBoost, to dynamically predict relapse risks at various stages of treatment, which enables pre-emptive, resource-efficient interventions. These insights have been extended by another study which introduced a conversational AI agent in virtual reality that integrated relapse prevention techniques with motivational interviewing especially for patients with substance use disorders (Yeo et al., 2024). Babu & Joseph (2024) emphasized upon the involvement of therapists in AI-augmented therapy to generate them with a balance of human insight and automation. These developments depict the transformative potential of AI in enhancing relapse prevention, offering predictive capabilities and personalized interventions that can significantly improve long-term treatment efficacy.

Delivery Methods of Cognitive Behavioural Therapy (CBT) using AI

The evolution of the delivery in therapy has been rapid ever since the involvement of technological advancements, expanding beyond traditional face-to-face interventions. Enhanced accessibility, increased engagement, and effectiveness in imparting CBT through innovative modalities. This part of the chapter delves into the key technological approaches through which CBT can be delivered to patients or users such as through chatbots, virtual reality, and real-time monitoring especially using wearable devices. Presently, the interventions are carried out in person, or with the assistance of a therapist. This is effective yet generates concerns regarding accessibility and cost. AI-enabled agents can mimic human-like dialogue still its accuracy is questionable and needs to be addressed (Gupta et al., 2022).

Chatbots in CBT

Chatbots trained in CBT interventions have been merged with AI to enhance psychotherapy. They instant, economical psychological assistance (Fitzpatrick et al., 2017). During a crisis period, such as the COVID-19 pandemic, a great number of the population, especially adolescents and young adults might have felt the necessity of counseling, when the rate of anxiety-depression increased, and in-person therapy became impossible to deliver. At that time, chatbots came to their rescue as a substitute which was similar to human interaction, marking a significant transition from face-to-face therapy to technology-assisted formats (Vilaza & McCashin, 2021). For instance, to offer sustenance, Woebot was designed in such a way that it could be used as an instant messenger app for both mobile phones and desktops. CBT is delivered in a brief, immediate first-aid format via everyday conversations and mood tracking (Fitzpatrick et al., 2017).

Additionally, a benefit of utilizing CBT-based chatbots is their capability to furnish a stigma-free space that would enable the users to strive for mental health support. Reluctance to share freely and a struggle in the expression of emotions are what many individuals, especially adolescents might be evident even at times with a therapist some of the reasons might be fear of judgment or societal stigma. Chatbots offer a safe space, free of judgment, and create a sense of anonymity where users can openly express themselves, practice solutions applied using CBT techniques, and receive immediate support. In the long run, this anonymity can promote early intervention and lessen the long-term effects of untreated mental health disorders by encouraging more people to participate in mental health interventions.

Real-Time Monitoring and Feedback

Barriers to getting help in the form of therapy can now be overcome using real-time monitoring via the internet, especially in very remote areas (Ruskin et al., 2004; Sorvaniemi et al., 2005). Stronger therapeutic relationships have been found to be formed between the client and the psychologist in remote therapy. The former felt content with the reception of treatment from time to time by real-time therapists via videoconferencing (Matsumoto et al., 2018). Ecological Momentary Interventions (EMI), for example, is an AI-enhanced tool that has been developed using machine learning algorithms that can enhance their database using the data provided and give forth personalized feedback (Heron & Smyth, 2010). As a consequence, it has the capability to benefit isolated areas, individuals with low socio-economic status, and adolescents (Matsumoto et al., 2018).

Furthermore, early detection of mental health problems can be facilitated by analysis of speech patterns, and physiological responses through biofeedback tech powered by AI and, text inputs (Fitzpatrick et al., 2017). Identification of the slightest changes in mood or behaviour that depict the deteriorating condition of the user by these technologies can prompt for timely interventions. Natural language processing (NLP) and sentiment analysis, for example, allow chatbots to identify distress signs in user discussions and offer prompt coping mechanisms or refer situations to human experts when needed (Inkster et al., 2018). By utilizing these developments, real-time feedback systems not only improve the efficacy of digital mental health interventions but also establish a safety net for vulnerable people, guaranteeing that assistance is accessible prior to a crisis.

Wearables in CBT

Biosensors and smartwatches blended with bio-markers such as heart rate variability, sleep patterns, and physical activities, etc., are some wearable devices that are easy to carry and, at the same time contribute significantly to detecting the biological contributors and thus, help with designing the therapies, especially in CBT (Bakker et al., 2016). Wearable technology helps control stress, anxiety, and panic disorders by enabling biofeedback-based cognitive behavioural therapy interventions (Kang et al., 2023). For example, to monitor stress reactions and promote behavioural activation, Fitbit and Apple Watch have been incorporated into digital cognitive behavioural therapy programs (Hekler et al., 2019). Data accuracy, adherence problems, and the requirement for customized feedback systems are obstacles that stand in the way of their potential (Baumeister & Montag, 2019).

Traditional versus AI-driven CBT:

The evolution of methods for delivering Cognitive Behavioral Therapy (CBT) has transitioned from face-to-face sessions to teletherapy, and currently to AI-integrated CBT via wearables, all of which demonstrate similar levels of effectiveness, participant adherence, and long-term results. This chapter segment focuses on a comparative review of conventional versus AI-driven CBT, examining factors such as accessibility, cost-effectiveness, diagnostic precision, cognitive restructuring and behavioral activation, real-time monitoring, relapse prevention, client engagement, and psychoeducation. It consolidates recent research findings and assesses the beneficial aspects along with the challenges linked to these modalities, highlighting their complementary functions in meeting the mental health requirements of the population.

Accessibility

Another key advantage of AI-infused CBT is that it can address output: geographic and resource-based barriers. Traditional CBT is effective for face-to-face interactions, with trained therapists providing individualized counseling sessions. However, But its limitations are particularly apparent in underserved and rural areas (Dobson & Dozois, 2021). Conversely, AI-enhanced CBT is delivered via mobile apps and chatbots. For example, Touchkin's Wysa helps with anxiety, energy, focu, relaxation, loss and conflicts, to develop mental toughness and mental output: well-being through a text-based conversational interface (Inkster, 2018). Another platform,Woebot is an automated conversational tool for delivering CBT, tracking mood, and facilitate daily conversations (Fitzpatrick et al., 2017). However, it is important acknowledge that access to this technology and the virtual world may still be a barrier for certain demographics, necessitating a hybrid approach.

Cost-Effectiveness

Another area in which AI-CBT has a clear advantage is the cost of therapy. Because traditional CBT requires many therapist–client interactions, it can be very expensive, which makes it inaccessible to individuals with limited resources (Beck, 2017). AI-CBT reduces the need for therapist involvement and increases automation, thereby lowering the cost of therapy. Research has shown that AI-driven interventions can be highly scalable, providing personalized therapy at a fraction of the cost (Beg & Verma, 2024). However, there is an upfront cost to developing and maintaining AI platforms, as well as a need for rigorous evaluation to ensure effectiveness.

Diagnostic Accuracy

Rather, standard CBT relies on therapist interpretation and client self-reports, which while effective may be subject to human variability and bias (Dobson & Dozois, 2021). AI-CBT increases diagnostic accuracy through the use of emerging technologies such as natural language processing (NLP) and machine learning. AI tools such as IlluminateAI use data-driven insights to more accurately identify patterns of depression or anxiety (Agarwal, 2024; Wanderley Espinola et al., 2022). However, current AI models often struggle to address comorbid conditions and rely heavily on training data that may not adequately represent diverse populations, raising questions about generalizability.

Cognitive Restructuring and Behavioral Activation

Cognitive restructuring and behavioral activation (BA) are essential elements of cognitive behavioral therapy (CBT). In traditional CBT, exercises guided by therapists and structured activity planning foster a strong therapeutic relationship but often demand significant time and effort (Mazzucchelli et al., 2009). AI-CBT streamlines these processes by utilizing natural language processing (NLP) and adaptive algorithms to identify cognitive distortions and suggest tailored activities. For instance, AI-based applications like HealMe and MUBS have demonstrated success in providing customized interventions and incorporating gamification into the therapeutic process to boost engagement (Xiao et al., 2024; Rohani et al., 2020). Nonetheless, the absence of human empathy and nuanced understanding in AI tools may hinder their ability to effectively tackle intricate emotional issues.

Real-Time Monitoring and Relapse Prevention

One significant benefit of AI-CBT is the ability to monitor and provide feedback in real-time. Wearable devices and mobile applications can consistently track symptoms and offer immediate feedback for making treatment adjustments as needed (Nepal et al., 2024; Garcia-Ceja et al., 2015). Furthermore, predictive analytics within AI systems can detect the risk of relapse and suggest preventive measures (Yeo et al., 2024). In contrast, conventional CBT often depends on scheduled check-ins and self-reported records, which may overlook early indicators of relapse. While AI tools show potential in this area, it is important to address ethical issues concerning data privacy and the risk of excessive reliance on automated solutions.

Client Engagement and Psychoeducation

Engagement from clients plays a vital role in the success of treatments. In conventional CBT, the level of engagement largely hinges on the therapist's interpersonal skills, which can differ significantly (Dobson & Dozois, 2021). AI-CBT addresses this challenge by providing interactive platforms, gamified activities, and instant feedback, which can be especially appealing to tech-savvy younger demographics (Liu et al., 2022). Furthermore, AI technologies facilitate scalable psychoeducation through tailored content delivery, which has shown effectiveness in addressing issues like ADHD and anxiety (Selaskowski et al., 2023; Tielman et al., 2017). However, sustaining engagement over the long term poses a difficulty, as users may lose interest in digital solutions that lack human interaction.

User Experience (UX) Design

User experience (UX) design plays a critical role in the effectiveness of AI-driven CBT platforms, especially those delivered via wearables or mobile applications. A well-designed interface enhances user engagement, satisfaction, and adherence to therapeutic tasks. Studies show that digital tools with intuitive navigation, minimal cognitive load, and personalized interaction significantly improve treatment outcomes (Mohr et al., 2017). For example, emotionally supportive design and interactive feedback loops can encourage users to consistently track symptoms or complete therapy modules, reducing dropout rates in self-guided interventions.

While the technological sophistication of wearable devices and mobile apps has grown exponentially, their efficacy in supporting mental health interventions especially in formats like iCBT hinges critically on how user-friendly these systems are (Chikersal, 2020). A seamless, empathetic, and intuitive user experience (UX) plays a pivotal role in bridging the perceived distance between human and digital care, especially when therapeutic alliance is traditionally built face-to-face(Hohman et al.,2019). Thoughtfully designed interfaces that offer clear feedback, accessible language, and personalized content can enhance a user's sense of being seen and understood, directly influencing their willingness to engage and persist with treatment. UX design also impacts how easily iCBT supporters—many of whom vary in clinical experience—can navigate, interpret, and act on AI-driven insights. As seen in participatory design studies like those by Hirsch et al.,2017 the interpretability and transparency of system interfaces directly affect the trust and usability of AI tools. These findings underscore a broader need for human-centered design that supports both patients and practitioners in dynamic care environments, where emotional resonance, timely feedback, and relational depth are essential for sustaining adherence and achieving therapeutic outcomes.

DISCUSSION

The application of artificial intelligence in CBT is expanding the possibilities of affordable solutions in the field of mental health care. The existing wearable devices and systems powered by AI for real-time monitoring of a person's well-being have shown a potential for enhancing patient engagement, tracking symptoms, and preventing relapse. Study highlights that AI enabled apps and chatbots can support people with symptoms of depression and anxiety and help them engage in structured therapy and get instant responses. For instance, the AI-based mental health monitoring through wearable technology and mobile sensing has enabled the autonomous and continuous sampling of mental health outcomes such as stress, anxiety, and depression, which in turn facilitates early identification of the need for help. Nonetheless, while these technologies have been found to be useful, they have some limitations in terms of accuracy, user engagement, and ethical issues. While AI systems can easily offer a set of structured therapeutic tasks, they may lack the ability to provide the crucial human element of empathy required to deal with more delicate emotional and psychological issues. (Sinha, 2024)

Cognitive behavioral therapy (CBT) based on artificial intelligence (AI) improves availability by removing spatial and economic barriers to treatment. There are not enough licensed mental health care givers in many countries; thus, the AI-powered chatbots are useful for those who cannot afford to pay for therapy or when it is not easily available. These digital solutions enable people to get mental health support when they need it most and maybe improve the use of therapeutic strategies. However, challenges like the digital divide, unavailability of the internet, and the cost of wearable devices are barriers, especially for individuals from low-income countries. Furthermore, it is still unclear if AI-based CBT solutions can remain financially sustainable over the long term, considering that these technologies demand considerable investment for both development and maintenance (Borghare et al., 2024).

Another possible additional benefit of using AI in CBT is that it can actually track people in real life and help prevent them from worsening. The current literature also shows that wearable devices and biosensors are capable of unconsciously collecting physiological and behavioral data including heart rate variability, sleep, and physical activity, which can help in making timely interventions during the time of distress. Research has shown that machine learning models are accurate in identifying the relapse of mental illness, which in turn leads to early intervention and better long term outcomes of the treatment process. However, these advancements pose ethical concerns including the privacy of data, security of data, and autonomy of the patient. The solutions for mental health based on AI depend on the use of personal data and thus, it is crucial for the developers as well as the policy makers to address the issue of data protection and prevent its misuse (Shajari et al., 2023).

AI models in CBT must be developed with attention to diversity and inclusion to prevent algorithmic bias. If these systems are trained predominantly on data from limited populations, they may fail to generalize across cultural, gender, or socio-economic differences. For instance, emotion recognition algorithms or cognitive restructuring tools may misinterpret expressions or patterns unique to certain groups, leading to inaccurate suggestions or poor outcomes (O'Neil, 2016; Obermeyer et al., 2019). Addressing these gaps requires inclusive datasets and fairness-aware model development.

However, in the delivery of CBT, i.e., the therapeutic interventions, there is still a role for humans in the process. The variety of human disorders we see in practice often forces them to use their judgment in many of the cases, including those that are comorbid or require urgent attention. There could be differences in the training data used for AI models that may not be consistent with the cultural, contextual, and individual variations in the psychological responses of individuals. A properly designed model will integrate both AI and human oversight to create a synergistic approach between the two fields of practice to ensure that the AI-based therapies actually improve and do not replace conventional mental health care practice (Jiang et al., 2024).

The upcoming stage for AI-augmented CBT focuses on refining existing technologies to boost their accuracy, personalization, and user interaction. Advancements in natural language processing (NLP) and deep learning may result in more effective conversational AI agents that can respond with greater empathy and relevance to the concerns of users. For AI to be seamlessly incorporated into mental health care systems, it is essential to establish regulatory frameworks that address ethical and legal issues. To foster public trust and guarantee the responsible use of AI in mental health treatment, issues related to informed consent, data privacy, and accountability must be addressed. The use of wearable devices and real-time monitoring could enhance the efficacy and safety of AI-driven CBT, making it more customized, accessible, and data-driven. While AI improves operational efficiency, it cannot replace human compassion and clinical supervision. A synergistic approach that integrates AI capabilities with human expertise may provide the most effective means of leveraging the benefits of AI-based CBT while mitigating its limitations. Enhancing AI's ability to comprehend complex emotional subtleties and ensuring these technologies comply with ethical and clinical standards for safe and effective mental health support are vital research areas.

Case Study: AI-Driven CBT in Action – Adapting Woebot for Substance Use Disorders Through the W-SUDs Intervention

A recent study (Prochaska, 2021) explored the adaptation of Woebot, an AI-driven digital mental health platform rooted in cognitive behavioural therapy (CBT), for addressing substance use disorders (SUDs). This version, called W-SUDs, was evaluated for feasibility, engagement, and preliminary efficacy among American adults aged 18–65 who screened positive for substance misuse. Participants were recruited between March 27 and May 6, 2020, through online advertisements and flyers, resulting in a final sample of 101 individuals with a mean age of 36.8 years. The app demonstrated a high registration rate (78.9%), with engagement peaking in the early weeks of use. A total of 1571 mood ratings were submitted by 90 participants (89.1%), with each providing an average of 17.5 entries, while 1399 craving and 1403 pain ratings were recorded by 87 participants. Craving scores declined significantly over time, particularly from weeks 4 through 9, with reductions reaching approximately 50% of initial ratings by the final two weeks. A McNemar test further confirmed a significant shift in reported craving severity, with more participants indicating minimal cravings post-treatment ($p < .001$). In contrast, pain ratings remained stable over the nine-week period (mean 2.3/10) and showed no significant change.

Importantly, participants who completed both pre- and post-treatment assessments demonstrated significant improvements in psychological outcomes. Confidence scores increased across all eight measured domains, and reductions were observed in substance use frequency, as well as in AUDIT-C (alcohol use), DAST-10 (drug use), PHQ-8 (depression), and GAD-7 (anxiety) scores (all $p < .05$). Correlational analyses revealed that greater reductions in alcohol-related risk (AUDIT-C) were associated with fewer substance use occasions ($r = 0.48$), lower depression ($r = 0.36$) and anxiety ($r = 0.34$) scores, and higher confidence levels ($r = -0.39$). Similarly, improvements in drug-related measures (DAST-10) were linked to reductions in depressive symptoms ($r = 0.40$), though not with other measured variables. Subgroup analysis showed that participants already engaged in therapy experienced greater reductions in depression symptoms compared to those not in therapy ($p = .01$), suggesting that W-SUDs may be particularly effective when used alongside ongoing clinical care. Overall, W-SUDs proved to be a feasible, engaging, and promising digital intervention for substance use support, with clinically meaningful improvements observed in mood, cravings, and psychological functioning. These findings support the growing potential of conversational agents like Woebot as scalable and accessible complements to traditional mental health and addiction services.

Future Trends

The future of AI-CBT lies in the integration of cutting-edge technologies such as virtual reality (VR), augmented reality (AR), and multi-modal AI systems. VR exposure therapy is already being used effectively for phobias and PTSD by simulating anxiety-inducing environments in a controlled manner (Freeman et al., 2017). Additionally, multi-modal AI systems that combine text, voice, facial expression, and physiological signals can offer a more holistic understanding of mental states, potentially improving diagnostic accuracy and personalized interventions (Cummins et al., 2018). These innovations are poised to redefine how digital mental health is delivered in the coming decade.

CONCLUSION

In the end, comparing CBT with AI-CBT reveals how they can complement each other in mental health treatment. While conventional CBT is still considered the benchmark for personalized and empathetic care, particularly in intricate situations that need deep emotional exploration, AI-CBT presents an innovative and scalable alternative to close the gaps in accessibility, affordability, and real-time monitoring. By merging AI-assisted tools with traditional methods, a combined approach can utilize the advantages of both strategies to deliver comprehensive mental health services.

REFERENCES

Agarwal, S., Jalan, M., Wilcox, H. C., Sharma, R., Hill, R., Pantalone, E., ... & Robinson, K. A. (2022). Evaluation of mental health mobile applications. PMID: 35675475

Aghdam, L. & Science Arena Publications. (2017). Effectiveness of cognitive-behavioral therapy on the reduction of marital disaffection of women filing for divorce. *International Journal of Philosophy and Social-Psychological Sciences, 3–6*, 1–7. https://sciarena.com/storage/models/article/jpN8UTmaxPHN2fkw0xyso AlVAXwJnoJHOJKsGOnRIy4oGoBVLmtQhwawN5Pu/effectiveness-of-cognitive-behavioral-therapy-on-the-reduction-of-marital-disaffection-of-women-fi.pdf

Andersson, G., Carlbring, P., Titov, N., & Lindefors, N. (2019). Internet Interventions for Adults with Anxiety and Mood Disorders: A Narrative Umbrella Review of Recent Meta-Analyses. *Canadian Journal of Psychiatry, 64*(7), 465–470. DOI: 10.1177/0706743719839381 PMID: 31096757

Assuncao, G., Patrao, B., Castelo-Branco, M., & Menezes, P. (2022). An overview of emotion in artificial intelligence. *IEEE Transactions on Artificial Intelligence, 3*(6), 867–886. DOI: 10.1109/TAI.2022.3159614

Baker, R., Owens, M., Thomas, S., Whittlesea, A., Abbey, G., Gower, P., Tosunlar, L., Corrigan, E., & Thomas, P. W. (2011). Does CBT facilitate emotional processing? *Behavioural and Cognitive Psychotherapy, 40*(1), 19–37. DOI: 10.1017/S1352465810000895 PMID: 21392415

Beck, A. T. (1970). Cognitive therapy: Nature and relation to behavior therapy. *Behavior Therapy, 1*(2), 184–200. DOI: 10.1016/S0005-7894(70)80030-2 PMID: 27993332

Beck, N. A. (2017). *Mobile apps for cognitive restructuring: A review and comparative analysis*. Pepperdine University.

Beg, M. J., & Verma, M. K. (2024). Exploring the Potential and Challenges of Digital and AI-Driven Psychotherapy for ADHD, OCD, schizophrenia, and Substance Use Disorders: A Comprehensive Narrative review. *Indian Journal of Psychological Medicine*, 02537176241300569. Advance online publication. DOI: 10.1177/02537176241300569 PMID: 39677516

Borghare, P. T., Methwani, D. A., & Pathade, A. G. (2024). A comprehensive review on harnessing wearable technology for enhanced depression treatment. *Cureus*. Advance online publication. DOI: 10.7759/cureus.66173 PMID: 39233951

Broderick, J. E., Keefe, F. J., Schneider, S., Junghaenel, D. U., Bruckenthal, P., Schwartz, J. E., Kaell, A. T., Caldwell, D. S., McKee, D., & Gould, E. (2016). Cognitive behavioral therapy for chronic pain is effective, but for whom? *Pain, 157*(9), 2115–2123. DOI: 10.1097/j.pain.0000000000000626 PMID: 27227692

Byrne, S. M., Fursland, A., Allen, K. L., & Watson, H. (2011). The effectiveness of enhanced cognitive behavioural therapy for eating disorders: An open trial. *Behaviour Research and Therapy, 49*(4), 219–226. DOI: 10.1016/j.brat.2011.01.006 PMID: 21345418

Chand, S., Ravi, C., Chakkamparambil, B., Prasad, A., & Vora, A. (2018). CBT for depression: What the evidence says. *Current Psychiatry, 17*(9), 14–23.

Chikersal, P., Belgrave, D., Doherty, G., Enrique, A., Palacios, J. E., Richards, D., & Thieme, A. (2020, April). Understanding client support strategies to improve clinical outcomes in an online mental health intervention. In *Proceedings of the 2020 CHI conference on human factors in computing systems* https://doi.org/DOI: 10.1145/3313831.3376341

D'Alfonso, S. (2020). AI in mental health. *Current Opinion in Psychology, 36*, 112–117. DOI: 10.1016/j.copsyc.2020.04.005 PMID: 32604065

Dobson, K. S. (2016). New frontiers in Cognitive-Behavioral Therapy for Depression. *International Journal of Cognitive Therapy, 9*(2), 107–123. DOI: 10.1521/ijct.2016.9.2.107

Dobson, K. S., & Dozois, D. J. (Eds.). (2021). *Handbook of cognitive-behavioral therapies*. Guilford Publications.

Eysenck, H. J. (1952). The effects of psychotherapy: An evaluation. *Journal of Consulting Psychology, 16*(5), 319–324. DOI: 10.1037/h0063633 PMID: 13000035

Fitzpatrick, K. K., Darcy, A., & Vierhile, M. (2017). Delivering cognitive behavior therapy to young adults with symptoms of depression and anxiety using a fully automated conversational agent (Woebot): A randomized controlled trial. *JMIR Mental Health, 4*(2), e7785. DOI: 10.2196/mental.7785 PMID: 28588005

Freeman, C. (1990). Cognitive Behavioural Therapy for Psychiatric Problems: A Practical Guide. Edited by Keith Hawton, Paul M. Salkovskis, Joan Kirk and David M. Clark. Oxford: Oxford University Press. 1989. 456 pp. £40 (hb), £15 (pb). *British Journal of Psychiatry, 157*(3), 468–468. DOI: 10.1192/S000712500006356X

Freeman, D., Reeve, S., Robinson, A., Ehlers, A., Clark, D., Spanlang, B., & Slater, M. (2017). Virtual reality in the treatment of persecutory delusions: Randomised controlled experimental study testing how to reduce delusional conviction. *The British Journal of Psychiatry, 210*(6), 429–436. DOI: 10.1192/bjp.bp.116.193564 PMID: 28302702

Goodwin, T. R., Maldonado, R., & Harabagiu, S. M. (2017). Automatic recognition of symptom severity from psychiatric evaluation records. *Journal of Biomedical Informatics, 75*, S71–S84. DOI: 10.1016/j.jbi.2017.05.020 PMID: 28576748

Gupta, M., Malik, T., & Sinha, C. (2022). Delivery of a mental health intervention for chronic pain through an artificial intelligence–enabled app (Wysa): Protocol for a prospective pilot study. *JMIR Research Protocols, 11*(3), e36910. DOI: 10.2196/36910 PMID: 35314423

Heron, K. E., & Smyth, J. M. (2010). Ecological momentary interventions: Incorporating mobile technology into psychosocial and health behaviour treatments. *British Journal of Health Psychology, 15*(1), 1–39. DOI: 10.1348/135910709X466063 PMID: 19646331

Hirsch, T., Merced, K., Narayanan, S., Imel, Z. E., & Atkins, D. C. (2017, June). Designing contestability: Interaction design, machine learning, and mental health. In *Proceedings of the 2017 Conference on Designing Interactive Systems* (pp. 95-99).https://doi.org/DOI: 10.1145/3064663.3064703

Hohman, F., Head, A., Caruana, R., DeLine, R., & Drucker, S. M. (2019, May). Gamut: A design probe to understand how data scientists understand machine learning models. In *Proceedings of the 2019 CHI conference on human factors in computing systems* (pp. 1-13). https://doi.org/DOI: 10.1145/3290605.3300809

Inkster, B., Sarda, S., & Subramanian, V. (2018). An empathy-driven, conversational artificial intelligence agent (Wysa) for digital mental well-being: Real-world data evaluation mixed-methods study. *JMIR mHealth and uHealth, 6*(11), e12106. DOI: 10.2196/12106 PMID: 30470676

Jauhar, S., McKenna, P. J., Radua, J., Fung, E., Salvador, R., & Laws, K. R. (2014). Cognitive–behavioural therapy for the symptoms of schizophrenia: Systematic review and meta-analysis with examination of potential bias. *The British Journal of Psychiatry, 204*(1), 20–29. DOI: 10.1192/bjp.bp.112.116285 PMID: 24385461

Jiang, M., Zhao, Q., Li, J., Wang, F., He, T., Cheng, X., . . . Fu, G. (2024). A Generic Review of Integrating Artificial Intelligence in Cognitive Behavioral Therapy. *arXiv preprint arXiv:2407.19422*. https://doi.org//arXiv.2407.19422DOI: 10.48550

Johnson, K. B., Wei, W. Q., Weeraratne, D., Frisse, M. E., Misulis, K., Rhee, K., Zhao, J., & Snowdon, J. L. (2021). Precision medicine, AI, and the future of personalized health care. *Clinical and Translational Science, 14*(1), 86–93. DOI: 10.1111/cts.12884 PMID: 32961010

Kabat-Zinn, J. (2003). Mindfulness-based interventions in context: Past, present, and future. *Clinical Psychology : a Publication of the Division of Clinical Psychology of the American Psychological Association, 10*(2), 144–156. DOI: 10.1093/clipsy.bpg016

Kendell, R. E. (1980). Diagnostic and statistical manual of mental disorders. *The American Journal of Psychiatry, 137*(12), 1630–1631. DOI: 10.1176/ajp.137.12.1630

Kodal, A., Fjermestad, K., Bjelland, I., Gjestad, R., Öst, L. G., Bjaastad, J. F., Haugland, B. S. M., Havik, O. E., Heiervang, E., & Wergeland, G. J. (2018). Long-term effectiveness of cognitive behavioral therapy for youth with anxiety disorders. *Journal of Anxiety Disorders, 53*, 58–67. DOI: 10.1016/j.janxdis.2017.11.003 PMID: 29195188

König, P., Zwiky, E., Küttner, A., Uhlig, M., & Redlich, R. (2024). Brain functional effects of cognitive behavioral therapy for depression: A systematic review of task-based fMRI studies. *Journal of Affective Disorders*. Advance online publication. DOI: 10.1016/j.jad.2024.09.084 PMID: 39299583

Koutsouleris, N., Hauser, T. U., Skvortsova, V., & De Choudhury, M. (2022). From promise to practice: Towards the realisation of AI-informed mental health care. *The Lancet. Digital Health, 4*(11), e829–e840. DOI: 10.1016/S2589-7500(22)00153-4 PMID: 36229346

Lazarus, A. A. (1977). Has behavior therapy outlived its usefulness? *The American Psychologist, 32*(7), 550–554. DOI: 10.1037/0003-066X.32.7.550 PMID: 889160

Matsumoto, K., Sutoh, C., Asano, K., Seki, Y., Urao, Y., Yokoo, M., Takanashi, R., Yoshida, T., Tanaka, M., Noguchi, R., Nagata, S., Oshiro, K., Numata, N., Hirose, M., Yoshimura, K., Nagai, K., Sato, Y., Kishimoto, T., Nakagawa, A., & Shimizu, E. (2018). Internet-based cognitive behavioral therapy with real-time therapist support via videoconference for patients with obsessive-compulsive disorder, panic disorder, and social anxiety disorder: Pilot single-arm trial. *Journal of Medical Internet Research, 20*(12), e12091. DOI: 10.2196/12091 PMID: 30559094

Matusiewicz, A. K., Hopwood, C. J., Banducci, A. N., & Lejuez, C. W. (2010). The effectiveness of cognitive behavioral therapy for personality disorders. *Psychiatria Clinica, 33*(3), 657–685. DOI: 10.1016/j.psc.2010.04.007 PMID: 20599139

Mitchell, M. D., Gehrman, P., Perlis, M., & Umscheid, C. A. (2012). Comparative effectiveness of cognitive behavioral therapy for insomnia: A systematic review. *BMC Family Practice*, *13*(1), 1–11. DOI: 10.1186/1471-2296-13-40 PMID: 22631616

Mohr, D. C., Tomasino, K. N., Lattie, E. G., Palac, H. L., Kwasny, M. J., Weingardt, K., Karr, C. J., Kaiser, S. M., Rossom, R. C., Bardsley, L. R., Caccamo, L., Stiles-Shields, C., & Schueller, S. M. (2017). IntelliCare: An eclectic, skills-based app suite for the treatment of depression and anxiety. *Journal of Medical Internet Research*, *19*(1), e10. DOI: 10.2196/jmir.6645 PMID: 28057609

Na, H. (2024) Cbt-llm: A chinese large language model for cognitive behavioral therapybased mental health question answering. arXiv preprint arXiv:240316008 https://doi.org/DOI: 10.48550

Niles, A. N., Axelsson, E., Andersson, E., Hedman-Lagerlöf, E., Carlbring, P., Andersson, G., ... & Ljotsson, B. (2021). Internet-based cognitive behavior therapy for depression, social anxiety disorder, and panic disorder: Effectiveness and predictors of response in a teaching clinic. *Behaviour Research and Therapy, 136*, 103767. https://doi.org/ rights and contentDOI: 10.1016/j.brat.2020.103767Get

O'Donnell, L. A., Brydon, D. M., & González-Prendes, A. (2022). Technological Advances in Cognitive-Behavioral Therapy and Clinical Practice: Challenges in an Evolving Field. *American Journal of Psychotherapy*, *75*(4), 186–190. DOI: 10.1176/appi.psychotherapy.20220003 PMID: 36052453

O'Neil, C. (2016). *Weapons of Math Destruction: How Big Data Increases Inequality and Threatens Democracy*. Crown Publishing Group.

Obermeyer, Z., Powers, B., Vogeli, C., & Mullainathan, S. (2019). Dissecting racial bias in an algorithm used to manage the health of populations. *Science*, *366*(6464), 447–453. DOI: 10.1126/science.aax2342 PMID: 31649194

Olatunji, B. O., Davis, M. L., Powers, M. B., & Smits, J. A. (2013). Cognitive-behavioral therapy for obsessive-compulsive disorder: A meta-analysis of treatment outcome and moderators. *Journal of Psychiatric Research*, *47*(1), 33–41. DOI: 10.1016/j.jpsychires.2012.08.020 PMID: 22999486

Prochaska, J. J., Vogel, E. A., Chieng, A., Kendra, M., Baiocchi, M., Pajarito, S., & Robinson, A. (2021). A therapeutic relational agent for reducing problematic substance use (Woebot): Development and usability study. *Journal of Medical Internet Research*, *23*(3), e24850. DOI: 10.2196/24850 PMID: 33755028

Proudfoot, J., Goldberg, D., Mann, A., Everitt, B., Marks, I., & Gray, J. A. (2003). Computerized, interactive, multimedia cognitive-behavioural program for anxiety and depression in general practice. *Psychological Medicine*, *33*(2), 217–227. DOI: 10.1017/S0033291702007225 PMID: 12622301

Rizzo, A., & Koenig, S. T. (2017). Is clinical virtual reality ready for primetime? *Neuropsychology*, *31*(8), 877–899. DOI: 10.1037/neu0000405 PMID: 29376669

Rodgers, M. M., Pai, V. M., & Conroy, R. S. (2014). Recent advances in wearable sensors for health monitoring. *IEEE Sensors Journal*, *15*(6), 3119–3126. DOI: 10.1109/JSEN.2014.2357257

Ruskin, P. E., Silver-Aylaian, M., Kling, M. A., Reed, S. A., Bradham, D. D., Hebel, J. R., Barrett, D., Knowles, F.III, & Hauser, P. (2004). Treatment outcomes in Depression: Comparison of remote treatment through telepsychiatry to In-Person Treatment. *The American Journal of Psychiatry*, *161*(8), 1471–1476. DOI: 10.1176/appi.ajp.161.8.1471 PMID: 15285975

Shajari, S., Kuruvinashetti, K., Komeili, A., & Sundararaj, U. (2023). The emergence of AI-based wearable sensors for digital health technology: A review. *Sensors (Basel)*, *23*(23), 9498. DOI: 10.3390/s23239498 PMID: 38067871

Sinha, R. (2024). The role and impact of new technologies on healthcare systems. *Discover Health Systems*, *3*(1), 96. Advance online publication. DOI: 10.1007/s44250-024-00163-w

Sorvaniemi, M., Ojanen, E., & Santamäki, O. (2005). Telepsychiatry in emergency consultations: A follow-up study of sixty patients. *Telemedicine Journal and e-Health*, *11*(4), 439–441. DOI: 10.1089/tmj.2005.11.439 PMID: 16149889

Vilaza, G. N., & McCashin, D. (2021). Is the automation of digital mental health ethical? Applying an ethical framework to chatbots for cognitive behaviour therapy. *Frontiers in Digital Health*, *3*, 689736. DOI: 10.3389/fdgth.2021.689736 PMID: 34713163

Wanderley Espinola, C., Gomes, J. C., M^onica Silva Pereira, J., & dos Santos, W. P. (2022). Detection of major depressive disorder, bipolar disorder, schizophrenia and generalized anxiety disorder using vocal acoustic analysis and machine learning: An exploratory study. *Research on Biomedical Engineering*, *38*(3), 813–829. DOI: 10.1007/s42600-022-00222-2

Yalom, I. D. (2019). *When Nietzsche wept: A novel of obsession*. Basic Books.

Yang, X., Li, Z., & Sun, J. (2020). Effects of cognitive behavioral therapy–based intervention on improving glycaemic, psychological, and physiological outcomes in adult patients with diabetes mellitus: A meta-analysis of randomized controlled trials. *Frontiers in Psychiatry, 11*, 711. DOI: 10.3389/fpsyt.2020.00711 PMID: 32848906

Ye, B. Y., Jiang, Z. Y., Li, X., Cao, B., Cao, L. P., Lin, Y., Xu, G.-Y., & Miao, G. D. (2016). Effectiveness of cognitive behavioral therapy in treating bipolar disorder: A n updated meta-analysis with randomized controlled trials. *Psychiatry and Clinical Neurosciences, 70*(8), 351–361. DOI: 10.1111/pcn.12399 PMID: 27177717

Young, Z., Moghaddam, N., & Tickle, A. (2020). The efficacy of cognitive behavioral therapy for adults with ADHD: A systematic review and meta-analysis of randomized controlled trials. *Journal of Attention Disorders, 24*(6), 875–888. DOI: 10.1177/1087054716664413 PMID: 27554190

Zamboni, L., Centoni, F., Fusina, F., Mantovani, E., Rubino, F., Lugoboni, F., & Federico, A. (2021). The effectiveness of cognitive behavioral therapy techniques for the treatment of substance use disorders: A narrative review of evidence. *The Journal of Nervous and Mental Disease, 209*(11), 835–845. DOI: 10.1097/NMD.0000000000001381 PMID: 34698698

Chapter 3
AI-Powered Wearables for Mental Health Monitoring

Moumita Mukherjee
https://orcid.org/0000-0003-2925-471X
Adamas University, Kolkata, India

ABSTRACT

Artificial intelligence (AI) would not only become indispensable in the future, but it has already become imperative for major walks of life, like education, healthcare, human resource management, etc. AI tools are utilised for mental health monitoring and assessing wellbeing in terms of stress, anxiety, depression, sleep, etc., AI and Internet of Things (IoT)-based wearable devices are used as cognitive support system for various mental health conditions, including Parkinson's, Autism, Depression, etc. However, as manufacturers compete to build monitoring devices, this chapter explores their accessibility and efficiency in practicality. In mental healthcare, three crucial areas are diagnosis, therapeutic intervention, and rehabilitation. Monitoring is valuable for diagnosis and prognosis evaluation for further treatment plans. This chapter discusses the implications of the confluence of AI technology in mental health monitoring and the subsequent diagnosis and treatment. Ethical concerns regarding AI algorithms in use for monitoring, diagnosis and intervention are also discussed.

INTRODUCTION

For individuals with chronic health conditions, be it physical or mental, continuous monitoring is often advised. However, from human frailty of memory lapses in the same, to limited support be it affordability of caregivers or inaccessible family

DOI: 10.4018/979-8-3373-0467-0.ch003

members, the task of monitoring could become taxing. Furthermore, the challenges of accessing traditional healthcare for the monitoring of treatment interventions during the Covid-19 period across the world perhaps highlighted the issue even more.

Limited healthcare access in rural areas globally, be it the U.S. or Africa stems from weak infrastructure and scarce resources. Digital tools—mobile apps, telemedicine, and e-learning—can enhance health literacy and outcomes by enabling remote consultations, self-monitoring, and education. In rural America, they bridge care gaps; in Africa, they provide vital health information and doctor connections. However, adoption of digital technology faces barriers like poor internet, low digital literacy, and cultural differences. Overcoming these requires better connectivity, training, and culturally tailored solutions. By leveraging digital health, underserved communities can gain improved healthcare access and decision-making power, leading to better well-being (Maha et al., 2024).

Progress in artificial intelligence (AI) offers groundbreaking possibilities for transforming healthcare, improving its efficiency, accessibility, and financial viability. Through strategic policy support for AI adoption, we can promote fairness, elevate care standards, and ensure that innovations in technology, therapies, and pharmaceuticals deliver widespread societal benefits. AI can optimize the distribution of medical resources. Predictive analytics can anticipate patient inflow and improve the management of hospital beds, personnel, and medical devices. This helps direct resources to areas of greatest need, minimizing inefficiencies and raising care standards. Additionally, AI can address critical healthcare issues, including escalating expenses, operational inefficiencies, and the growing need for superior care. By automating routine administrative duties—such as appointment booking, invoicing, and maintaining digital health records—AI can cut costs and enhance workflow efficiency. This allows medical staff to dedicate more time to patient interaction.

In medical diagnostics, AI improves precision and facilitates earlier disease detection, often resulting in less aggressive and more affordable treatment pathways. Furthermore, AI-powered customized treatment strategies can enhance conventional methods by delivering more precise and efficient care, boosting patient recovery rates while alleviating financial strain on healthcare infrastructures.

Löchner et al (2025) present a comprehensive report of research evidence highlighting the advantages and challenges in applications of digital technology in the mental health care sector. Besides critical evaluation of the evidence-based research findings on AI driven technology already available, Löchner et al (2025) also speculate on future potential applications of AI in mental healthcare, concluding with a recommended model of TEQUILA, each letter representing a key feature of the model, which highlights key factors and obstacles in shaping a positive and ethical future. The model proposes a conscientious path forward in digital mental health which depends on cultivating *Trust* through strong safeguards for personal information, clear AI processes, and transparency, alongside the need for reliable evidence and precise and well-defined rules of law to uphold Quality. Attention to effective interface design that meets varied requirements, as well as ethical consistency with users' Interests, will be vital. At the same time, establishing clear Liability measures and consistent Accreditation will help ensure responsibility as the field progresses.

CONCEPTUAL FRAMEWORK

The use of AI enabled devices is increasingly becoming feasible for not only monitoring mental and physical health parameters, but also using machine learning (and deep learning) protocols to better diagnose and predict prognosis, as well as use AI to better design interventions for enhanced health outcomes. Larger hospitals and other medical establishments are increasingly using machine learning for big data analysis. In such set ups, huge number of patients may be subjected to digital monitoring with variety of AI enabled devices, some even generic such as strategically placed cameras to monitor patients at risk for falling from their hospital beds when attempting to stand without help, to specific devices placed on individual patient's bodies, to generate tremendous volumes of data, which are thereafter analysed via machine learning to generate patterns of predictable behaviour, on some occasions even particular forecasts of potentially preventive measures for vulnerable patients (Bajwa et al, 2021; Maleki Varnosfaderani & Forouzanfar, 2024).

There is a revolutionary change thus happening in the healthcare sector spanning both medical and mental health and allied domains, with AI presenting ground-breaking solutions for diagnosis, prediction, management and treatment of mental health conditions in larger hospital and clinical settings, though caution is advisable to incorporate the final decision making power remaining in human brains, albeit informed well by information, facilitated by AI and experience gathered by generations of medical and mental health professionals.

In the light of such work, the larger conceptual framework must incorporate the integration of AI in different areas, not merely limited to monitoring mental health parameters, with the ultimate goal of better mental health care outcomes for human beings. A schematic presentation of the framework is given in Figure 1. However, the focus of the present chapter remains on mental health monitoring devices.

Figure 1. Conceptual Framework Connecting AI Integration in Aspects of Mental Health Care

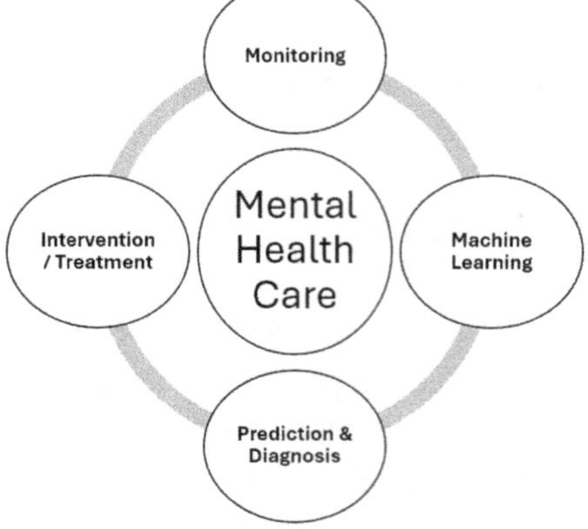

In the increasing trend of smartening up lives and livelihoods, the use of AI (artificial intelligence) enabled devices, be it a smart watch, or portable heart monitoring device, has seen an unforeseen surge. The product manufacturers medical technology industry are routinely designing devices to monitor health parameters, some even incorporating AI technology to create alarms for action which facilitate self-care for persons with varied health conditions that require regular tracking.

The World Health Organization (WHO, 2024) recommends use of AI technology, not limited to smartphones and wearables, to especially expand the accessibility and affordability of health care to the margins of the populations, consisting of individuals and groups, who are often located beyond the scope of traditional medical care. The strategic salience of ethical governance and defined guidelines for validation and regulation backed by the resources of WHO paint a hopeful picture of AI enabled digital technology making impactful change in the overall healthcare sector (WHO, 2024).

The present chapter presents a scoping review of the research literature available with respect to mental health monitoring devices powered by artificial intelligence, alongside note of the emerging and growing market in the domain. The chapter is divided into two broad segments comprising an overview of the review of the research papers and chapters published, and a segment focusing on applications and market implications.

RESEARCH REVIEW

The primary focus of the present work is to ascertain the extent of research already available on wearable devices, with particular emphasis on research generated and published in the last five years, 2019-2024, and further in 2025. Search of published research in the given timeline particularly focused on mental health monitoring with the help of AI-enabled devices was conducted with multiple databases, including Google Scholar and EBSCO Connection, and shortlisted based on criterion of monitoring using known psychophysiological parameters. The major focus of the published research was found to be anxiety and depression, recognized as two of the most common mental health conditions.

Characteristics of Included Studies

Majority of the studies included in the review shared one major characteristic – that of the patient population, though variation of different groups of patient population were considered for greater generalizability and universality of the wide scope of applications of AI enabled devices. Demographic data also varied across different studies owing to inclusion of studies across different geographical locations of the world, spanning different countries and ethnicities, as well as age groups. The next segment begins with one of the most prevalent areas of mental health problems – that of depression and anxiety.

The World Health Organization (WHO, 2023a) estimates depression prevalence rate at about 5% for adults globally whereas marginally less for India at approximately 4.5% (Arvind et al., 2019; WHO, 2023a). The prevalence of anxiety across the world is around 4% (WHO, 2023b). However, in India prevalence of anxiety appears to have risen especially among the younger population since Covid-19 times, as reported in multiple surveys, including a report by the Lancet (Alonso et al., 2018; India State-Level Disease Burden Initiative Mental Disorders Collaborators, 2020; Manjunatha, et al., 2022; Santomauro, 2021). Therefore, it stands to reason that the majority of the research published has explored these conditions. However, additional published research was found which had explored cognitive impairments especially in elderly population, persons with developmental disabilities, such as intellectual disability, Autism Spectrum Disorders (ASD), as well as more recent work in an area of notable clinical interest focused on Bipolar disorders, etc.

Summaries of a few select research from the most recent five-year span are presented in the next segment to give a coherent note of the investigations done to observe impact and effectiveness of use of mental health monitoring devices, powered by artificial intelligence.

Table 1 provides a brief overview of majority of the articles reviewed with further elaborations following the same.

Table 1. Summary Table of Review of the Research Literature

Citation	Research Objectives	Key Findings
Sequeira et al. (2020)	To review evidence on identifying and predicting mood disorders in adolescents and children using wearable and mobile devices.	Found wearable devices useful for tracking depression and predicting its progression; smartphone apps dominated research focus.
McGorry (2020)	To discuss the shift in healthcare approaches during COVID-19 in Australia, especially digital services.	Patients preferred digital/telehealth solutions; regional setups are valuable when supported by primary care.
Mirmomeni et al. (2021)	To analyze the role and development of wearable devices in clinical and non-clinical settings.	AI in wearables shifting from monitoring to predictive use; emphasized ethical, privacy, and regulatory concerns.
Jansen et al. (2022)	To assess AI-based wearable sensors for aging and physical health monitoring.	AI wearables enhance elderly care; real-time analysis possible; future looks promising with interdisciplinary advances.
Herath et al. (2024)	To examine vision-based AI technologies for assessing elderly mobility.	Vision-powered devices can aid in screening age-related conditions like Parkinson's and Sarcopenia.
Abd-alrazaq et al. (2023a)	To review wearable AI devices diagnosing anxiety and depression.	Smartwatches detect symptoms via activity, sleep, and heart rate data; promising but not yet clinically independent.
Abd-alrazaq et al. (2023b)	(Follow-up from previous)	Stressed combined use with traditional tools for accurate diagnosis.
Szabo et al. (2023)	To evaluate assistive tech's impact on people with disabilities.	Potential noted, but limited by cost, transparency, and specificity of conditions.
Almufareh et al. (2023)	To explore AI's role in assisting persons with disabilities.	Live captioning, sign language translation, and brain-computer interfaces offer inclusion and autonomy.
Sokołowska et al. (2025)	To analyze digital tools for diagnosing/treating ASD.	Virtual environments offer new possibilities in ASD care; compensates for real-life social challenges.
Wankhede et al. (2024)	To examine AI in diagnosis/intervention for ASD.	AI enables accurate detection, VR-based training, robot therapies; future potential promising despite data privacy issues.
Wang et al. (2024)	To assess AI-powered IoT wearables for depression detection.	AI improved detection accuracy by 10% over traditional methods, especially through sleep data analysis.
Zafar et al. (2024)	To highlight AI's contribution to mood-based mental health management	AI personalizes care plans, improves diagnostic and treatment efficiency, especially in psychiatric settings.

continued on following page

Table 1. Continued

Citation	Research Objectives	Key Findings
Parekh (2023)	To explore AI in management of Bipolar Disorder (BD).	AI supports personalized interventions based on lifestyle, genetics, and medical data.
Milic et al. (2025)	To review AI's effectiveness in BD care using PICO methodology.	AI improves prediction and care quality; promising for early detection and intervention, but still developing.

Sequeira et al (2020) conducted a scoping review of published evidence based research involving identification and prediction of mood disorders in adolescents and children, using wearable monitoring devices and mobiles, reported or hosted on five research databases. Sequeira et al (2020) analysed evidence from 30 empirical articles spanning twelve years from 2008-2019, shortlisted from a preliminary yield of over 800 publications, to reveal more use of smartphone applications recording a vast array of data ranging from monitoring and observations of physical and mental health indicators of mood problems, along with considerations of feasibility. Overvall, their review indicated the utility of monitoring devices for detecting and tracing the trajectory of depression, with implications for prediction.

McGorry (2020) in an editorial on the integrative approach of medical and health care sectors in Australia in the Covid-19 pandemic times, noted that telehealth and digital services were backed by evidence and received clear preference from patients and their families, compared to traditional emergency and medical establishments, to the extent feasible. McGorry (2020) also recorded the value of regional community linked health care set-ups provided they were adequately supported by primary medical set-ups, should that be needed.

Mirmomeni et al (2021) noted the widely increasing applications of wearable digital machines in clinical settings ranging from diabetes to Parkinson's, to non-clinical set-ups of athletes engaged in sporting activities. They also recorded the changing trajectory of wearable devices from mere monitoring to identifying potential triggers for seizures or fits, based on predictive algorithms powered by AI. Advanced monitoring devices could also offer feedback in real-time to sportspersons on risks of injury and performance. Mirmomeni et al (2021) further commented on the challenges in adopting AI enabled technology, especially in terms of protection and transparency of recorded data, as well as the necessity of regulations for the same. They also emphasized that all stakeholders from medical practitioners to regulatory bodies to biomedical technology experts, pharma industry and of course patients, must find common ground of communication and cooperate to develop frameworks for most judicious use and efficient outcomes.

Jansen et al., (2022) summarised several research published in a special issue of Sensors, covering the potential of wearable sensors based on artificial intelligence technology for age -related physical ability and activity parameter analysis. While challenges remain in checking these technologies and optimizing data analysis, advancements of biotechnology, artificial intelligence and materials science can overcome this problem. Wearable sensors can start the artificial intelligence algorithm in real time and provide an analysis idea. Wearable sensors are integrated into various devices to monitor body movement, heart rate and blood pressure. Such technology are being integrated into devices to improve medical services and allow people to function more independently. Jansen et al., (2022) concluded that AI enabled wearable devices are emerging as most promising tools to support healthy aging in the present, and likely to result in notable progress and health care management for the elderly in future.

Herath et al (2024) elaborate the uses and challenges of vision-oriented technology in assessment and facilitation of elderly individuals for functional mobility. Their book particularly focused on screening of functional impairments for age related diseases such as Sarcopenia and Parkinson's, and extrapolates on how devices powered by AI may assist in overcoming such challenges.

Abd-alrazaq et al (2023a) reviewed research articles exploring unique properties of wearable devices powered by AI to diagnose anxiety and depression. Their review largely revealed noticeable use of wrist worn smart devices for detection of symptoms of anxiety and depression in the 18-65 years age range. Frequency of use of three types of psychophysiological data were noted, viz, physical activity, sleep and heart rate. The review identified the promising potential of AI enabled devices in wearable form facilitating preliminary screening and identification of anxiety and depression, among the common mental health conditions (Abd-alrazaq et al., 2023a). Wearable devices powered by AI promises outcomes of detection and prediction of depression; however, the technology is still nascent, and not ready for independent use in clinical practice. Wearable AI needs to be used in concurrence with other tools and techniques for accuracy in diagnosis and prediction of depression (Abd-alrazaq et al., 2023b).

Szabo et al., (2023) reviewed 23 studies to conclude that while tech-based applications developed, offer hope, their prospects remain limited owing to high cost, partial transparency, and dependence on particular disabilities. Szabo et al., (2023) discussed potential implications of such assistive AI enabled technologies in assisting individuals with disabilities manage their unique tribulations.

Almufareh et al., (2023) discussed the promise of AI powered technology to enhance quality of life for persons with disabilities, with real-time assistance provided to them in a variety of forms such as live captioning, translation of languages including sign language, and more advanced brain computer interfaces to facilitate their empowerment via inclusion and independent functioning.

Sokołowska et al., (2025) discussed the use of modern technology for the diagnosis and treatment of developmental disorders described by autism spectrum disorders (ASDs). The complex characteristics of ASD are ideal for explaining the advantages and disadvantages of the digital world. The possibilities of the virtual environment (VE) are not yet known, VE adds to the traditional approach by providing expanded diagnosis and treatment options for ASD. Unlike real-life obstacles related to social problems and overwhelming environments, these new technologies provide unique compensation for deficits related to ASD. Sokołowska et al., (2025) concluded that incorporation of digital technology would facilitate catering to the specific needs of persons with ASD.

Wankhede et al., (2024) reviewed research on uses of artificial intelligence (AI) based technology in diagnosis and intervention for ASD. AI, via analysis of multitude of sources of data ranging from patterns of behaviour and electronic health records, to neuroimaging, and genetics, facilitates expeditious detection and individualised assessment of ASD. Machine learning algorithms produce an acceptable degree of accuracy in differentiating ASD from other developmental disorders as well as the typically developing brain, laying the foundations for prompt intervention planning and execution. AI-powered intervention programmes consisting of enhanced systems of communication, training based in virtual reality, and robot-facilitated therapies provide the opportunity of improvement in communication skills and social interactions for persons with ASD. Wankhede et al., (2024) concluded that despite concerns for privacy of data, AI-enabled assessment approaches and intervention techniques offer a promising future for individuals with ASD, with greater precision, and more individualised care. Done properly, mental health professionals can utilize AI to improve the quality of delivery of care for persons with ASD.

Wang et al (2024) researched wearable Internet of Things powered by AI for detection of depression, and presented a comparative picture of AI enabled against traditional measures. Depression may be reflected in a multitude of physiological characteristics such as gloomy countenance, abnormal bearing, variations in electrocardiogram (ECG) signals, and irregular sleep patterns, to list a few. With AI enabled monitoring of psychophysiological data of sleep, Wang et al (2024) showed 10% improvement in baselines. Their results have implications for potential applications of wearable AI devices for monitoring and detection of mental conditions.

Artificial intelligence (AI) enables consistent data-driven health-care, supporting medical and psychiatric practitioners to better envisage changes in the mood, customise intervention plans and execute timely treatment regimen for the patients (Zafar et al, 2024). Use of AI also facilitates the design of individualised intervention programmes based on requisite analysis of hereditary, medical and past lifestyle, with a pragmatic approach for more improved management of BD (Parekh, 2023). Milic et al (2025) reviewed recent studies on AI applications in management of bipolar disorders (BD), focusing on mood tracking, episode prediction, and personalized treatment. Their selected research met PICO criteria, exclusively concerning BD patient population, who received customized treatment based on AI-driven mood monitoring, including conventional therapies where applicable, with outcome facets of prediction accuracy, treatment efficacy, and care quality. Milic et al (2025) found that AI tools show promise in enabling proactive, data-driven care through wearable, smartphone, and social media analytics, enhancing early detection and dynamic interventions. Their conclusion was that despite being nascent, AI could revolutionize BD management, though further development is needed to maximize its clinical impact.

TECHNOLOGY IN APPLICATION

The technology at work for mental health care monitoring typically involves machine learning and deep learning protocols, as delineated in the few published research summarised as follows. Both have indicated fair amount of success, though other considerations such as cost, feasibility, and authenticity are yet to be ascertained at a large scale, for mass uses.

Heart rate variability (HRV), electroencephalogram (EEG), galvanic skin response (GSR), skin temperature (ST), blood pressure (BP), sleep, salivary cortisol, and blood oxygen saturation (SpO2) are currently the most often employed signals in mental health detection. In order to precisely assess a person's mental health, researchers typically employ a variety of markers. The wearable gadget may simultaneously gather information including posture, activity trajectory, activity categorization, and steps. The devices may be worn on various body areas depending on the demands of the users and the functions executed. Next set of steps typically involve data preprocessing, followed by feature identification which are then subjected to machine learning protocols (Long et al, 2022).

Naeem and Coronato, (2022) studied infrastructure powered by artificial intelligence (AI) to help patients with cognitive impairments adhere to their treatment regimen at home and minimize medication errors. The AI-based system chooses a suitable monitoring procedure after evaluating the patients' disabilities. The barcode technique, optical character recognition, and a Deep Learning (DL)-based classifier are available ways to track the medication process. The GUI shows that the system can help patients avoid medication errors and take the right medication. Because it increases the likelihood of accurate detection and reduces the likelihood of medication errors, this integration of three distinct tools to monitor the medication process has benefits. When a patient has mild cognitive impairment, this approach works better.

Garriga et al (2022) studied machine learning models using health records electronically for uninterrupted monitoring of patients at risk for psychological crisis for a 28-day period. mental health, based on the ROC (receiver operating characteristic) curve. The design included a six-month follow up to assess their algorithm's utility in clinical prediction and practice, reflecting crisis alleviation in more than sixty percent of cases.

Domains of Application of AI enabled health monitoring devices

The healthcare industry is undergoing a transformation with AI health monitoring devices in the picture. Continuous, real-time insights into health conditions with primarily non-invasive technology could well be a game changer for the medical and psychiatric community, in the future. Domains of key applications of AI in health monitoring include:

- Wearable Health Devices: Smartwatches, or fitness trackers, that are also a common type of wearables, use AI algorithms for tracking physical activity, sleep cycles, and stress levels, and also for identifying early symptoms of health problems. Such wearables can provide data on daily habits and facilitate early detection of issues such as arrhythmias, abnormal heart rhythms or respiratory issues.
- Remote Patient Monitoring: AI-based devices facilitate remote monitoring of patients, empowering health practitioners to monitor vital statistics like heart rate, BP blood glucose, oxygen saturation, etc., without office visits. This is important for managing chronic conditions such as diabetes, heart disease and hypertension.
- Predictive Analytics for Early Diagnosis: The data collected from health tracking devices is analyzed by AI algorithms to make predictions about potential health risks. In fact, by identifying patterns in vital signs, AI can detect when conditions like heart attacks, strokes or complications from diabetes will rear their heads. This facilitates early intervention and personalized treatment plans.
- AI in Sleep Tracking: AI sleep monitoring devices track users' sleep cycles and quality, provide insights into sleep disorders such as insomnia, sleep apnea and restless leg syndrome. Important insights are obtained from these devices regarding attributes which influence sleep outcomes, thereby having

implications for superior plan of management and intervention for sleep related health conditions.

- AI based Fall Detection and Emergency Alerts for Elderly: AI services can be embedded in home monitoring systems or even worn in the form of a device that helps detecting falls of elderly patients. In the event of a fall, an automatic emergency alert is sent to healthcare providers or family members, providing fast response times and minimizing the potential consequences of a prolonged period of immobility.
- Chronic Disease Management Using AI Devices: Devices with AI capabilities allow individuals who suffer from chronic diseases to keep track of their conditions by monitoring symptoms and vital signs. These devices can aid patients in taking their medication, scheduling doctor appointments, and noticing patterns that diverge from their normal health state that may indicate a worsening condition.
- Telemedicine and Virtual Health Assistance: Telemedicine makes use of devices that patients can use at home to send data to their health providers in real-time during virtual consultations. With the aid of Artificial Intelligence, the data can be analyzed instantly and gives the physicians much deeper insight into the condition to improve patient-focused decisions.
- Personalized Health and Wellness: Health monitoring devices provide data that can be analyzed by AI to develop customized health strategies like specific workouts and meal plans. AI modifies these plans as time progresses and as the person's data and overall health improves or worsens.
- Post-Surgery and Rehabilitation Monitoring: Another area of application of AI-powered devices is monitoring health parameters of patients who are recovering from injury or surgery. From abnormalities of movement patterns assisting in evaluating improvement from physiotherapy and the overall outcome of treatment regimen, AI-based devices can help record authentic data that might assist in planning and continuing or discontinuing or changing planned interventions for rehabilitation.
- Nutrition Monitoring: Artificial Intelligence health devices that monitor food consumption, metabolic processes, and nutritional information are helping people to understand their eating patterns better. With the aid of Artificial Intelligence, these devices give recommendations to enhance the nutrition of a person and identify diet-related health risks such as food allergies or intolerances.
- AI in Diagnostics. Traditional devices are undergoing transformation to advanced digital versions enabled with artificial intelligence for more sensitive and better identification of abnormalities in respiratory patterns to body temperature, which in turn are assisting doctors make faster diagnosis, and

on occasion even second opinions in the form of preliminary identification of conditions.

- AI in Mental Health Monitoring: Besides currently experimental and potentially more established applications in medical healthcare domains, AI enabled devices are routinely also finding applications in mental health monitoring, especially in areas like stress levels, cognitive task performance, mood state, via wearable technology as well as smartphone apps. For conditions that may have situational triggers, wearable devices may be notably helpful by recording significant deviations of parameters, raise alarm, give real-time feedback, and even call for help, should that be necessary, reducing risk of dire consequences for the individual.

AI Enabled Wearables: Real-Life Applications

Over the last few years several products to be used for mental health monitoring and assessment have been designed, developed, and manufactured that incorporate AI technology in wearable forms. A few are described briefly as follows:

- Fitbit: Primarily designed for fitness monitoring, Fitbit at a later stage incorporated technology to analyse variations in heart rate for stress management. A higher heart rate may elicit a suggestion for a quick session of meditation, thus effecting positive de-stressing change.
- Apple Watch (or other smart watches): A sophisticated piece of wearable technological device, the Apple Watch is fitted with ECG functioning that can track rhythms of the heart, and raise an alarm to users, on identifying irregularities of the heart rhythm which could be indicative of stress. This smart device can also prompt breathing which is often an utterly simple yet life-affirming action, helping with stress management, occasionally even helping reduce heart palpitations.
- Oura Ring: A lustrous ring, which tracks sleep, while simultaneously providing insights into the ways in which one's lifestyle affects their mental health. If someone has had a difficult night of sleep, the oura ring might advise following that up with a relatively a lighter day.
- Moodbeam: A smart wearable device that tracks moods of the wearer day long, yielding data that can then be analysed to detect patterns, thus helping users identify triggering agents for their affective.
- Spire Stone: This tiny smart device, which clips on outfits, monitoring breathing patterns, can provide reminders for deep breathing on detecting stiffness and anxiety.

A NEW AGE: THE 'CHATTY' THERAPIST

The application domain of AI-powered technology presently is not restricted to merely monitoring but also now involves providing counselling and psychotherapeutic help to individuals via 'chat bots'. Various options are available on the market, and some have earned credibility to the extent of coverage from health insurance and employee welfare programmes paid for by organizations for their staff for their mental health support, just one click away. One of the most popular chat-bot therapists is 'Woebot Health' designed by Stanford post-doctoral researcher Alison Darcy in 2017 (Mental health needs have multiplied, *n.d.*). Another entrant in the therapeutic chatbot sphere is Earkick, featuring a cartoon panda in a bandana, who's like your friendly neighbourhood therapist with a kick, which may be especially endearing to younger clients who may share their thoughts and feelings either in speech or text, and the panda may engage in conversation for better understanding of the problem (reflecting back), and/ or offering an activity suggestion such as deep breathing to help one calm down. (AI chatbots are here to help, 2024; Earkick, Your Free Personal AI Therapist, n.d.)

Chat-bots like Woebot Health, Marlee, Wysa, Replika, Earkick, and Youper use the basic premise of providing a 'listening ear' using techniques of 'active listening' and 'reflecting' for preliminary level of intervention, especially for stress and low moods, and also as a resource tool for information. These can be used as a first ground of mental health support for generic mental health support, not acute in nature. However, as the Chatbot manufacturers also caution, these should not be taken as an equal alternative to human psychotherapy with a professionally trained therapist, and are not designed or equipped at the present time to provide crisis intervention. In addition, for deeper intervention and acute crisis, the technological platforms are designed to suggest professional support, including helpline contact information particularly for those indicating imminent thoughts of suicide or self-harm.

The present-day demand for mental health care and the inadequacy of the traditional infrastructure, and insufficiency of qualified and trained mental health care practitioners (MHP) increases the burden on both sides of the aisle - the professionals in practice, have workloads much higher than ideal, and yet non-availability of therapists for those in need. The scarcity also raises the cost of mental health care, where each therapy session with a trained MHP costs well above what many are able to afford, especially if not covered in the purview of health insurance. Therefore, despite questions remaining about degree of authenticity of chatbots being a primary therapeutic stop, chatbots are cheaper, often even free of cost, easily accessible at the effort of one click, and available 24/7, which put them at a distinct advantage of an obvious choice, especially for younger people versed in technology.

Effectiveness of AI enabled devices and their applications

Dehbozorgi et al., (2025) reviewed multitude of published research evidence to conclude varied degrees of effectiveness of AI in advancing mental health diagnosis, monitoring and interventions, resulting in marked progress for studying and treating diverse set of mental health conditions. Previously published studies have shown notable improvement in scores on depression upon frequent use of Wysa app (Inkster et al, 2018), enhanced performance on diagnosis and treatment of mental health issues with machine learning (Li, 2023), as well as efficiency in psychiatric evaluation, based on live symptom monitoring and customised delivery of therapeutic help (Liang, 2022).

The discussion on effectiveness must also include a note on cost-efficiency. In the Indian AI enabled smart device market, one of the most popular forms of wearables is smart watch. While adaptive smartphone screens and AI sentient backpacks exist (Jeevanandam, 2023), the lure of smartwatches is wide and far reaching, transcending age brackets from young school going adolescents to middle aged office goers. A quick search yields lists of popular smartwatches with varying features including widely ranging battery life of 18 hours to 20 days, depending on active to standby modes, as well as prices (as listed on Amazon) ranging from around 1800 INR (approx. $21) for a local brand such as Fire-Boltt or Noise to 85000 INR (approx. $995) for an Apple Series 10 device. Smart watches also offer a notable range of features from step counting to help towards fitness management, to heart rate and BP monitoring, core health monitoring elements, as well as more customised tracking of periods for women, to other sophisticated sports modes for sports enthusiasts. Researchers have also studied user satisfaction for AI enabled wearable smart watches. User experience and trust were found to have been notably influenced by product quality, service quality, perceived convenience, and perceived ease of use for smart watch wearers. In turn, both user experience and trust significantly impacted user satisfaction and served as partial mediators in the relationship between the influencing factors and user satisfaction (Uzir et al., 2021).

Masoumian Hosseini et al (2023) caution researchers and readers alike about reliability of health readings from mass market smart watches, based on their review of studies published that have looked into effectiveness and consistency of readings of such wearables. Masoumian Hosseini et al (2023) note that variations in detection capabilities between smartwatches and their underlying algorithms carry significant consequences for healthcare applications. Ensuring both the precision of these algorithms and the devices' ability to accurately identify shifts in health conditions is essential. This highlights the need for the advancement of medically certified smartwatches and the integration of AI-based assistants in clinical settings.

Challenges of Mental Health Monitoring

Digital phenotyping (the process of using data retrieved from personal digital devices and wearables for measuring observable traits, characteristics and behaviours of individuals) for mental health is expanding quickly, yet it faces significant and complex hurdles. These include determining whether and when the data collected can be effectively applied in real-world settings, incorporating the views and experiences of users and patients into research, addressing potential data biases, and questioning the assumption that digital phenotyping is inherently more objective than traditional psychiatric evaluation methods (Birk & Samuel, 2022). Common issues of novel digital phenotyping approaches include validation techniques, requisite approvals from pertinent regulatory bodies, and their judicious and prudent incorporation in medical and psychological research and patient care (Abbas et al., 2021).

Recent research indicates tremendous potential of Artificial Intelligence (AI) enabled technology and its applications in the health sector. These prospects, though, are not free from multitude of challenges, including but not limited to ethical and confidentiality and privacy related matters, insufficient awareness, inconsistencies of technology and professional accountability (Wubineh, 2023).

Ethical Considerations of AI in Mental Health Monitoring and Care

In use of AI enabled devices in therapy and mental healthcare, ethical considerations deserve attention. On a necessary cautionary note, manufacturers often present a disclaimer, be it for AI powered technology-based tools ranging from monitoring tools for diagnosis or treatment, to chatbots for clickable text-based psychotherapy in a nutshell, urging users not to rely on devices for complete medical / psychiatric assessment and treatment support. The devices are designed and intended to be used as auxiliary tools for diagnosis and intervention, under the supervision of trained professional experts, and not as a final destination for either diagnosis or treatment.

Besides the legal and cognitive disclaimer, first ethical concern is that of legal consent (Gerke et al, 2020). Consent must be explicit and so must be the terms of the statements the patients are expressing consenting to, or it becomes moot to obtain the same. From larger hospitals and clinics using cameras or other psychophysiological recording instruments, to routine digital smart devices such as smart watches, that people pick up without a second thought could be obtaining our consent, with those pesky terms and conditions written in miniscule font, that we often "agree" or "consent" to without fully reading or comprehending them. Thus, immersed with the consent factor is the issue of limited cognizance regarding the degrees of privacy there were signing away (Gerke et al, 2020; Jain, 2023).

The second issue is that of data privacy and access (Gerke et al, 2020; McGorry, 2020). In a world often subject to data fraud, especially where technological infrastructure has not yet caught up to the demands of cyber security networks, with unprecedented rise in demands especially in the post-Covid 19 era, where does the accountability of protection of all data obtained through AI enabled technology lie and in what form?

Another question of ethical concern is to what extent are the manufacturers responsible, and what recourse is available to individual consumers, should there be violations? Here comes the genuine issue of regulatory commissions if any that may oversee the commercial trading and consumer protection aspects of the service. In the USA the responsibility of regulation of AI enabled and other digital technology in mental and medical health care partly lies with the Food and Drug Administration (FDA) and partly with Federal Trade Commission (FTC, Löchner et al, 2025). Closer to home, in India, while the Indian Council of Medical Research (ICMR) defines certain guidelines for ethical use of AI enabled technology in healthcare, it also records the need for governance that must be sensitive to the nuances of use and abuse or misuse of the same. The legal provisions are restricted to Medical Device Rules, 2017, and the 2020 amendment to a certain degree (Naik & Reddy, 2025). The Health Data Management Policy (2020) and its amendment (2022) are also a reasonable step towards clearer guidelines (Jain, 2023). However, despite the guidelines and policies being there, there is lack of clarity at this time regarding which government or autonomous body has the regulatory authority in the matter, in India. India's existing technology regulations do not specifically cover telehealth, and the country lacks comprehensive data protection laws to handle the vast amounts of data that widespread digital healthcare would produce. There are also worries about patient privacy breaches in digital healthcare systems. Beyond privacy-related legal issues, there is also no established system to guarantee informed consent in the context of digital healthcare (Jain, 2023).

Besides data fraud, another ethical grey area is the degree of transparency regarding biases of algorithms used in AI powered technology for mental health monitoring. Besides deep learning and machine learning based technological biases (interchangeably referred to as algorithmic biases), cultural biases of language and regional nuances, as well as sociological implications of the same could be areas of concern. Researchers have also recorded genuine technical limitations of machine learning algorithms from sensitivity to missing data to slow prediction speed and low generalizability (Long et al., 2022).

Further concerns lie regarding the anthropomorphic connection to chat bots and devices in the absence or far less presence of human physicians, with which a relationship could potentially be the platform for a dystopian sphere where human connectedness is even less than its current form. Where children and adolescents are being routinely cautioned to reduce screen time, and engage in face-to-face interactions and relationships, what are the ethical implications for adults, relying more on technology and artificial intelligence for mental health care, something that is fundamentally individual, personal and private in nature?

Global to Local - Indian Scene

To put things in perspective, for India, a short overview of the mental health landscape is necessary. The Ministry of Health and Family Welfare (MOHFW, February 7th, 2025) website with reference to The National Mental Health Survey (NMHS) 2015-16, reports that 10.6% Indian adults suffer from mental illnesses, whereas prevalence across the lifetime is 13.7%. Research indicates that 15% of Indian adults facing mental health problems needing intervention, where urban prevalence rate is almost double (13.5%) that of rural areas (6.9%). The alarming note is that of the gap in treatment access for the people. According to the MOHFW, between 70% to 92% Indians with mental illnesses do not get proper medical and psychiatric treatment and care, due to severely limited awareness, stigma of insanity, and scarcity of professionally trained experts to provide the requisite treatment interventions. India reportedly has 0.75 psychiatrists per 100,000 (one lakh) individuals, in stark contrast to WHO recommendation of at least 3 psychiatrists per lakh people.

Meghrajani et al., (2023) writes of the heavy burden of mental health related problems on individual citizens, their families and the society in a larger context, significantly damaged functioning, diminishing quality of life, often with notable negative economic and social outcomes. Sociocultural factors like stigma of discrimination towards persons with mental illnesses, poverty, gender inequalities, fast urbanization, make the task of addressing mental health problems tougher. The researchers also highlighted the lack of access to mental healthcare with notable scarcity of qualified mental health practitioners, affecting quality of care one may receive, particularly in rural regions, with insufficient infrastructure. Lacuna in awareness and consequently acceptance of mental health problems, and inadequacy of rural citizens' integration into primary healthcare systems worsen the potential for adequate access to mental health care facilities (Meghrajani et al., 2023).

While the Government is making some effort in addressing the mental health care needs of the citizens, such as the Tele-Manas program and funding to National Institute of Mental Health and Neuro-Sciences (NIMHANS), but the share of allotment for mental health from the total health budgetary allocation is less than 2% (Keshav Desiraju India Mental Health Observatory, IMHO, March 7th, 2024).

India is also seeing an upsurge in the availability and use of AI-enabled apps and platforms especially among the younger population. Digital platforms as well as mobile apps have been launched as alternatives to traditional approaches for mental health tracking. From questionnaires to tracing mood changes, as well as conversation (text based input and output), avenues of data gathering have been incorporated into the designs, which shall be subjected to algorithms powered by artificial intelligence for mental health and wellness tracking, identification of problems or tiggers (if applicable) and provide customised advice, as needed. Mobile apps available include Wysa, YourDOST, and others to list a few.

To monitor psychophysiological correlates such as activity levels, heart rate, sleep patterns, blood pressure, etc., AI powered wearable devices are also being used, most commonly in the form of smart wrist watches. These devices record patterns of regular and irregular parameters, some even designed to alert the consumers to irregularities such as heightened stress levels, suggesting actions such as breathing exercises to regulate the same.

Other devices with AI-powered back up involve analysis of facial expressions, speech patterns and intonations, to help diagnose mental health conditions, prior to a formal diagnosis by a professional. In a country such as India, where for financial or social stigma, whichever reason, mental health care access may be limited, such preliminary diagnosis may encourage consumers to seek help, where they did not before.

AI-enabled equipments are also being used in hospital set-ups to record and analyze patient data to flag apprehensible psychological health issues, assisting informed prediction for faster and often life saving interventions.

Telehealth and virtual modes of therapy are also becoming a viable alternative if not the norm, at least in urban sectors, for those not affected by the digital divide. There is also a gradually increasing pattern of adoption of artificial intelligence for internet based platforms which dispense mental health related support via chats or video calling. In the Indian mental health scene, where human resource is majorly lacking, AI-based technology could help prioritise patients seeking help, in degrees of criticality in some cases even recommending preliminary steps of intervention, prior to a human therapist stepping in. Tele Counselling has particularly taken flight in India since the lockdown during Covid19 pandemic, and has continued post reopening of institutions and organizations alike, highlighting the convenience and cost saving.

Artificial intelligence powered technology has also been noted to show some success in early identification of mental health related problems. Faster detection of warning signs for anxiety, depression and post traumatic stress disorder (PTSD), with the help of AI, can facilitate timely assistance and potential prevention of undesirable outcomes especially for cases who are experiencing emotional turmoil. Further, Indian startups, working in the domain of mental health are also incorporating artificial intelligence in devising models of prediction of psychological health conditions, based on text, speech or behavioural data, with potential implications for timely intervention.

Like other nations, issues of AI enabled mental health care solutions remain inundated with their challenges in the Indian mental health landscape as well, including lack of assurance of privacy and confidentiality of data, and potential misuse of the same, as well as insufficient representation of diverse communities in base data sample used to generate the algorithms to facilitate openness and inclusivity in mental health care.

AI-enabled mental health monitoring in India is an emerging field with significant potential to enhance mental health care access, facilitate early intervention, and deliver personalized support. However, it is essential to balance technological progress with ethical considerations, privacy issues, and the importance of human empathy in mental health care. While challenges remain, reason to hope do as well. The Government of India backed National Mental Health Programme is gradually incorporating AI powered digital mental health tools into its repertoire to address the upsurge of mental health care needs of the country, a sign of the acceptance and utilisation of AI in the judicious way it would prove beneficial.

LIMITATIONS AND FUTURE DIRECTIONS

The present work sets a preliminary platform with limited resources to navigate the increasingly fascinating sphere of AI enabled devices for mental health, be it in early screening and diagnosis or predicting outcomes or planning interventions. Research published were investigated within select time span and select data bases and could be expanded in a more comprehensive review. Further work could move the research to an application platform or even action policy platform for better connect with the chief stake holders of the entire process, the people, i.e., the patients and assess actively the degree of efficacy of such devices.

Policy Recommendations

The present chapter highlights certain areas of concern regarding AI powered wearables for mental health, especially those related to ethical dilemmas. From data privacy and confidentiality, legal and moral elements of consent to biases in algorithms, concerns remain for judicious use of AI enabled wearable devices. Therefore, governments must ensure patients' rights are upheld, by empowering them with suitable policies, and enforcing the same via direct or indirect avenues of regulatory bodies and Institutional Ethical Committees, where the latter must be proactive in treatment execution for patients, rather than merely upholding research participants' rights. Policies and guidelines must also be devised and applied for some regulation of costs and durability of devices, and enhancing their accessibility by incorporating these within essential commodities with lowered burdens of taxation and consequently lower prices for consumers, without compromising on quality or accuracy of the products.

CONCLUSION

The present world is one of rapidly changing technological know-how. Small to large scale manufacturers are rivalling in a race to produce tech-based and artificial intelligence i.e., AI-powered devices, machines for physical and mental health monitoring with varied degree of success. While technological advancements are inevitable, it is important to note that AI is based on large data, and health care at the individual level, must of necessity be customized. The age-old notion of 'bed-side manner' of physicians is not a mere term, but rather has connotations for building trust, and rapport building, which is of even greater significance for psychological health issues. Be it diagnosis, or treatment interventions, the nuanced uniqueness of lived experience and even the vagaries of daily life for individual patients/ clients, makes it essential that the reliance is not entirely upon AI-enabled devices but rather at the reasoned understanding and expertise of a trained human professional.

REFERENCES

Abbas, A., Schultebraucks, K., & Galatzer-Levy, I. R. (2021). Digital Measurement of mental health: Challenges, promises, and future directions. *Psychiatric Annals, 51*(1), 14–20. DOI: 10.3928/00485713-20201207-01

Abd-alrazaq, A., AlSaad, R., Aziz, S., Ahmed, A., Denecke, K., Househ, M., Farooq, F., & Sheikh, J. (2023a). Wearable Artificial Intelligence for Anxiety and Depression: Scoping Review. *Journal of Medical Internet Research, 25*, e42672. Advance online publication. DOI: 10.2196/42672 PMID: 36656625

Abd-Alrazaq, A., AlSaad, R., Shuweihdi, F., Ahmed, A., Aziz, S., & Sheikh, J. (2023b). Systematic review and meta-analysis of performance of wearable artificial intelligence in detecting and predicting depression. *NPJ Digital Medicine, 6*(1), 84. Advance online publication. DOI: 10.1038/s41746-023-00828-5 PMID: 37147384

AI chatbots are here to help with your mental health, despite limited evidence they work. (March 24th, 2024). *The Hindu.* Retrieved from https://www.thehindu.com/sci-tech/technology/ai-chatbots-are-here-to-help-with-your-mental-health-despite-limited-evidence-they-work/article67984641.ece

Almufareh, M. F., Tehsin, S., Humayun, M., & Kausar, S. (2023). Intellectual disability and technology: An artificial intelligence perspective and framework. *Journal of Disability Research, 2*(4), 58–70. DOI: 10.57197/JDR-2023-0055

Alonso, J., Liu, Z., Evans-Lacko, S., Sadikova, E., Sampson, N., Chatterji, S., Abdulmalik, J., Aguilar-Gaxiola, S., Al-Hamzawi, A., Andrade, L. H., Bruffaerts, R., Cardoso, G., Cia, A., Florescu, S., de Girolamo, G., Gureje, O., Haro, J. M., He, Y., de Jonge, P., & Thornicroft, G.WHO World Mental Health Survey Collaborators. (2018). Treatment gap for anxiety disorders is global: Results of the World Mental Health Surveys in 21 countries. *Depression and Anxiety, 35*(3), 195–208. DOI: 10.1002/da.22711 PMID: 29356216

Arvind, B. A., Gururaj, G., Loganathan, S., Amudhan, S., Varghese, M., Benegal, V., Rao, G. N., Kokane, A. M., B S, C., P K, D., Ram, D., Pathak, K., R K, L. S., Singh, L. K., Sharma, P., Saha, P. K., C, R., Mehta, R. Y., & T M, S. (2019). Prevalence and socioeconomic impact of depressive disorders in India: Multisite population-based cross-sectional study. *BMJ Open, 9*(6), e027250. DOI: 10.1136/bmjopen-2018-027250 PMID: 31253618

Bajwa, J., Munir, U., Nori, A., & Williams, B. (2021). Artificial intelligence in healthcare: Transforming the practice of medicine. *Future Healthcare Journal, 8*(2), e188–e194. DOI: 10.7861/fhj.2021-0095 PMID: 34286183

Birk, R. H., & Samuel, G. (2022). Digital Phenotyping for Mental Health: Reviewing the challenges of using data to monitor and predict mental health problems. *Current Psychiatry Reports*, *24*(10), 523–528. DOI: 10.1007/s11920-022-01358-9 PMID: 36001220

Dehbozorgi, R., Zangeneh, S., Khooshab, E., Nia, D. H., Hanif, H. R., Samian, P., Yousefi, M., Hashemi, F. H., Vakili, M., Jamalimoghadam, N., & Lohrasebi, F. (2025). The application of artificial intelligence in the field of mental health: A systematic review. *BMC Psychiatry*, *25*(1), 132. Advance online publication. DOI: 10.1186/s12888-025-06483-2 PMID: 39953464

Earkick, Your Free Personal AI Therapist. (n.d.). *Earkick*. Retrieved from https://earkick.com/

Garriga, R., Mas, J., Abraha, S., Nolan, J., Harrison, O., Tadros, G., & Matic, A. (2022). Machine learning model to predict mental health crises from electronic health records. *Nature Medicine*, *28*(6), 1240–1248. DOI: 10.1038/s41591-022-01811-5 PMID: 35577964

Gerke, S., Yeung, S., & Cohen, I. G. (2020). Ethical and Legal Aspects of Ambient Intelligence in Hospitals. *Journal of the American Medical Association*, *323*(7), 601–602. DOI: 10.1001/jama.2019.21699 PMID: 31977033

Herath, H. M. K. K. M. B., Jayasekara, A. G. B. P., & Madhusanka, B. G. D. A. (2024). *Vision Attentive Technology: AI Approaches for Functional Mobility Assessments in Elderly Healthcare* (1st ed.). CRC Press., DOI: 10.1201/9781032622088

Inkster, B., Sarda, S., & Subramanian, V. (2018). An Empathy-Driven, Conversational Artificial Intelligence Agent (Wysa) for Digital Mental Well-Being: Real-World Data Evaluation Mixed-Methods Study. *JMIR mHealth and uHealth*, *6*(11), e12106. DOI: 10.2196/12106 PMID: 30470676

Jain, D. (2023). Regulation of Digital Healthcare in India: Ethical and Legal Challenges. *Healthcare (Basel)*, *11*(6), 911. DOI: 10.3390/healthcare11060911 PMID: 36981568

Jansen, C.-P., Gordt-Oesterwind, K., & Schwenk, M. (2022). Wearable Motion Sensors in Older Adults: On the Cutting Edge of Health and Mobility Research. *Sensors (Basel)*, *22*(3), 973. DOI: 10.3390/s22030973 PMID: 35161719

Jeevanandam, N. (2023). Five interesting Generative AI-powered smart wearables in 2023. IndiaAI. Retrieved from https://indiaai.gov.in/article/five-interesting-generative-ai-powered-smart-wearables-in-2023

Keshav Desiraju India Mental Health Observatory. (March 7th, 2024). The interim Union Budget FY 2024-25: where does mental health stand? Retrieved from https://cmhlp.org/imho/blog/the-interim-union-budget-fy-2024-25-where-does-mental-health-stand/

Li, X. (2023). Evaluation and Analysis of Elderly Mental Health Based on Artificial Intelligence. *Occupational Therapy International*, *7077568*, 1–11. Advance online publication. DOI: 10.1155/2023/7077568 PMID: 36817324

Liang, L., Zheng, Y., Ge, Q., & Zhang, F. (2022). Exploration and Strategy Analysis of Mental Health Education for Students in Sports Majors in the Era of Artificial Intelligence. *Frontiers in Psychology*, *12*, 762725. DOI: 10.3389/fpsyg.2021.762725 PMID: 35308079

Löchner, J., Carlbring, P., Schuller, B., Torous, J., & Sander, L. B. (2025). Digital interventions in mental health: An overview and future perspectives. *Internet Interventions : the Application of Information Technology in Mental and Behavioural Health*, *100824*, 100824. Advance online publication. DOI: 10.1016/j.invent.2025.100824 PMID: 40330743

Long, N., Lei, Y., Peng, L., Xu, P., & Mao, P. (2022). A scoping review on monitoring mental health using smart wearable devices. *Mathematical Biosciences and Engineering*, *19*(8), 7899–7919. DOI: 10.3934/mbe.2022369 PMID: 35801449

Maha, C. C., Kolawole, T. O., & Abdul, S. (2024). Revolutionizing community health literacy: The power of digital health tools in rural areas of the US and Africa. *GSC Advanced Research and Reviews*, *19*(02), 286–296. DOI: 10.30574/gscarr.2024.19.2.0189

Maleki Varnosfaderani, S., & Forouzanfar, M. (2024). The Role of AI in Hospitals and Clinics: Transforming Healthcare in the 21st Century. *Bioengineering (Basel, Switzerland)*, *11*(4), 337. DOI: 10.3390/bioengineering11040337 PMID: 38671759

Manjunatha, N., Jayasankar, P., Suhas, S., Rao, G. N., Gopalkrishna, G., Varghese, M., & Benegal, V. (2022). Prevalence and its correlates of anxiety disorders from India's National Mental Health Survey 2016. *Indian Journal of Psychiatry*, *64*(2), 138–142. DOI: 10.4103/indianjpsychiatry.indianjpsychiatry_964_21 PMID: 35494323

Masoumian Hosseini, M., Masoumian Hosseini, S. T., Qayumi, K., Hosseinzadeh, S., & Sajadi Tabar, S. S. (2023). Smartwatches in healthcare medicine: Assistance and monitoring; a scoping review. *BMC Medical Informatics and Decision Making*, *23*(1), 248. DOI: 10.1186/s12911-023-02350-w PMID: 37924029

McGorry, P. (2020). Mental health and COVID-19: Are we really all in this together? *The Medical Journal of Australia, 213*(10), 454–455. DOI: 10.5694/mja2.50834 PMID: 33107074

Meghrajani, V. R., Marathe, M., Sharma, R., Potdukhe, A., Wanjari, M. B., & Taksande, A. B. (2023). A Comprehensive Analysis of Mental Health Problems in India and the Role of Mental Asylums. *Cureus, 15*(7), e42559. DOI: 10.7759/cureus.42559 PMID: 37637646

Mental health needs have multiplied. Support hasn't. (n.d.). Until now. *Woebot Health*. https://woebothealth.com/faq/

Milic, J., Zrnic, I., Grego, E., Jovic, D., Stankovic, V., Djurdjevic, S., & Sapic, R. (2025). The Role of Artificial Intelligence in Managing Bipolar Disorder: A New Frontier in Patient Care. *Journal of Clinical Medicine, 14*(7), 2515. DOI: 10.3390/jcm14072515 PMID: 40217964

Mirmomeni, M., Fazio, T., von Cavallar, S., & Harrer, S. (2021). From wearables to THINKables: artificial intelligence-enabled sensors for health monitoring. In *Wearable Sensors* (pp. 339–356). Academic Press. DOI: 10.1016/B978-0-12-819246-7.00012-7

Naeem, M., & Coronato, A. (2022). An AI-Empowered Home-Infrastructure to Minimize Medication Errors. *Journal of Sensor and Actuator Networks, 11*(1), 13. DOI: 10.3390/jsan11010013

Naik, S., & Reddy, B. (2025). A framework for the governance of AI in healthcare for India: An exploratory model. *Indian Journal of Medical Ethics*. Advance online publication. DOI: 10.20529/IJME.2025.037

Parekh, A. E., Shaikh, O. A., Simran, N., Manan, S., & Hasibuzzaman, M. A. (2023). Artificial intelligence (AI) in personalized medicine: AI-generated personalized therapy regimens based on genetic and medical history: short communication. *Annals of Medicine and Surgery (London), 85*(11), 5831–5833. DOI: 10.1097/MS9.0000000000001320 PMID: 37915639

Santomauro, D. F., Herrera, A. M. M., Shadid, J., Zheng, P., Ashbaugh, C., Pigott, D. M., & Ferrari, A. J. (2021). Global prevalence and burden of depressive and anxiety disorders in 204 countries and territories in 2020 due to the COVID-19 pandemic. *Lancet, 398*(10312), 1700–1712. DOI: 10.1016/S0140-6736(21)02143-7 PMID: 34634250

Sequeira, L., Perrotta, S., LaGrassa, J., Merikangas, K., Kreindler, D., Kundur, D., Courtney, D., Szatmari, P., Battaglia, M., & Strauss, J. (2020). Mobile and wearable technology for monitoring depressive symptoms in children and adolescents: A scoping review. *Journal of Affective Disorders, 265*, 314–324. DOI: 10.1016/j.jad.2019.11.156 PMID: 32090755

Sokołowska, E., Sokołowska, B., Chrapusta, S. J., & Sulejczak, D. (2025). Virtual environments as a novel and promising approach in (neuro)diagnosis and (neuro) therapy: A perspective on the example of autism spectrum disorder. *Frontiers in Neuroscience, 18*, 1461142. DOI: 10.3389/fnins.2024.1461142 PMID: 39886337

Szabó, P., Ara, J., Halmosi, B., Sik-Lanyi, C., & Guzsvinecz, T. (2023). Technologies Designed to Assist Individuals with Cognitive Impairments. *Sustainability (Basel), 15*(18), 13490. DOI: 10.3390/su151813490

The Ministry of Health and Family Welfare. (MOHFW, February 7th, 2025). *Advancing Mental Healthcare in India.* Retrieved from https://mohfw.gov.in/?q=pressrelease-206

Uzir, M. U. H., Al Halbusi, H., Lim, R., Jerin, I., Abdul Hamid, A. B., Ramayah, T., & Haque, A. (2021). Applied Artificial Intelligence and user satisfaction: Smartwatch usage for healthcare in Bangladesh during COVID-19. *Technology in Society, 67*, 101780. DOI: 10.1016/j.techsoc.2021.101780 PMID: 34697510

Wang, W., Chen, J., Hu, Y., Liu, H., Chen, J., Gadekallu, T. R., Garg, L., Guizani, M., & Hu, X. (2024). Integration of Artificial Intelligence and Wearable Internet of Things for Mental Health Detection. *International Journal of Cognitive Computing in Engineering, 5*, 307–315. DOI: 10.1016/j.ijcce.2024.07.002

World Health Organization. (2023a). Depressive Disorder (Depression). Retrieved from https://www.who.int/news-room/fact-sheets/detail/depression

World Health Organization. (2023b). Anxiety disorders. Retrieved from https://www.who.int/news-room/fact-sheets/detail/anxiety-disorders

World Health Organization. (2024). Leading the Future of Global Health with Responsible Artificial Intelligence. Retrieved from https://cdn.who.int/media/docs/default-source/digital-health-documents/who_brochure_ai-vision_web.pdf?

Wubineh, B. Z., Deriba, F. G., & Woldeyohannis, M. M. (2023). Exploring the opportunities and challenges of implementing artificial intelligence in healthcare: A systematic literature review. *Urologic Oncology, 42*(3), 48–56. DOI: 10.1016/j.urolonc.2023.11.019 PMID: 38101991

Zafar, F., Fakhare Alam, L., Vivas, R. R., Wang, J., Whei, S. J., Mehmood, S., Sadeghzadegan, A., Lakkimsetti, M., & Nazir, Z. (2024). The Role of Artificial Intelligence in Identifying Depression and Anxiety: A Comprehensive Literature Review. *Cureus*, *16*(3), e56472. DOI: 10.7759/cureus.56472 PMID: 38638735

Chapter 4
Efficacy of Wearable AI Driven Mindfulness Practices in the Management of Stress:
Systematic Review

Sanchita Ghosh
https://orcid.org/0000-0002-1410-0024
West Bengal State University, Kolkata, India

Debangshu Chanda
West Bengal State University, Kolkata, India

Chirantani Mukhopadhyay
West Bengal State University, Kolkata, India

ABSTRACT

Mindfulness is increasingly being recognized for reducing stress, where wearables incorporating artificial intelligence offer innovative avenues for mindfulness practices & stress relief. This review was registered at PROSPERO & it examined the effectiveness of AI-based wearables for mindfulness practices & stress management. Following the review question, elements of PICO (Population, Intervention, Comparison, Outcome) were identified. Using the PRISMA model, & following the Joanna Briggs Institute & the Indian Council of Medical Research guidelines, 800 studies were identified, with 7 meeting the inclusion criteria. 5 core themes were identified through thematic analysis, including: AI and personalization in mindfulness interventions, physiological monitoring and biofeedback, accessibility and

DOI: 10.4018/979-8-3373-0467-0.ch004

usability, psychological and behavioral outcomes, & future directions in wearable technology. Limitations include dearth of relevant studies & potential keyword gaps in study selection. Further research is needed to validate reliability and effectiveness.

INTRODUCTION

The benefits of mindfulness are now well acknowledged. It is an interesting and effective introspective technique with ancient roots, often used for stress relief and general well-being. During this practice, one intentionally focuses on the present, and tries to remain as focused as possible, without any form of judgment. The ancient roots referred here are deeply rooted into Eastern spiritual traditions, especially Buddhism, Taoism (LaFargue, 2010) and Hinduism, for example in Patanjali Yoga Sutras (Bryant, 2015). Starting from Buddha's eightfold path teachings, related to "being in the present", to the Satipatthana Sutta, which outlines focusing on "breath, body and emotions", the ancient roots of mindfulness is like a tapestry woven in search of the true present. According to classical Buddhist philosophy, mindfulness is basically one of the 52 different functions of the mind (Olendzki, 2009). As per the traditional languages of Buddhism, that is, Sanskrit and Pali, the closest word indicative of mediation is *Samadhi*. As per etymology, the term *Samadhi* means concentrating our mind and putting it upon an object. According to the Buddhist understanding of the operating principle of mind, experience comes from the conjunction of three primary elements, which includes our consciousness, an organ and an object. Any object is capable of being cognized through an organ; on the contrary, an organ cognizes any object. This can also be constructed as "our consciousness about such an object arises through the usage of an organ". Hence, it is clear that there are different ways of depicting this same phenomenon or occurrence. When one places his/her/their mind upon a "sensory object" it is just as much meditation as when someone places one's mind upon some kind of "conceptual object", and it is not possible to do both at once.

Ekaggataa, literally meaning one pointedness, which can also be broken down as "one (*ek*) pointed (*agga*) ness (*taa*)", is the main attribute of meditation and the term most frequently used to describe it. The basic yet necessary goal of meditation is to unify the mind, focus it on a single thing, and place it there. This occurs spontaneously in every mental moment to some degree; otherwise, mental experience would be seriously lacking in coherence. Buddhist theories of mind hold that consciousness arranges different supporting mental processes around a single object at a time. One way to interpret this would be as a single episode of consciousness, in reaction to a stimulus, lasts for a very short time, and then dies nearly instantly. In reaction to another stimulus, another mental moment instantly occurs, and this one also ends instantly. Subjective experience appears to us as just like a modern livestream during one's moments of consciousness. The continuity and stability of subject and object are projected onto such streams, in a manner similar to how a story is built from quickly shown movie frames. Every frame has a single focus, which is known as one-pointedness. However, concentration based meditation focuses on extending this singularity of focus throughout several subsequent mental moments.

Until now we discussed more on meditation, but now we aim to describe what mindfulness really means.Described as the capacity to stay completely aware and present in the present moment, devoid of any judgment, "mindfulness" has successfully emerged as a truly potent strategy for fostering resilience. Resilience, in this context, is our ability to cope, bounce back and withstand tough situations, places, people and more. Mindfulness is not exactly equal to what we call meditation. According to Thich Nhat Hanh (2006), when one is directed towards the core of reality, one feels truly free, since it offers wisdom, help and understanding. Meditation, as per Hanh, is "looking face-to-face" instead of fleeing away or ignoring realities of life. Unlike true Buddhist meditation, this is not a mere belief on the principle of non-duality, not a battlefield, but it is about acceptance of the fact that, where there is love, there is suffering as well, where there is peace, there is conflict as well. The task of a meditation practitioner is not to suppress hard emotions like anger, disgust, fear etc., but rather to invite other emotions to "take care of" the problematic emotion. This concept is beyond any discrimination, battle, conflict or anything similar. The goal of meditation is to bring forth freedom of acceptance and understanding the true meaning of love. This "energy", according to Hanh, has the power to heal. In fact, if mindfulness is the energy of Buddha, the seed of mindfulness is nothing but the "baby Buddha", present within all of us. If one is aware of what one is doing at that present moment, mindfulness is present. This includes things like drinking water, being angry, feeling sad, feeling joyful and much more. Mindfulness, for Hanh, is an energy as well. It enhances consciousness by encouraging a nonjudgmental, present-centered mindset (Chiovato et al., 2021). According to Olendzki (2009), mindfulness meditation is a tool which aims to bring a transformation from suffering to well-being. Though other forms of focus-based meditations have indispensable advantages, mindfulness meditation is a relevant tool for a complex mind like ours. Unlike a restricted and sharply pointed focus, it opens up a broader arena, just like a flashlight instead of a laser.

Without knowing oneself and awareness about one's capacities and limits, practicing mindfulness can become a double-edged sword. Because, despite the ultimate aim of mindfulness practice is to be comfortable with all forms of experiences, it can be really unworthy and next to impossible to do this all at once. Just as our capacity to tolerate various forms of experiences varies from day to day, it also varies during different periods of our lifetime. The ability to tolerate supposedly difficult experiences also varies greatly from one person to the other. Certain people are born with neural networks and patterns within the nervous systems which are more intensely reactive to any form of changes or threats, while others are born with systems which are comparably less reactive than the former set of people. Human development is extremely complex hence it is important to take mindfulness practices in accordance with one's own capabilities and capacities. There are some forms of practices, like sitting calmly and holding one's breath for long periods of time, or even participating in a completely silent meditation retreat. Many forms of meditations are present which tend to bring up relatively difficult thoughts and feelings stored within oneself. If an individual can bear these difficult mental contents without feeling fully flooded, it tends to be a useful practice to free oneself up. However, if they make them feel very overwhelmed, the experience may make them more frightened of their own mind, and hence be counterproductive (Siegel, 2010).

From the Buddhist era (around 500 BCE) to 2025, much has changed. The same has occurred in the field of mindfulness meditation. As a result of the recent AI revolution and the advent of wearables, future-focused mindfulness training using wearable technology, such as biometric-measuring smartwatches, seems to be really promising (Wrede et al., 2023). Some of the smart applications available nowadays are made exclusively for self-help purposes without any feedback, while others act as wearable mechanisms targeted towards providing feedback to the users.

The increasing integration of wearables and artificial intelligence (AI) into mindfulness and stress management practices has garnered significant attention in recent years. This growth is fueled by the accelerating awareness of the need for mental health care and the potential of newer technology to improve traditional mindfulness practices, which have long been clinically validated as effective in alleviating stress. Pairing of AI (in some or the other form) with wearables has a dual benefit. Wearables give us real-time physiological feedback and AI tries to personalize the mindfulness interventions, allowing them to reach not only a wider audience, but also people who truly face difficulties during conventional mindfulness sessions. Hence, in the subsequent paragraphs here, we have seen some of the devices along with how AI is used in mindfulness and stress management, how they can be helpful, as well as its applications and implications on mental health according to latest research.

There are several reasons why wearables have been utilized for mindfulness and stress management. A few of them are explained here. With respect to physiological monitoring, wearable devices have been used to keep track of various physiological parameters, including heart rate and breathing patterns, allowing for immediate feedback on stress levels. When it came to intervention effectiveness, a study even reported (Smith et al., 2020) that participants experienced a 15.8% decrease in occurrence and 28.2% less days with anxiety due to the use of these wearables. Not just individual-based but even community-based programs have been using wearables, especially in community stress management programs, showing significant improvements in mindfulness skills and physiological responses after those interventions (Carter et al., 2018).

AI's Role in enhancing mindfulness is gaining popularity in recent days. Be it personalized interventions where AI can tailor mindfulness practices to individual needs, enhancing user engagement and effectiveness (Ramana & Peri, 2024), or for monitoring progress over time, providing insights into the effectiveness of mindfulness practices and therapies (Borthakur, 2021), or in alignment to healthcare, where AI has shown potential in managing pain and reducing stress, thereby improving overall health outcomes (Ramana & Peri, 2024).

Nowadays, when it comes to bridging technology and mental health, wearable technology, such as the Apple Watch, offers portable solutions for anxiety management, making mindfulness practices more accessible to users, empowering individuals to manage their mental well-being, fostering resilience in stressful environments (Muthuraman, 2024). In fact, the combination of wearables and AI promotes a holistic approach to mental health, integrating physiological data with mindfulness practices for comprehensive stress management (Carter et al., 2018).

In conclusion, however the merging of wearable devices and AI into mindfulness practice offers many positive outcomes, yet there are also certain concerns to think about and rectify in the near future, especially the ones which may not be beneficial for stress management, or the ones which might bring up a number of other issues. Primarily, there is a danger with wearable technology that solely relies on artificial intelligence as it may lead to the reduction of internal motivation for engaging in mindfulness practices or other related stress exercises. A second concern is domestic data breaches that arise from the cloud based storage, along with other biometric related privacy issues which may envelop users within a shroud of fear that these tools might be misused. This consideration regarding privacy concerns should be incorporated, in order to be at par with traditional mindfulness approaches, critical for maximizing the mental health benefits while using technology. Another consideration might be accessibility related challenges, including pricing, hardware requirements, potential for cybersickness, and issues with digital literacy and access to technology. In this context, cybersickness is a collection of symptoms like oculomotor issues, disorientation, and nausea, caused by the prolonged use of computers, TV, smartphones, or other digital devices with a screen that involves being stationary for a long time. This systematic review aims to explore the overall effectiveness of mindfulness practice through AI based wearable devices meant for stress reduction.

BACKGROUND

Concept and Definitions

In today's fast-paced times, the world is crammed up with several new-age technologies which although are impressive, can pose a serious threat to our focus and concentration. Not only children, but adults too are affected by the same. This is because these technological advancements tend to distract us and eventually lead to stress, making us yearn for mental peace in our daily lives, which by then we start to lack. This is where mindfulness practice can assist to provide us an easy but effective way to combat stress. Mindfulness is technically defined as "the awareness that arises through paying attention, on purpose, in the present moment and non-judgmentally" (Kabat-Zinn, 1990). Its objective is to increase psychological flexibility and decrease habitual stress responses by sustaining an open, accepting awareness of both external and internal experiences (thoughts, feelings, and bodily sensations).

Defining stress is not so difficult. In simple terms, stress is one of those phenomena, which the world has always been worried about in the last decade. Why? Because stress seems to be one of the biggest contributors to mental health problems, not just within the boundaries of workspaces but also at home. "Stress can be technically defined as a state of worry or mental tension caused by a difficult situation" (World Health Organization, 2023). In other words it's a psychological & physiological response of an individual to external or internal demands that are perceived as overwhelming or exceeding the person's coping resources. One of the most classical definitions was given by Lazarus and Folkman (1984). According to them, stress has been defined as a type of relationship between the individual and the environment, which is evaluated by that individual as exhaustive or exceeding one's resources and posing a threat towards mental well-being. Hans Selye (1936), on the other hand, defined stress as a nonspecific response of one's body, to any demand or stressor. This approach was the first to focus on stress as a biological function. A stressor is defined as a stimulus which acts as a threat to an individual's homeostasis, whereas a stress response is defined as the reaction of the living organism in order to regain one's homeostasis (Chrousos, 2009). Cannon (1932) was the first to coin the term homeostasis and expressed various physiological factors like blood glucose level, blood pressure etc. have certain preferred ranges and any deviations lead to certain physiological responses, for restoring back to the preferred level. Selye (1936,1950) stressed upon the adaptive nature related to such stress responses, when iconically formulating the General Adaptation Syndrome (GAS). In 1976 Selye also introduced two terms: distress (what we call non-productive and bad stress) and *eustress* (the type of stress which is productive and/or useful in some manner). This was done in order to cast a contrast between the maladaptive versus the adaptive consequences as a result of stress response. However, even after emphasizing both the maladaptive and adaptive sides of stress response, it seems to be problematic to separate these two.

Speaking from a biological point of view, stress is usually defined in terms of a threat to the optimal balance of the body, i.e., our homeostasis (Levine, 2005; Romero et al., 2009). According to Levine & Ursin (1991) stress not only includes the process, but also includes the perceptual processing of the stimulus or *input* (whatever our brains process and perceive), as well as the behavioural outcome or *output* (whatever we do) (Levine, 2005). Physiologically, it seems that stress might be marked by either the absence of any unpredictable and anticipatory response of the neuroendocrine reaction (Koolhaas et al., 2011). Stress reduction refers to a set of techniques designed to mitigate the physiological and psychological effects of stress, often by inducing relaxation, promoting awareness, mindfulness practices and enhancing coping strategies. In simple words, stress reduction is the procedure through which stress can be reduced.

Historical Context

Mindfulness based stress reduction programme is a widely researched intervention that combines mindfulness meditation and yoga to reduce stress. There is a quite long historical context of technology in Mindfulness Based Stress Reduction (MBSR) and other meditation practices involving mindful breathing. The key components involved within it may include yoga, mindfulness meditation, body awareness, group discussions, home practices and retreat programs. The first record which is found is of 1979, when the Mindfulness-Based Stress Reduction Clinic was founded by Jon Kabat-Zinn, at the University of Massachusetts Medical Center (Brantley, 2005). It was followed by the Center for Mindfulness in Medicine, Health Care, and Society at the University of Massachusetts Medical School, almost two decades later. Researchers have been working on how mindfulness can be used by literally anyone in everyday life (Kabat-Zinn, 1994) since it is evident that one need not be spiritual or a Buddhist to practice MBSR (Brantley, 2005).

There are a number of outcomes which were found recurrently in various studies which includes high degree of conformance with various meditation techniques, maintenance of an overall enhanced mental status over time, as well as a high degree of significance attributed to the meditation training modules (Kabat-Zinn et al, 1986), increased vigor, improved physiological markers of vagal tone both at rest and during exposure to a cognitive stressor (Balconi et al, 2019). Apart from the ameliorating effects stated above, there were certain other factors which were noted to significantly decrease. These include decreased stress and anxiety measures and decreased mental fatigue (Balconi et al, 2019). Studies have also shown that MBSR improves psychological health, reduces depression, and enhances emotional regulation (Kabat-Zinn, 1990; Goyal et al., 2014). Hölzel et al. (2011) found that specific mindfulness practices, such as MBSR, can even change brain structure, specifically increasing density of gray matter of the brain, within those regions of the brain which are associated with emotional regulation and self-control.

Mindfulness Techniques

There are various types of mindfulness techniques. Some are related to breathing, while the others are related to posture, paying attention to the body, or observation. For example, in Mindful Breathing, people focus on their breath, thereby noticing the subtle sensation of their breath entering and leaving their body, and sometimes also the rise and fall of the chest or the stomach. It is instructed that whenever the mind wanders, they should gently redirect their entire attention back to their breath only, without any judgment. Another extremely popular technique is Body Scan, where the participants systematically bring their attention to different parts of their body (as instructed to them or self-instructed), just noticing any kind of sensations without judgment. This process usually starts with focusing on the toes or the feet and eventually, the focus moves up to the head. It is a bottom-to-top meditation technique. Mindful Walking is yet another form of mindfulness practice, where the users focus on the sensations of their feet touching the ground, their moving body, as well as their breath. Sometimes it also involves observing the surroundings i.e., the sights, sounds, and smells around them. This technique is sometimes referred to as walking meditation. Similar to this technique is another technique called Mindful Eating, where the participants focus on eating without any distractions, and by paying attention to the taste, texture, and smell of the food which is being eaten. Journaling is also a technique in which the person uses writing as the medium to reflect on daily experiences, thoughts, and emotions in a journal. Loving-kindness meditation, also known as metta meditation, is also a mind based practice that cultivates positive emotions and compassion by extending goodwill towards oneself and others. It involves mentally sending kind and warm wishes, like "May you be happy," "May you be well," and "May you be at peace," to various individuals, starting with oneself and gradually expanding to others, even those who are difficult to deal with. Many experts also focus on Gratitude Practice. It involves expressing gratitude for the things in one's life, which can help cultivate a more positive and mindful outlook towards life. There are a number of formal and informal practices of mindfulness as well. Classically formal practices involve four postures including sitting, standing, walking and lying down. Apart from these, practices like Retreats (or Retreat Practices) are also growing in popularity. Retreats are special vacations dedicated entirely to mindfulness cultivation. There are many more techniques or ways, however, the most important ones have been summarized here (Siegel, 2010).

The Brain and Cognition

Now it is essential to understand how the brain is related to all the various kinds of mindfulness based practices mentioned above. According to Davidson, people who are distressed have more activity in the right prefrontal cortex. In lay terms this is on the frontal areas of the brain associated with planning, maintaining focus during speech. It also plays a role in inhibiting feelings, including potential consequences (Goleman, 2003). The left prefrontal cortex, on the other hand, helps to establish positive feelings, including the rewards of certain behaviors and actions. It has been observed that among healthy patients, the activity on both sides are balanced. It has been observed that monks who have practiced thousands of hours of meditation practices, had dramatic shifts to left prefrontal cortex activation. In another study by Kabat-Zinn, workers showed more left prefrontal activation, after being trained in an eight-week meditation course. Shockingly, even the immune response seemed to be better among people who meditated (Davidson et al., 2003). Additionally, Sara Lazar found that meditators had thickened cerebral cortex in the anterior insula, sensory cortex, and prefrontal cortex (Lazar et al., 2005).

Current Development

In the previous section we understood the benefits of mindfulness and meditation, as far as scientific evidence is concerned. Now, we will delve into how devices and computing can be used to benefit people through the route of mindfulness. There have been certain recent developments of AI and wearables in the field of Mindfulness Based Stress Reduction and Mindful Breathing meditation, but it seems to be extremely limited and non-specific in nature, not at all covering the wide varieties of mindfulness practices. The interaction of AI and wearable technology is revolutionizing the field of mindfulness and meditation, particularly in the field of Mindfulness-Based Stress Reduction (MBSR) and Mindful Breathing.

Some of the important developments include programs, feedback systems, and trackers. For example, Tailored meditation programs deal with AI algorithms that can evaluate individual data, such as heart rate, breathing patterns, and sleep quality, to make individualized meditation plans for diverse individuals. These plans can adjust in real-time based on the user's current state and needs. On the other hand, usage of AI apps have also been used to provide real-time feedback to individuals, based on breathing, type of posture, and level of focus/distraction. This current feedback system helps users to modify their meditation practice and achieve a deeper state of relaxation. Additionally, in the field of adaptive mindfulness techniques, AI has been used to constantly work on adjusting the duration of meditation sessions based on the user's attention span, stress levels, and other factors. This ensures that the practice remains engaging and effective.

There has also been evolution of wearable technology for enhanced mindfulness, which includes a number of devices including *biofeedback devices* (wearable devices which are able to check physiological data including heart rate, skin conductance, and breathing rate, in order to provide real-time feedback on stress levels and relaxation, used to guide mindfulness practices and check the progress time to time); *smartwatches* and *fitness trackers* (devices which can track behavioral data like physical activity, sleep patterns, and levels of stress); and *neurofeedback headbands* (devices that can measure real-time brainwave activity and provide real-time feedback to help users train their minds to achieve a state of relaxation and focus).The integration of AI and wearable technology in mindfulness can have the capability to significantly improve mental health and well-being. Integrating the power of technology with ancient wisdom,humans can create a more mindful and compassionate future. By integrating such assisted mindfulness practices into daily routines, users can improve their overall well-being.

The body of research on self-compassion and mindfulness has grown significantly in recent years. Self-compassion refers to the process of treating oneself with the same kind of kindness, humility and care which one usually displays to very close friends and other near and dear ones. As per various researchers, self-compassion and mindfulness-based therapies are two distinct approaches that can significantly reduce burnout within professional settings (Neff et al., 2020; Suleiman-Martos et al., 2020). This is because they can help people with particular training needs, like those with impairments, access serious mental health training programs of high quality as well as various personalized interventions. Digital mental health tools are becoming more and more popular. In the near future, people will have unrestricted access to excellent training programs that are tailored to their own requirements and preferences, which is one of the major benefits that smart technology will offer. As expected, AI already has shown potential and proved to be having a significant role in a number of digital mental health programs (D'Alfonso, 2020; Boucher et al., 2021). AI apparel is expected to drastically change how mental health interventions have been perceived till date, especially in those cases which are addressed to persons with special training needs (Shaban-Nejad et al., 2018). Extended Reality (ER) including Virtual Reality (VR), Augmented Reality (AR), and Mixed Reality (MR) technologies are slowly gaining their positions in the field of mental health interventions (Flavián et al.,2019 Goharinejad S et al.,2022). In VR, a person interacts with the digitally created environment as if the environment is real. On the other hand, in AR, the user experience of the real world around us is enhanced by the incorporation of images, videos or holograms. Finally, MR blends our real world with the virtual world, allowing us to interact with both digital (non-real) as well as physical (real) elements. A recent experimental study investigated the role of immersive technologies in such intervention based programs which aim to guide relaxation and develop self-regulation skills among people with mental health conditions (Riches et al., 2023). The increasing interest in neurocognitive advancement has also directed the research focus on neurofeedback technologies with a special focus on non-invasive smart brain-sensing headbands (Ghosh et al.,2023). Mindfulness programs which are digitally assisted, have been said to be helpful for individuals to develop their meditation practice more swiftly and effectively (Brandmeyer et al.,2013). Numerous uses of AI are now available to support mental health training. For example, chatbots, sometimes referred to as conversational agents, are a common application of artificial intelligence. In lay terms, AI based chatbots are certain interactive digital platforms which create an artificial sense of real-time interaction with real human beings. These unique human–machine interaction interfaces enable a computer program to have a meaningful text-based or speech-based conversation with human users (Boucher et al., 2021). During training, AI chatbots can give users advice, encouragement, and positive reinforcement in addition to teaching them new skills. Users' motivation to participate in activities aimed at enhancing mental health can be raised by AI chatbots. They can also serve as a self-management strategy developer, a mood tracker, and a tool for mental health assessments (Haque et al., 2023).

Thanks to immersive technologies, users can now feel completely present in the virtual reality environment (Schuemie et al., 2001). By immersive technologies, we mean to indicate such technological advancements which provide a sense of absolute or near-absolute realness, blurring the line between what is real (physical) and what is not (digital) (e.g. VR, AR, MR). Having intentional and open-hearted presence constitutes a key training component in almost every mindfulness practice in general (Kabat-Zinn, 2003). Moreover, VR technologies have been said to be promising in providing users with newer, elaborate, and even some transcendental experiences, motivating them to closely observe and reflect upon these experiences as well as the mental and emotional states that these experiences tap into (Asad et al.,2021). VR input/output devices produce remarkable sensory experiences. Furthermore, these gadgets give consumers the exceptional chance to control and recognize sensory flow (Seabrook et al., 2020; Bruggeman et al., 2018). In order to help users become even more conscious of their own mental and emotional states, attention, anxiety, as well as their general preparedness to be fully involved in training tasks. These neurofeedback devices are not very expensive, quite portable, user friendly, and absolutely non-invasive. Setting up such devices is easy and can be easily connected to a smartphone application which can be downloaded for free (Millstine et al., 2019; Pawade et al., 2022).

Mindfulness apps (AI-powered) which incorporate a curriculum based on emotional intelligence, were also found effective among students for ameliorating social awareness, emotional awareness as well as increasing resilience combined with emotion regulation skills. In the field of mental health, SERMO is a relatively new mobile application which has a chatbot (AI based). It was found to be really effectively executed in mindfulness practices, in combination with cognitive behavior therapy (CBT), in order to support people living with various kinds of mental health problems especially those which are related to regulating emotions and effectively handling one's thoughts and feelings. Users also reported positively about the efficacy and attractiveness of this mobile application (Denecke et al.,2021).

Ajivar is yet another AI conversation platform. It uses AI as well as machine learning in order to achieve individualized mindfulness practices for training skills related to emotional intelligence. This is delivered using text-oriented conversations, various visuals, and related positive affirmations, a technique used for self-help. Ajivar was reported to be able to effectively respond to users' emotions. At the same time, it can identify users' underlying beliefs and appropriately adjust responses. *Wysa* is another, conversational AI based agent, which is empathy-driven in nature. It aims to deliver digital mindfulness training. Specifically, *Wysa*, through written as well structure-based conversations can respond to users' emotions and suggest therapeutic self-regulation strategies based on mindfulness, cognitive behavioral therapy (CBT), and techniques based on positive psychology. Inkster et al. (2018) investigated in a study, whether *Wysa* could actually help 129 users in order to develop their self-management skills. The results obtained were indicative of the fact that mindfulness training with *Wysa* is quite effective and acceptable. These findings also revealed that this app significantly helped its users to develop affirmative self-expression as well as resilience skills related to one's emotion.

Another example is *Youper*, which is basically an AI chatbot. Chatbot, as described earlier too, is a human-computer interface providing real-time feel. *Youper* is developed in order to train its users for detecting, apprehending, and managing their thoughts and feelings. This AI based app not only involves mindfulness training, but also combines elements of cognitive behavioral therapy (CBT) as well as acceptance and commitment therapy (ACT). Mehta et al. (2021) seems to claim that mindfulness training based on artificial intelligence, can be helpful for users, in order to provide systematic training on regulating one's emotions, which in turn may result in long-term and successful reduction in symptoms.

METHOD

Study Design

We chose the instructions of the Joanna Briggs Institute's (JBI) guidelines for systematic reviews based on effectiveness as well as the Beginner's Guide for Systematic Reviews published by the Indian Council of Medical Research (ICMR). The framework offered a comprehensive and transparent approach to examining the efficacy of wearable AI driven Mindfulness practices in the management of stress.

Inclusion Criteria

After the identification of the review question, elements of PICO (Population, Intervention/Exposure, Comparison, and Outcome) were identified as followed:

Population

Individuals who are 15 to 60 years of age, including but not limited to students, working professionals, healthcare workers, and individuals with high-stress lifestyles.

Intervention/Exposure

Smart wearable devices designed to support mindfulness and stress management.

Comparison

Comparable conventional methods of mindfulness or mindfulness techniques.

Outcome

Reduction in stress levels measured using physiological and psychological indicators.

Studies involving persons between 15 to 60 years of age were included, while studies involving children up to 15 years of age and persons above 60 years of age, were excluded. Only those studies were included which involve the usage of wearable devices aimed for stress management and mindfulness. Studies which referred to wearable devices aimed for any domain other than stress management and mindfulness or which include technologies based on stress management and mindfulness, other than wearable devices, were also excluded.

Search Strategy

We focused upon the PICO framework while conducting the search. Here, population referred to "adults", while exposure referred to "wearable devices" or "wearables", while no specific criteria were used for describing the comparison, while "reduced stress", "better mindfulness", "stress relief" referred to the outcomes.

Search Procedure

The literature search started through the registration of the topic of the systematic review at PROSPERO, with the registration number CRD42024618977. After receiving approval, the screening was conducted from December 2024 to January 2025 by the authors after deciding upon the search strategy mentioned previously, using Google Scholar, PubMed, ResearchGate, PsycINFO, Science Direct, and Web of Science. For Google Scholar "mindfulness+mindful+stress+wearable+devices+AI" was used, for PubMed the following were utilized: "(mindfulness) AND (wearable)", "(mindful) AND (wearable)", "(stress) AND (wearable)", "(mindfulness) AND (AI)", "(mindful) AND (AI)", "(stress) AND (AI)", "(mindfulness) AND (wearable devices)", "(mindful) AND (wearable devices)", "(stress) AND (wearable devices)", and for ResearchGate "mindfulness mindful stress wearable devices AI" was utilized.

Search Records

The citations used in our review were not managed using any software. Instead, the citations were manually generated in accordance with APA 7the edition standard. Screening was manually conducted to minimize bias (Shamseer et al., 2015). Using the inclusion and exclusion criteria, the papers were shortlisted. Full text was available for all the 7 papers. The findings were elaborated using the PRISMA flow diagram to ensure transparency as well as adherence to expected standards of a systematic review. The same is presented in Figure 1. Studies from any country were considered for this review, provided that they were written in English. Only those papers were included which addressed the usage and/or effectiveness of wearable devices as related to mindfulness and/or stress. Disagreements among individual judgements were resolved through mutual discussion.

Quality Appraisal

The study being qualitative in nature was assessed (Figure 1) using the Joanna Briggs Institute's (JBI) guidelines. "Yes" was coded green, "no" was coded red, "unclear" was coded yellow and "not applicable" was coded blue. Regardless of the bias discovered, all of the 7 studies were included for the systematic review, since this enhanced the overall synthesis, findings and discussion related to the objective of the review. The JBI standardized appraisal tool based on study design appropriate for JBI reviews of effectiveness was selected as per guided (see Appendix 3.2 of JBI manual regarding the JBI standardized appraisal tools).

Figure 1. Systematic review critical appraisal using JBI critical appraisal checklist for randomized controlled trials

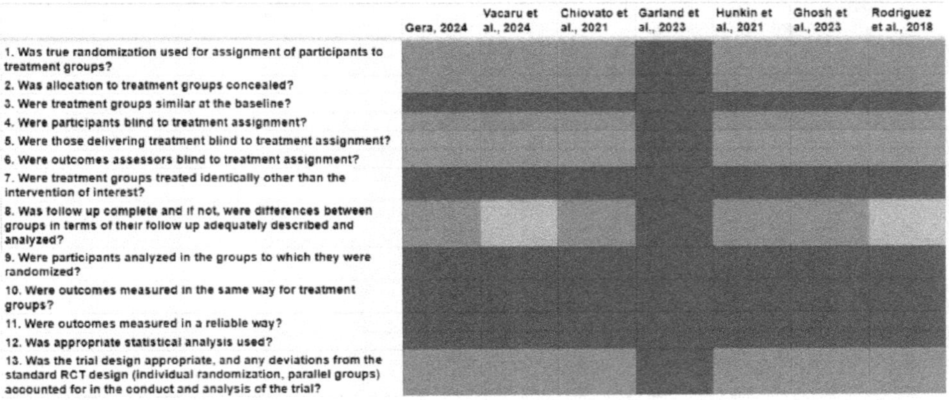

Figure 2. Prisma flow diagram

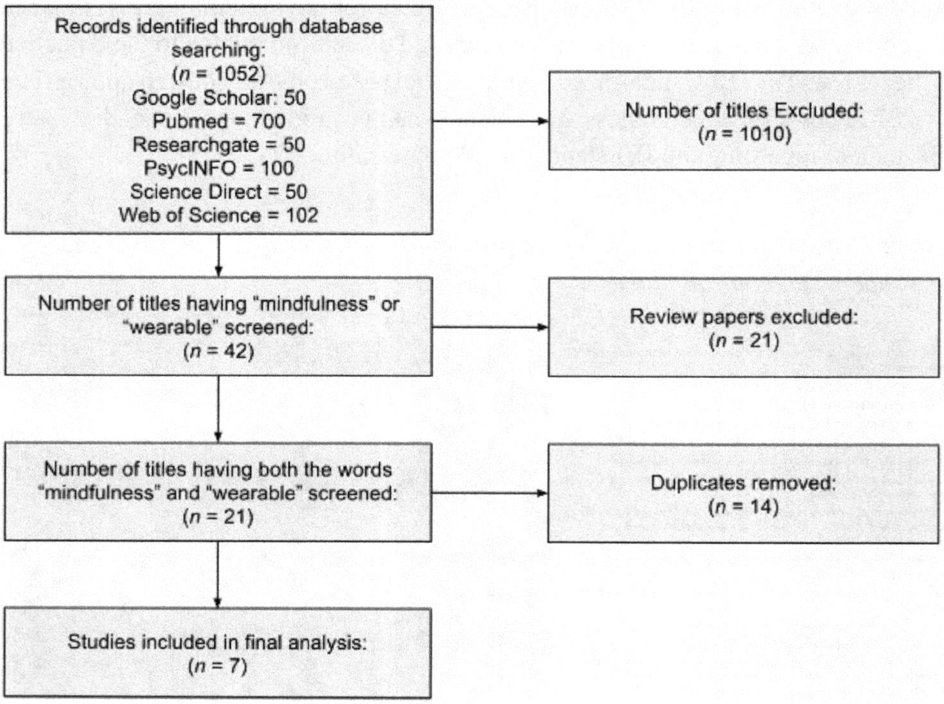

Synthesis

In our analysis of the seven studies, we extracted key information pertaining to the study aims, types of methods used, results, and other notable information. We then employed Braun and Clark's (2006) qualitative thematic analysis framework to identify and code salient concepts and ideas. Through a process of iterative refinement, these codes were organized into emergent themes. To ensure the trustworthiness and validity of our findings, we conducted a rigorous review of the themes, verifying their coherence and alignment with the original data. This involved a systematic comparison of the obtained themes with the coded extracts, followed by a re-examination of the data, in order to confirm that the themes captured the essential insights. Throughout this process, we were alert of the possibility of bias. Therefore, we took steps to mitigate this risk by incorporating data from a number of sources and methodologies, without being rigid. This approach enhanced the credibility and applicability of our conclusions.

RESULTS

Search Results

1052 records from across six databases were obtained as a result of the search, as mentioned. 1010 records were remaining after irrelevant papers were removed. 42 of those records were selected based on their titles, while 21 review papers were excluded leading to a complete text review of the remaining 21 articles. After the exclusion of 14 duplicate studies, 7 articles were finalized for inclusion. The studies which were excluded involved those which did not focus on individuals (below 15 years & above 60 years of age), studies not involving wearable technology and studies that did not report on stress and/or mindfulness. A total of 7 articles were identified for inclusion in the systematic review.

Characteristics of Included Studies

The 7 studies which were included were published during the period of 2018 to 2024. Study types included quantitative studies (n = 7). The studies investigated well-being, various wearable sensors, pain relief, postures and much more. Participants included students, patients, meditators, healthcare professionals and other adults. The types of wearables included AI voice assistants, EEG headbands etc. Detailed study characteristics are presented in Table 1.

Table 1. Study characteristics

Study	Year	Author	Type	Sample
Integrating Mindfulness: Enhancing Art and Design Student Well-being through AI Voice Assistants and Wearable Technology	2024	Gera, P.	Quantitative	13 Art and Design student (Mean age 35.6)
Selecting optimal wearables for measuring physiological arousal in robot-delivered mindfulness-based exercises	2024	Vacaru, S., Fekete, K., Luyten, L. P., Saldien, J., & Bosse, T.	Quantitative	28 healthy Dutch-speaking adults (50% female, mean age 23.68).
Evaluation of mindfulness state for the students using a wearable measurement system	2021	Chiovato, A., Demarzo, M., & Notargiacomo, P.	Quantitative	13 Brazilian high school students (8 female, 5 male; mean age 16.1 years) in a controlled, remote classroom setting.

continued on following page

Table 1. Continued

Study	Year	Author	Type	Sample
Zoom-Based Mindfulness-Oriented Recovery Enhancement Plus Just-in-Time Mindfulness Practice Triggered by Wearable Sensors for Opioid Craving and Chronic Pain	2023	Garland, E. L., Gullapalli, B. T., Prince, K. C., Hanley, A. W., Sanyer, M., Tuomenoksa, M., & Rahman, T.	Quantitative	63 Opioid-treated chronic pain patients
Evaluating the feasibility of a consumer-grade wearable EEG headband to aid assessment of state and trait mindfulness	2021	Hunkin, H., King, D. L., & Zajac, I. T.	Quantitative	68 adults (mean age 22.66 ± 7.35), with 29 completing 14 days of home practice.
Mindfulness Using a Wearable Brain Sensing Device for Health Care Professionals During a Pandemic: A Pilot Program	2023	Ghosh, K., Nanda, S., Hurt, R. T., Schroeder, D. R., West, C. P., Fischer, K. M., Bauer, B. A., Fokken, S. C. Ganesh, R. Hanson, J. L., Lindeen, S. A., Pruthi, S., & Croghan, I. T.	Quantitative	40 healthcare professionals (HCPs; 85% female, 87.5% White) employed at a Midwestern U.S. healthcare facility.
Wearable Sensors for Measuring Movement in Short Sessions of Mindfulness Sitting Meditation: A Pilot Study	2018	Victor H. Rodriguez, Carlos T. Medrano, Inmaculada Plaza	Quantitative	31 participants (28 novices, 3 experienced meditators)

Summary of Findings

AI-driven wearable devices have emerged as innovative tools in stress reduction and mindfulness interventions. Firstly, we realized that there is a dearth of studies related to mindfulness based wearable devices, especially as related to stress. The seven studies analyzed in our systematic review provide insights into their efficacy, mechanisms, and potential improvements, with special reference to AI used (if any) as well as stress relief and mindfulness. Using thematic analysis, five core themes were identified as a result of the qualitative thematic analysis, including: (1) AI and Personalization in Mindfulness Interventions, (2) Physiological Monitoring and Biofeedback, (3) Accessibility and Usability, (4) Psychological and Behavioral Outcomes, and (5) Future Directions in Wearable Technology.

AI and Personalization in Mindfulness Interventions

User engagement has been truly revolutionized via the integration of AI into mindfulness applications, providing real-time personalized experiences while using wearables. AI-driven systems help individuals to manage their stress through adaptive mindfulness exercises tailored to their physiological and psychological states.

This is a very essential element since humans have a diverse range of capacities, intelligence, talent, interests, skills and emotional needs. In one of the studies, Gera (2024) discusses how an AI-powered voice assistant named SATI, helped to tailor mindfulness exercises for art and design students, focusing on the role of AI in adaptive intervention strategies. Here, it is interesting to note that the choice to name "SATI," was theoretically based on Buddhism's concept of *Sati* which translates to awareness, and not a random abbreviation.This singular word itself, encapsulated the essence of the chatbot's function, implied a nurturing and encouraging identity, which was best portrayed by a standard female voice. This was not a decision based on personal taste, but a reaction to society and cultural norms that equate caregiving and support with femininity. The very interactive nature of this assistant, enabled students to customize their mindfulness exercises based on their creative workflow, helping them manage stress while enhancing creativity. Likewise, in a study by Garland et al. (2023) which dealt with the exploration of Just-in-Time Adaptive Interventions (JITAI), AI based algorithms were used as a tool to trigger mindfulness-based exercises. These practical triggers worked on indicators which were alarmed through physiological stress and such were effectively detected using certain wearable sensors. This approach seemed to have a dual benefit. Not only did it facilitate real-time based intervention, it also demonstrated the future potential of AI in delivering mindfulness strategies precisely whenever needed.

AI-driven mindfulness interventions are open to user feedback due to their advanced complexity. Feedback is very important because it helps the system to learn, improve and grow, just like human beings who receive their feedback for similar purposes. AI-enhanced mindfulness applications examine biometric data during the usage of the application and hence, deliver exercises that are meant to enhance relaxation and focus. A study by Garland et al. (2023) indicated that using the MORE+JITAI system, stress levels can be predicted and machine learning algorithms can start guided interventions proactively. Here MORE means Mindfulness-Oriented Recovery Enhancement. As users have specific stress and relaxation patterns, AI recommendation systems enable refinement, thereby making the device more effective in promoting mindfulness.

AI-based applications seem to struggle with a number of issues despite the noticeable recent advancement in wearable technology. One main drawback which we noticed is the ability to guarantee accurate, error free and context-specific AI responses. The study by Gera (2024) reported that the background noise in the environment sometimes caused the SATI voice assistant to function ineffectively. This was believed to contribute to unauthentic mindfulness experiences. Furthermore, AI based applications need to be enhanced to identify deep stress signals or even the slightest changes in heart rate variability (HRV) or EEG signals. Future work needs to emphasize improving AI's contextual awareness as well as incorporating additional biometric measures for better personalization and stress alleviation.

Physiological Monitoring and Biofeedback

Our analysis revealed that wearable devices are able to contribute significantly to mindfulness interventions by providing real-time physiological data, allowing users to track their own stress levels as well as their progress in mindfulness. For example, Vacaru et al. (2024) compared two devices, the Empatica E4 wristband and the Shimmer sock, for detecting electrodermal activity (EDA) variations. Both of these devices seemed to effectively track stress-based arousal and relaxation responses as a result of the robot-delivered mindfulness interventions. However, discrepancies in sensitivity between the two anatomical locations (wrist and foot) suggest that the placement of the device on a person's body also influences data accuracy.

The study by Chiovato et al. (2021) extended the territory above context, by utilizing heart rate variability (HRV) as a factor to measure mindfulness states among students. It was reported that there is a correlation between increased HRV and heightened mindfulness. It suggests that biofeedback can be used to enhance self-awareness when it comes to mindfulness meditation practices. In a similar manner, Hunkin et al. (2021) conducted a study using EEG metrics embedded within the Muse headband and examined its applicability for self-reports of mindfulness. All these studies emphasize the potential and usefulness of wearable technology.

Among many advantages, one of the major features of wearable biofeedback technology is its ability to provide objective and empirical measurements of mindfulness and calmness. Unlike self-reported subjective mindfulness experiences, these biofeedback devices can effectively track even subtle physiological changes, offering comprehendible evidence of stress responses and progress to the users. According to Vacaru et al. (2024) biological measures of stress-reduction were much more reliable and valid for consideration, than user-reported experiences. This is because participants often tend to underestimate the levels of their own physiological arousal, which may lead to improper reports. Hence, for providing more accurate stress assessments, AI-driven analytics can be utilized.

Though there are many benefits of AI-driven wearables, yet several limitations also emerge as a concern. For example, sensor accuracy, device comfort, and real-time data processing emerge as challenges in ensuring seamless physiological monitoring, since certain wearable devices require calibration and proper placement to deliver accurate readings. We recommend that future research should focus on optimizing AI-integrated sensor technology in wearables, to enhance reliability and usability in stress and mindfulness applications.

Accessibility and Usability

Gera (2024) raised issues related to different interfaces, especially readability within the app, and problems of consistency in performance if run in a noisy environment setup. The issue of wearability was highlighted by Vacaru et al. (2024) reporting that the Shimmer and the Empatica E4 devices received comparable ratings regarding comfortable usage. This indicates that wearability is a critical determinant for ongoing engagement in stress management and mindfulness practices. Hence accessibility and tactile comfort in AI-enabled wearable mindfulness devices are crucial to encourage widespread adoption.

The research by Ghosh et al. (2023) explored the feasibility of the Muse-S headband for healthcare professionals during the COVID 19 pandemic, indicating high adherence rates (92.5%) despite minor usability issues related to fit and sensor placement. These findings highlight the importance of designing devices which not only balances technological complexity with user-friendly interfaces, but are also relevant. Future research has the scope to focus on improving the ergonomic design and ensuring the ease of accessibility across diverse populations to support the adoption of mindfulness especially in stress-prone environments.

Additionally, when it comes to postures, Rodriguez et al. (2018) investigated the role of wearable sensors in improving mindfulness posture stability. Zafus (a round, firm meditation cushion primarily used in Zen meditation to support a comfortable and upright posture) and meditation benches were reported to be better at providing support for a stable meditation posture than chairs. The importance of designing AI-enhanced mindfulness wearables that accommodate different user preferences and postural needs were hence highlighted through this study. Hence, we suggest that, as such technologies evolve, priority should be given towards user feedback for ensuring that the design of the wearable devices are inclusive and adaptable for all users who are seeking stress relief and mindfulness enhancement, not just able-bodied individuals.

Psychological and Behavioral Outcomes

The psychological benefits of AI-driven mindfulness interventions are well-documented in certain studies which we explored. AI-driven wearable technologies have been found to improve stress resilience, emotional regulation, and overall well-being of the participants. For instance, Garland et al. (2023) found out that participants in the "MORE+JITAI" (Mindfulness-Oriented Recovery Enhancement plus Just-in-Time Adaptive Intervention) group reported significant reductions in opioid cravings and stress, along with increased positive affect. These findings align with the study by Ghosh et al. (2023), where healthcare professionals using the Muse-S device experienced reductions in stress along with improvements in resilience as well as their quality of life.

A study by Rodriguez et al. (2018) examined the impact of wearable sensors on meditation posture stability, as mentioned previously. Their study concluded that posture stability could enhance mindfulness experiences. This clearly suggests that AI-driven wearable technology can complement traditional mindfulness practices by providing feedback on physical alignment (proper or improper), thereby improving the efficacy of meditation. Hence it is evident that these studies highlight the psychological and behavioral benefits of AI-driven mindfulness interventions, reinforcing their positive role in mental health and stress management.

Further, Hunkin et al. (2021) found that EEG-based feedback in meditation improved self-regulation, helping users maintain focus and reduce distractions. The ability of AI-enhanced wearables to provide neurophysiological feedback represents a promising avenue for enhancing mindfulness practices. Future research should explore how different psychological profiles respond to AI-driven mindfulness tools to ensure that interventions are tailored to individual needs.

Therefore, our systematic review of AI-driven wearable mindfulness interventions reveals their substantial potential in stress reduction and mental well-being. The themes which we identified, definitely underscore the multi-faceted role of wearable technology in mindfulness and stress-based research. As further advancements continue, ensuring better and user-friendly designs, refining capabilities of AI, and expanding research to diverse populations will be pivotal in maximizing the efficacy of these AI-powered interventions in stress and mindfulness applications.

DISCUSSION

Innovative methods of stress management have been made possible by the combination of technology and mindfulness practices. Hence, our discourse below explores how the usage of various wearable technologies, a number of adaptive systems, and the integration of AI in such devices are constantly revolutionizing the field of mindfulness practices. We tried to examine the interactions among physiological monitoring, adaptive interventions, user engagement, and AI-assisted mindfulness training, through thematic analysis.

Physiological Insights through Wearable Technology

Biological parameters associated with mindfulness and stress can now be monitored with wearable technology. In comparison to traditional self-assessment techniques, metrics like electrodermal activity (EDA) and heart rate variability (HRV) offer a substantial benefit through real time data of body responses (Chung et al., 2023; Rodriguez et al., 2018). EDA basically measures the changes in the electrical conductivity of our skin, reflecting sweat gland activity and hence, primarily indicating sympathetic nervous system activity. HRV, on the other hand, measures the variation in time between heartbeats, providing various insights about the balance between the sympathetic and parasympathetic branches of the Autonomic Nervous System (ANS). It is also essential to know that the sympathetic and parasympathetic nervous systems are two divisions of the ANS, working together to maintain homeostasis by regulating involuntary bodily functions like heart rate, digestion, and breathing. As far as researchers and practitioners are concerned, due to the objective evaluation of stress levels and mindfulness states, more objective analysis is made possible through these physiological indicators. At the forefront of this change are gadgets like the Empatica E4 wristband and Muse EEG headband. For example, research has shown a positive correlation between mindfulness and HRV value, specifically root mean square of successive differences (rMSSD) over 88 ms (Hickey et al., 2021). According to the above mentioned results, wearable technology can be used to detect how mindfulness-based practices affect the physiological stress response, offering substantial evidence.It's fascinating to note that various forms of meditation have differential effects on biomarkers. Reduced heart rate and increased rMSSD, which indicate improved relaxation, have been associated with focused attention meditation, Conversely, open monitoring meditation, which means actively watching thoughts and feelings, has been linked to lower cortisol levels and active HRV ratios (Vacaru et al., 2024). These outcomes show how mindfulness practices can be used to reduce stress in a variety of ways, which are biologically linked.

Wearable technology permits physiological conditions of the users to be continuously monitored during their daily lives, which in turn is especially useful for reducing stress levels. Rodriguez et al. (2018), for instance, highlighted how wearable sensors may identify minute variations in stress markers during a range of activities, showing how people react to different situations.

Adaptive Interventions: Tailoring Mindfulness in Real-Time

A significant development in personalized support of mindfulness is the introduction of Just-in-Time Adaptive Interventions (JITAIs). JITAIs provide exercise depending on one's stress level measured by using real-time data from wearable technology. It has been particularly beneficial for treating a variety of health issues, such as chronic pain, opioid addictions, and overall stress (Hickey et al., 2021; Vacaru et al., 2024). Predicting stress level through the use of biological data, machine learning algorithms are important for improving JITAIs. These predictions have sensitivity rates between 60% and 76%, which provide space for accurate timing of interventions (Chung et al., 2023). For example, a JITAI system may encourage the user to interact when a wearable gadget detects raised heart rate or heightened EDA levels, which are signs of stress. This immediate feedback loop is essentially helpful at the time of stress.

However, the experiences of the users differ regarding the effectiveness of JITAI. Vacaru et al. (2024) pointed to a number of factors such as baseline stress level, personality of the individual as well as previous experience of mindfulness practice affect those experiences. In response to the growing need of individualized attention, researchers now begin to look at increasingly complex algorithms that can adapt to the distinct physiological and psychological characteristics of each individual in order to overcome these difficulties. Combining situational information into JITAI systems is another interesting approach to adaptive interventions. According to the study done by Rodriguez et al.,(2018), to improve the prediction and availability of the system, we can integrate biological data with details about a user's location, time of day, or level of activity.

Enhancing Engagement through Technology

Sustaining user's attention over time is one of the key challenges in mindfulness practices. Through making mindfulness exercises more engaging, interactive and accessible, technology has shown great potential. Increasing participation rates among diverse populations through the use of AI-powered voice bots can be highly effective as they provide personalized guidance and encouragement (Vacaru et al., 2024). Cultivating a sense of connection and assistance that can motivate users to incorporate mindfulness into their daily routines, these digital companions are quite impressive.

Wearable technology helps to enhance user engagement as they are convenient and provide real-time feedback. This capability helps to integrate mindfulness practices in daily lives, as they can engage in anywhere, at any time, and get real-time progress reports. According to studies, people who use mindfulness wearables have encouraging adherence rates; on average, they practice mindfulness 23.8 times a month (Rodriguez et al., 2018). This degree of involvement highlights the technological advancement in daily lives.

Connecting social relatedness into mindfulness based wearable devices is an evidence based strategy as it offers opportunities for peer support and accountability. The importance of fostering social connections as part of technology-based mental health interventions, was emphasized by Vacaru et al. (2024), stressing upon the fact that users are capable of engaging to virtual communities through some apps where they can share their progress, giving each other feedback and participate in group meditation sessions.

Despite these advancements maintaining user's engagement is quite challenging. As per the study by Hunkin et al. (2021), wear-off in novel manner can sometimes be the reason for reducing users' compliance. Developers are combining tactics like components of gamification, tailored objectives, and social relatedness features to address this problem. Here, gamification refers to adding game-like features, such leaderboards, challenges, or rewards to enhance the quality of user engagement over time. Letting users have a sense of autonomy in their mindfulness journey by giving particular objectives in accordance with their requirements and preferences is essential.

With rapid technological advancement, mindfulness training through the use of AI is giving us endless opportunities for increasing the engagement of users. Devices such as EEG neurofeedback devices and robotic mindful coaches provide users the space for getting real-time feedback of their effectiveness as well as their progress. Using advanced algorithms, these devices analyze brain activity and other physiological signals during the session and provide scope for improvement in the near future (Hickey et al., 2021; Vacaru et al., 2024). Research has also provided evidence regarding the significant benefits associated with AI-driven mindfulness training. For example, a 90-day study done by Mitesa et al. (2023) found that as a result of mindfulness training, the subjects saw significant reduction in stress level, increased resilience and enhanced quality of life, highlighting the potential for AI-based approaches to deliver significant output for mental health and well-being.

It is essential to tailor these solutions in accordance with the needs of each individual. Because if we do not consider these individual differences such as age group or cultural background, robotic mindful coaches give inconsistent results (Vacaru et al., 2024). This highlights the importance of considering diversity while developing AI based solutions. To ensure that AI-based devices meet the needs of various user populations, while also upholding the effectiveness of mindfulness, research have to develop increasingly complex customized algorithms, enhancing AI voice assistant's natural language processing capacity and most importantly finding harmony between traditional mindfulness practice and technological advancement (Hunkin et al., 2021).

Limitations

Certain limitations exist while conducting literature review. Due to the limited number of available research, no qualitative or mixed methods data were available to be analyzed. Limitations also include the exclusion of virtual reality and potential keyword gaps in study selection. Development and execution of VR-based mindfulness therapies depend much on cultural adaptability and diversity, which guarantees their inclusiveness, accessibility, and efficacy across various communities. Mental health stigma within any specific cultural context is essential to be addressed in order to promote engagement and adherence to the technology and the devices. This might require additional outreach programs, involving educating community members about mindfulness and its benefits, especially in this technological era. Diversity issues like age should be kept in mind, since members of all age groups might not easily understand and/or require the same kind and duration of mindfulness-based practices. Research on wearables involving mindfulness still does not adequately represent diverse racial and ethnic groups, leading to a lack of culturally relevant interventions and potentially reinforcing biases.

Future Scope

Qualitative or mixed methods research needs to be conducted along with more empirical research in order to validate the findings and bring forth more insights related to any side effects or long term negative effects of any wearable devices.

LONG-TERM IMPLICATIONS

According to our review findings, wearable AI-driven mindfulness practices for stress management show promising long-term implications, though they present both opportunities and challenges that require careful consideration:

Health Outcomes

The first point would definitely be "Sustained Mental Wellness". Studies suggest AI-enhanced mindfulness can improve self-regulation skills (attention, emotional control) and psychological well-being, particularly when integrated into daily routines (Hunkin,2021). Long-term use may reinforce neuroplasticity, leading to lasting cognitive and emotional benefits. Another factor is "Preventive Care". Continuous biometric monitoring could enable early detection of stress patterns, allowing proactive interventions to prevent chronic mental health issues.

Ethical and Societal Challenges

Firstly, "Data Privacy Risks" is the biggest challenge. Long-term storage of sensitive biometric data increases vulnerability to breaches or misuse, necessitating robust encryption and transparent data policies. Apart from this, due to "Algorithmic Bias" prolonged reliance on AI systems trained on non-representative datasets may perpetuate health disparities, particularly for marginalized groups. Over-Reliance on

Technology is yet another issue. Extended use might reduce intrinsic self-regulation abilities, creating dependency and undermining traditional mindfulness practices.

Workplace and Economic Impact

Productivity Gains is a major impact which needs to be discussed. Organizations adopting these tools report improved employee engagement and profitability, but long-term success depends on balancing corporate interests with genuine well-being support. Commercial exploitation, without proper regulation, monetization of user data could prioritize profit over user health, eroding trust in digital mental health solutions

CONCLUSION

The integration of AI and wearable technology in mindfulness and stress reduction interventions has shown considerable promise. Key themes emerging from the literature suggest that physiological monitoring, real-time adaptive interventions, and AI-based mindfulness training can enhance user engagement and efficacy. However, further research is needed to optimize these technologies for broader application and sustained effectiveness.

Ethical Issues

Data Privacy

Privacy is the key to future cyber security and data security. Wearable devices collect sensitive health and biometric information. If this kind of data is not securely stored by the operators or shared externally without any user consent, it may lead to privacy breaches and misuse.

Algorithmic Bias

AI algorithms may be trained on non-representative data, leading to inaccurate stress detection for certain populations (e.g., minority communities or people with atypical physiological responses), which can perpetuate health inequalities and inequities.

User Autonomy

In case of heavy reliance on AI recommendations, users' ability to self-reflect and make independent decisions about their mental health might get diminished, potentially eroding personal agency and control.

Psychological Harm

In case of over usage, such constant monitoring and feedback about stress levels can inadvertently heighten users' anxiety or stress, especially if they become overly fixated on their biometric data. Also, such over usage if guided by non-experts, can lead to unwelcome damage, which is completely against what the AI driven devices were meant for.

Informed Consent

Users may not fully understand what data is being collected, how it is used, or who has access to it. Lack of transparency can undermine truly informed consent. Such a consideration is especially true in an era where almost none of the user data is truly safe from even the simple mobile games and mobile applications, including social media.

Commercial Exploitation

Developers might use collected biometric data for commercial gain (such as targeted advertising or selling data to third parties) without providing clear benefits to the user. Such exploitation would lead to a technological disaster as well as a health based catastrophe.

Transparency

The functioning of AI algorithms is often opaque. For-profit organizations most often do not share, or do not want to share the real algorithms. Users should have access to clear explanations of how decisions are made and how their data influences outcomes. Users should also have clear knowledge of the adverse effects or side effects (if any) of the devices if used or overused.

Well-Being Versus Profit

There is also an ethical concern when institutions prioritize profit over user well-being, potentially leading to the development of features that benefit the business rather than genuinely supporting users' mental health. Such toxic-capitalism is not only an alarming fact for the end users, but also for humanity in general.

Assistance of Therapist

A therapist often plays a crucial role in guiding clients (especially the ones who have moderate to severe mental health issues), adapting techniques, and ensuring safe and effective experiences, just as mindfulness. Such tailored experience is especially necessary for the users who have a history of mental health issues or those users who are unsure of the devices. Moreover, a therapist can ensure that what degree of usage is suitable for a person, and to what extent should be increased or decreased. Not every person will have the same subjective experience during their mindfulness practice, hence, having an expert advisor is not just a choice but seems to be a necessity. For organizations, this would not only increase the credibility but also ensure that there is a backup ready for emergency situations, including a human touch.

ACKNOWLEDGMENT

Competing Interests:

The authors of this publication declare there are no competing interests.

Funding Agency:

This research received no specific grant from any funding agency in the public, commercial, or not-for-profit sectors. Funding for this research was covered by the authors of the article.

REFERENCES

Asad, M. M., Naz, A., Churi, P., & Tahanzadeh, M. M. (2021). Virtual reality as a pedagogical tool to enhance experiential learning: A systematic literature review. *Education Research International*, *2021*(1), 7061623. DOI: 10.1155/2021/7061623

Balconi, M., Fronda, G., & Crivelli, D. (2019). Effects of technology-mediated mindfulness practice on stress: Psychophysiological and self-report measures. *Stress (Amsterdam, Netherlands)*, *22*(2), 200–209. DOI: 10.1080/10253890.2018.1531845 PMID: 30472901

Borthakur, R. (2021). U.S. Patent Application No. 16/630,451.

Boucher, E. M., Harake, N. R., Ward, H. E., Stoeckl, S. E., Vargas, J., & Stoeckl, S. M. (2021). Artificial intelligence and cognitive behavioral therapy: An integrative approach to address challenges in mental health care delivery. *Journal of Medical Internet Research*, *23*(5), e26980. DOI: 10.2196/26980

Brandmeyer, T., Delorme, A., & Wahbeh, H. (2013). The neuroscience of meditation: Classification, phenomenology, correlates, and mechanisms. *Progress in Brain Research*, *244*, 1–29. DOI: 10.1016/bs.pbr.2018.10.020 PMID: 30732832

Brantley, J. (2005). Mindfulness-based stress reduction. In *Acceptance and mindfulness-based approaches to anxiety: Conceptualization and treatment* (pp. 131–145). Springer US., DOI: 10.1007/0-387-25989-9_5

Braun, V., & Clarke, V. (2006). Using thematic analysis in psychology. *Qualitative Research in Psychology*, *3*(2), 77–101. DOI: 10.1191/1478088706qp063oa

Bruggeman, K. J., & Wurster, S. W. (2018). The hiatus system: virtual healing spaces: low dose mindfulness based stress reduction virtual reality application. In *ACM siggraph 2018 appy hour* (pp. 1-2). https://doi.org/DOI: 10.1145/3213779.3213785

Bryant, E. F. (2015). *The yoga sutras of Patanjali: A new edition, translation, and commentary*. North Point Press.

Cannon, W. B. (1932). The Wisdom of the Body WW Norton & Company. Inc., New York.

Cannon, W. B. (1932). *The wisdom of the body*. W.W. Norton & Company, Inc.

Carter, R.III, Carter, K. S., Holliday, J., Holliday, A., & Harrison, C. K. (2018). A Community-based Stress Management Program: Using Wearable Devices to Assess Whole Body Physiological Responses in Non-laboratory Settings. *Journal of Visualized Experiments*, *55816*(131), •••. DOI: 10.3791/55816 PMID: 29443019

Chiovato, A., Demarzo, M., & Notargiacomo, P. (2021). Evaluation of mindfulness state for the students using a wearable measurement system. *Journal of Medical and Biological Engineering*, *41*(5), 1–14. DOI: 10.1007/s40846-021-00658-4 PMID: 34608381

Chrousos, G. P. (2009). Stress and disorders of the stress system. *Nature Reviews. Endocrinology*, *5*(7), 374–381. DOI: 10.1038/nrendo.2009.106 PMID: 19488073

Chrousos, G. P. (2009). Stress and disorders of the stress system. *Nature Reviews. Endocrinology*, *5*(7), 374–381. DOI: 10.1038/nrendo.2009.106 PMID: 19488073

Chung, K., Cho, H. Y., & Lee, D. (2023). Wearable devices for stress monitoring: A review of the current state and future prospects. *Journal of Medical Internet Research*, *25*, e42672. DOI: 10.2196/42672

D'Alfonso, S. (2020). AI in mental health. *Current Opinion in Psychology*, *36*, 112–117. DOI: 10.1016/j.copsyc.2020.04.005 PMID: 32604065

Davidson, R. J., Kabat-Zinn, J., Schumacher, J., Rosenkranz, M., Muller, D., Santorelli, S. F., Urbanowski, F., Harrington, A., Bonus, K., & Sheridan, J. F. (2003). Alterations in brain and immune function produced by mindfulness meditation. *Psychosomatic Medicine*, *65*(4), 564–570. DOI: 10.1097/01.PSY.0000077505.67574.E3 PMID: 12883106

Denecke, K., May, R., & Deng, Y. (2021). SERMO: An AI-based mobile application to improve emotion regulation through mindfulness and cognitive behavioral therapy. *Frontiers in Artificial Intelligence*, *4*, 654789. DOI: 10.3389/frai.2021.654789

Flavián, C., Ibáñez-Sánchez, S., & Orús, C. (2019). The impact of virtual, augmented and mixed reality technologies on the customer experience. *Journal of Business Research*, *100*, 547–560. DOI: 10.1016/j.jbusres.2018.10.050

Garland, E. L., Gullapalli, B. T., Prince, K. C., Hanley, A. W., Sanyer, M., Tuomenoksa, M., & Rahman, T. (2023). Zoom-based Mindfulness-Oriented Recovery Enhancement plus just-in-time mindfulness practice triggered by wearable sensors for opioid craving and chronic pain. *Mindfulness, 14*(6), 1329–1345. DOI: 10.1007/s12671-023-02137-0 PMID: 37362184

Gera, P. (2024). Integrating Mindfulness: Enhancing Art and Design Student Well-being through AI Voice Assistants and Wearable Technology (Doctoral dissertation, OCAD University).

Ghosh, K., Nanda, S., Hurt, R. T., Schroeder, D. R., West, C. P., Fischer, K. M., Bauer, B. A., Fokken, S. C., Ganesh, R., Hanson, J. L., Lindeen, S. A., Pruthi, S., & Croghan, I. T. (2023). Mindfulness using a wearable brain sensing device for health care professionals during a pandemic: A pilot program. *Journal of Primary Care & Community Health, 14*, 21501319231162308. DOI: 10.1177/21501319231162308 PMID: 36960553

Ghosh, K., Singh, A., & Ghosh, S. (2023). Neurofeedback: A comprehensive review on system design, methodology and clinical applications. *Biomedical Signal Processing and Control, 78*, 104024. DOI: 10.1016/j.bspc.2022.104024

Goharinejad, S., Alizadehsani, R., Khosravi, A., & Nahavandi, S. (2022). A comprehensive review of applications of EEG and fNIRS in brain-computer interface and neurofeedback. *Neuroscience and Biobehavioral Reviews, 138*, 104712. DOI: 10.1016/j.neubiorev.2022.104712

Goleman, D. (2003). Finding happiness: Cajole your brain to lean to the left. *New York Times, 4*(3).

Goyal, M., Singh, S., Sibinga, E. M., Gould, N. F., Rowland-Seymour, A., Sharma, R., Berger, Z., Sleicher, D., Maron, D. D., Shihab, H. M., Ranasinghe, P. D., Linn, S., Saha, S., Bass, E. B., & Haythornthwaite, J. A. (2014). Meditation programs for psychological stress and well-being: A systematic review and meta-analysis. *JAMA Internal Medicine, 174*(3), 357–368. DOI: 10.1001/jamainternmed.2013.13018 PMID: 24395196

Hanh, T. N. (2006). *True love: A practice for awakening the heart.* Shambhala Publications.

Haque, M. D. R., Rahman, M. M., & Islam, M. R. (2023). AI chatbots in mental health: A comprehensive review. *Journal of Medical Systems, 47*(1), 5. DOI: 10.1007/s10916-022-01810-3 PMID: 38127210

Hickey, S., Kavanagh, E., & McGrath, D. (2021). Wearable technology in the mindfulness space: A systematic review. *PLoS One*, *16*(11), e0259623. PMID: 34793519

Hölzel, B. K., Lazar, S. W., Gard, T., Schuman-Olivier, Z., Vago, D. R., & Ott, U. (2011). How does mindfulness meditation work? Proposing mechanisms of action from a conceptual and neural perspective. *Perspectives on Psychological Science*, *6*(6), 537–559. DOI: 10.1177/1745691611419671 PMID: 26168376

Hunkin, H., King, D. L., & Zajac, I. T. (2021). Perceived acceptability and feasibility of digital interventions for reducing problematic smartphone use: A qualitative study. *BMC Psychiatry*, *21*(1), 1–12. DOI: 10.1186/s12888-021-03420-2 PMID: 33388029

Inkster, B., Sarda, S., & Subramanian, V. (2018). An empathy-driven, conversational artificial intelligence agent (Wysa) for digital mental well-being: Real-world data evaluation mixed-methods study. *JMIR mHealth and uHealth*, *6*(11), e12106. DOI: 10.2196/12106 PMID: 30470676

Kabat-Zinn, J. (1990). *Full catastrophe living: Using the wisdom of your body and mind to face stress, pain, and illness*. Delta.

Kabat-Zinn, J. (1994). *Wherever you go, there you are: Mindfulness meditation in everyday life*. Hyperion.

Kabat-Zinn, J. (2003). Mindfulness-based interventions in context: Past, present, and future. *Clinical Psychology : a Publication of the Division of Clinical Psychology of the American Psychological Association*, *10*(2), 144–156. DOI: 10.1093/clipsy.bpg016

Kabat-Zinn, J., Lipworth, L., Burncy, R., & Sellers, W. (1986). Four-year follow-up of a meditation-based program for the self-regulation of chronic pain: Treatment outcomes and compliance. *The Clinical Journal of Pain*, *2*(3), 159–774. DOI: 10.1097/00002508-198602030-00004

Koolhaas, J. M., Bartolomucci, A., Buwalda, B., de Boer, S. F., Flügge, G., Korte, S. M., Meerlo, P., Murison, R., Olivier, B., Palanza, P., Richter-Levin, G., Sgoifo, A., Steimer, T., Stiedl, O., van Dijk, G., Wöhr, M., & Fuchs, E. (2011). Stress revisited: A critical evaluation of the stress concept. *Neuroscience and Biobehavioral Reviews*, *35*(7), 1743–1761. DOI: 10.1016/j.neubiorev.2011.02.003 PMID: 21316391

LaFargue, M. (2010). *The Tao of the Tao Te Ching: A translation and commentary*. State University of New York Press.

Lazar, S. W., Kerr, C. E., Wasserman, R. H., Gray, J. R., Greve, D. N., Treadway, M. T., ... & Fischl, B. (2005). Meditation experience is associated with increased cortical thickness. *neuroreport, 16*(17), 1893-1897. https://doi.org/DOI: 10.1097/01.wnr.0000186598.66243.19

Lazarus, R. S., & Folkman, S. (1984). *Stress, Appraisal, and Coping*. Springer Publishing Company.

Levine, S. (2005). Developmental determinants of sensitivity and resistance to stress. *Psychoneuroendocrinology, 30*(10), 939–946. DOI: 10.1016/j.psyneuen.2005.03.013 PMID: 15958281

Levine, S., & Ursin, H. (1991). What is stress? In Brown, M. R., Koob, G. F., & Rivier, C. (Eds.), *Stress: Neurobiology and Neuroendocrinology* (pp. 3–21). Marcel Dekker, Inc.

Levine, S., & Ursin, H. (1991). What is stress? In Brown, M. R., Koob, G. F., & Rivier, C. (Eds.), *Stress: Neurobiology and neuroendocrinology* (pp. 3–21). Marcel Dekker, Inc.

Mehta, U. M., Venkatasubramanian, G., & Chandra, P. S. (2021). The "mind" of the mindful therapist: Mindfulness meditation, neuroscience, and therapeutic presence. *Asian Journal of Psychiatry, 56*, 102547. DOI: 10.1016/j.ajp.2020.102547

Millstine, D. M., Chen, C. Y., & Bauer, B. A. (2019). Complementary and integrative medicine in the management of headache. *BMJ (Clinical Research Ed.), 364*, k5235. DOI: 10.1136/bmj.k5235 PMID: 28512119

Muthuraman, P. (2024, August). Anxiety Monitoring and Digital Meditation Aids. In *2024 International Conference on Electrical Electronics and Computing Technologies (ICEECT)* (Vol. 1, pp. 1-6). IEEE.

Neff, K. D., Tóth-Király, I., & Colosimo, K. (2020). Self-compassion is best measured as a global construct and is overlapping with but distinct from neuroticism: A response to Pfattheicher, Geiger, Hartung, Weiss, and Schindler (2017). *European Journal of Personality, 34*(5), 810–823. DOI: 10.1002/per.2248

Olendzki, A. (2009). Mindfulness and meditation. In *Clinical handbook of mindfulness* (pp. 37–44). Springer New York., DOI: 10.1007/978-0-387-09593-6_3

Pawade, D., Karia, S., & De Sousa, A. (2022). Neurofeedback: A comprehensive review on system design, methodology and clinical applications. *Asian Journal of Psychiatry, 67*, 102937. DOI: 10.1016/j.ajp.2021.102937

Ramana, Y. V. S., & Peri, P. (2024). AI-Based Mindfulness: The Future of Healthcare and Customer Service. In *Comparative Analysis of Digital Consciousness and Human Consciousness: Bridging the Divide in AI Discourse* (pp. 54-64). IGI Global. DOI: 10.4018/979-8-3693-2015-0.ch003

Riches, S., Schlosser, M., & Gaebler, M. (2023). Immersive virtual reality in the treatment of mental disorders: A systematic review of its efficacy. *Frontiers in Psychology*, *14*, 1122334. DOI: 10.3389/fpsyg.2023.1122334

Rodriguez, V. H., Medrano, C. T., & Plaza, I. (2018). Wearable sensors for measuring movement in short sessions of mindfulness sitting meditation: A pilot study. *Journal of Healthcare Engineering*, *2018*(1), 7275049. DOI: 10.1155/2018/7275049 PMID: 29854363

Rodriguez, V. H., Perez, C. J., & Latorre, J. M. (2018). Wearable sensors for measuring movement in short sessions of mindfulness sitting meditation: A pilot study. *Journal of Healthcare Engineering*, *7275049*, 1–9. Advance online publication. DOI: 10.1155/2018/7275049 PMID: 29854363

Romero, L. M., Dickens, M. J., & Cyr, N. E. (2009). The reactive scope model: A new model integrating homeostasis, allostasis, and stress. *Hormones and Behavior*, *55*(3), 375–389. DOI: 10.1016/j.yhbeh.2008.12.009 PMID: 19470371

Schuemie, M. J., van der Straaten, P., Krijn, M., & van der Mast, C. A. (2001). Research on presence in virtual reality: A survey. *Cyberpsychology & Behavior*, *4*(2), 183–201. DOI: 10.1089/109493101300117884 PMID: 11710246

Seabrook, E., Kelly, R., Foley, F., Theiler, S., Thomas, N., Wadley, G., & Nedeljkovic, M. (2020). Understanding how virtual reality can support mindfulness practice: Mixed methods study. *Journal of Medical Internet Research*, *22*(3), e16106. DOI: 10.2196/16106 PMID: 32186519

Selye, H. (1936). A syndrome produced by diverse nocuous agents. *Nature*, *138*(3479), 32–32. DOI: 10.1038/138032a0

Selye, H. (1950). *Stress: The physiology and pathology of exposure to stress*. Acta Medica Publication.

Selye, H. (1976). *Stress in health and disease*. Butterworths Scientific.

Shaban-Nejad, A., Michalowski, M., & Buckeridge, D. L. (2018). Health intelligence: How artificial intelligence transforms population and personalized health. *NPJ Digital Medicine*, *1*(1), 53. DOI: 10.1038/s41746-018-0058-9 PMID: 31304332

Shamseer, L., Moher, D., Clarke, M., Ghersi, D., Liberati, A., Petticrew, M., Shekelle, P., & Stewart, L. A. (2015). Preferred reporting items for systematic review and meta-analysis protocols (PRISMA-P) 2015: Elaboration and explanation. *BMJ (Clinical Research Ed.), 349*(jan02 1), g7647. Advance online publication. DOI: 10.1136/bmj.g7647 PMID: 25555855

Siegel, R. D. (2010). *The mindfulness solution: Everyday practices for everyday problems*. Guilford Press.

Smith, E. N., Santoro, E., Moraveji, N., Susi, M., & Crum, A. J. (2020). Integrating wearables in stress management interventions: Promising evidence from a randomized trial. *International Journal of Stress Management, 27*(2), 172–182. https://psycnet.apa.org/doi/10.1037/str0000137. DOI: 10.1037/str0000137

Suleiman-Martos, N., Albendín-García, L., Gómez-Urquiza, J. L., Vargas-Román, K., Ramirez-Baena, L., Ortega-Campos, E., & Cañadas-De la Fuente, G. A. (2020). Prevalence and predictors of burnout in midwives: A systematic review and meta-analysis. *International Journal of Environmental Research and Public Health, 17*(2), 641. DOI: 10.3390/ijerph17020641 PMID: 31963831

Vacaru, S., Fekete, K., Luyten, L. P., Saldien, J., & Bosse, T. (2024). Selecting optimal wearables for measuring physiological arousal in robot-delivered mindfulness-based exercises. *Advanced Robotics, 38*(1), 1–15. DOI: 10.1080/01691864.2024.2369797 PMID: 39507502

World Health Organization (WHO). (2023, February 21). *Stress*. https://www.who.int/news-room/questions-and-answers/item/stress

Wrede, S. J., Esch, T., & Michaelsen, M. M. (2023). Mindfulness in the Digital Workplace: an explorative study of the compatibility of mindfulness and technology. https://doi.org/DOI: 10.21203/rs.3.rs-2459776/v1

Chapter 5
Virtual Reality Exposure Therapy (VRET) for Phobia Treatment Enhancing Traditional Techniques Through Technology

Vijendra Nath Pathak
https://orcid.org/0000-0003-0731-7003
Parul University, Vadodara, India

K. M. Anjalee
Gurukul Kangari University Haridwar, India

Jotika Judge
https://orcid.org/0000-0002-9357-4040
Chitkara University, Rajpura, India

ABSTRACT

Virtual Reality Exposure Therapy (VRET) has revolutionized phobia treatment by combining traditional therapy with advanced technology. It immerses patients in virtual environments tailored to simulate specific fears, offering a safe, customizable, and accessible alternative to conventional exposure therapy. VRET is particularly beneficial for those with severe phobias or physical limitations. Its flexibility allows therapists to adjust exposure intensity, enhancing engagement and accelerating desensitization. Challenges include high equipment costs, cybersickness, and spe-

DOI: 10.4018/979-8-3373-0467-0.ch005

cialized training needs. As VR technology becomes more affordable, VRET has the potential to transform the treatment of anxiety disorders, expanding the accessibility and effectiveness of psychological care. This chapter explores VRET's principles, applications, challenges, and future directions.

1. INTRODUCTION

The chapter will explore Virtual Reality Exposure Therapy (VRET) as a transformative and technology-enhanced approach to treating phobias. Traditional exposure therapy, a proven method within Cognitive Behavioural Therapy (CBT) is highly effective for phobia treatment but faces challenges in accessibility, feasibility, and patient compliance. VRET addresses these limitations by gradually creating safe, immersive, and controlled environments where clients can confront their fears, fostering effective desensitization. This chapter will examine the principles behind VRET, its applications for various phobias, a comparative analysis of VRET versus traditional exposure therapy, and the future of virtual reality (VR) in mental health treatment.

Phobias, as a subset of anxiety disorders, represent one of the most pervasive psychological conditions, impacting daily functioning and quality of life for millions of individuals worldwide. These irrational and often overwhelming fears can be triggered by various stimuli, ranging from specific objects and animals to abstract scenarios such as public speaking or enclosed spaces. According to the *Diagnostic and Statistical Manual of Mental Disorders* (DSM-5), phobias are divided into three broad categories: specific phobias, social anxiety disorder (also known as social phobia), and agoraphobia (American Psychiatric Association, 2013). Phobia-related research has consistently demonstrated the debilitating effects of these disorders, highlighting the need for effective, scalable treatments.

1.1 Understanding Phobias

Phobias fall under the category of anxiety disorders and are defined by persistent, irrational fears that provoke immediate anxiety responses (Craske & Stein, 2016). The DSM-5 classifies phobias into three primary types:

1. **Specific Phobias:** Characterized by an irrational fear of specific objects or situations, such as animals (e.g., dogs, spiders), heights, flying, or enclosed spaces (American Psychiatric Association, 2013).

2. **Social Anxiety Disorder (Social Phobia):** An overwhelming fear of social interactions or situations where individuals fear judgment or humiliation (Hofmann, Asnaani, Vonk, Sawyer, & Fang, 2010).
3. **Agoraphobia:** Anxiety triggered by situations where escape may be difficult or help unavailable, such as open spaces, public transport, or crowded places (Gloster et al., 2011).

Phobias can lead to significant disruptions in daily life, including impaired functioning at work, school, or in social settings. For instance, a person with aviophobia may avoid traveling for professional opportunities, while those with social anxiety may withdraw from important events (LeBeau et al., 2010). Traditional exposure therapy aims to reduce these anxieties through repeated exposure to fear-inducing stimuli, gradually helping individuals reframe their fears. However, real-world implementation presents several challenges, including the availability of specific stimuli, patient willingness, and logistical concerns (e.g., arranging a flight for aerophobia treatment), which make traditional exposure therapy less accessible. This chapter will discuss these issues and put forth the advantages offered by VRET in comparison to traditional exposure therapies.

Phobias originate from a combination of psychological, biological, and environmental factors. Cognitive theories suggest irrational fears stem from maladaptive thought processes, including overgeneralization and catastrophic thinking (Beck, 1976). Biologically, phobias are linked to hyperactivity in the amygdala, the brain region responsible for processing fear and threat detection (Etkin & Wager, 2007). Neuroimaging studies have shown that individuals with specific phobias exhibit heightened activity in the amygdala and insular cortex when exposed to phobic stimuli, suggesting a neurological basis for exaggerated fear responses (Hahn et al., 2010).

2. THE PSYCHOLOGICAL BASIS OF PHOBIAS

The development and maintenance of phobias can be explained through various psychological theories, including cognitive, behavioral, and psychodynamic perspectives. These theories help explain the formation of irrational fears and provide the foundation for evidence-based therapeutic interventions.

2.1. Cognitive Theories of Phobia Development

Cognitive models attribute phobia development to maladaptive thought patterns, catastrophic thinking, and biased information processing (Beck, 1976). Individuals with phobias often overestimate the danger posed by a feared stimulus and underestimate their ability to cope with the perceived threat. This distorted thinking reinforces the fear response, creating a cycle of avoidance and heightened anxiety. One such distorted way of thinking is selective attention bias, wherein individuals with phobias display an exaggerated focus on threat-related stimuli, making them hypersensitive to perceived dangers. For example, a person with arachnophobia is likelier to notice spiders in their environment and interpret neutral stimuli as threatening (Teachman et al., 2007). Catastrophic misinterpretation is another cognitive distortion in which people with phobias frequently misinterpret benign stimuli as dangerous. For instance, an individual with social anxiety may interpret a brief pause in conversation as evidence of judgment or disapproval (Clark & Wells, 1995). Moreover, cognitive biases can manifest as selective memory, where individuals recall threatening encounters while ignoring non-threatening ones, reinforcing their fear (Beck & Clark, 1997).

2.2 Behavioral Theories of Phobia Acquisition

Behavioral theories, mainly classical and operant conditioning, play a central role in explaining how phobias are acquired and maintained. The work of Ivan Pavlov on classical conditioning and B.F. Skinner on operant conditioning forms the basis for understanding learned fear responses. As explained through classical conditioning phenomena, phobias often develop by pairing a neutral stimulus with a traumatic or aversive experience. Watson and Rayner's "Little Albert" experiment (1920) demonstrated how an infant could be conditioned to fear a previously neutral stimulus (a white rat) by associating it with a loud noise. This process leads to the creation of an automatic fear response to the conditioned stimulus. Operant conditioning, on the other hand, states that once a fear response is established, it is maintained through negative reinforcement. Avoidance behaviors, such as staying away from elevators due to claustrophobia, reduce anxiety in the short term. However, this avoidance prevents the individual from confronting and overcoming their fear, leading to the long-term persistence of the phobia (Mowrer, 1947). Another learning, namely observational Learning, states that individuals can acquire phobias by observing others' fearful responses. Bandura's (1977) social learning theory suggests that children and adults can develop fears by witnessing others display fear or distress toward a particular object or situation.

2.3. Psychodynamic Perspectives on Phobias

Psychodynamic theories, rooted in the work of Sigmund Freud, propose that phobias arise from unresolved unconscious conflicts and repressed emotions. According to Freud (1920), phobias may represent the symbolic displacement of anxiety-provoking thoughts onto a specific object or situation. For example, a person with agoraphobia may unconsciously fear losing control, which manifests as a fear of open spaces where they believe they may be vulnerable.

However, while psychodynamic perspectives have historical significance, they are less commonly used in contemporary phobia treatment due to limited empirical evidence supporting their efficacy compared to cognitive-behavioral approaches (McLeod, 2015).

2.4 Cognitive-Behavioral Models: Integration of Cognitive and Behavioral

Modern cognitive-behavioral models combine the strengths of cognitive and behavioral theories to provide a comprehensive explanation of phobia development and maintenance. The fear response combines distorted thoughts and learned avoidance behaviors that reinforce each other (Salkovskis, 1991). CBT interventions target both components, addressing negative thought patterns and gradually exposing individuals to feared stimuli to promote desensitization and cognitive restructuring.

Additionally, genetic predispositions and early life experiences contribute to the development of phobias. Twin studies have revealed that genetic factors account for approximately 30-40% of the variance in phobia susceptibility (Kendler et al., 1992). Moreover, classical conditioning plays a significant role in acquiring phobias, as demonstrated by Watson and Rayner's (1920) famous “Little Albert” experiment, where a young child was conditioned to fear a white rat through repeated pairings with a loud, aversive noise. This foundational study laid the groundwork for understanding how phobias can be learned and maintained through associative mechanisms.

3. THE BIOLOGICAL FACTORS OF PHOBIA DEVELOPMENT

Psychological or environmental factors do not solely explain phobias; biological mechanisms play a significant role in their development and maintenance. Research has shown that genetic predispositions, neurobiological processes, and the dysregulation of certain neurotransmitters contribute to heightened fear responses and the persistence of phobias.

3.1 Genetic Predisposition

Genetic factors significantly contribute to the development of phobias, with heritability estimates ranging between 30% and 40% (Kendler et al., 1992). Twin studies provide strong evidence for a genetic component, showing that monozygotic (identical) twins are likelier to share specific phobias than dizygotic (fraternal) twins.

- **Family and Twin Studies:** Studies suggest a familial link exists for specific phobias, such as animal and blood-injection-injury phobias (Hettema et al., 2001). Children of individuals with phobias are more likely to develop similar fears, supporting the role of genetic vulnerability.
- **Gene-Environment Interaction:** While genetic predisposition increases vulnerability, environmental triggers such as traumatic experiences or observational learning are often necessary to activate the development of phobias. This concept is known as the diathesis-stress model, where genetic vulnerability interacts with environmental stressors to trigger a disorder (Eley et al., 2015).

3.2 Neurobiological Mechanisms

Neuroimaging studies have identified several brain structures involved in the processing of fear and anxiety, including the **amygdala**, **prefrontal cortex**, and **hippocampus**. Dysfunctions within these regions contribute to the exaggerated fear responses seen in individuals with phobias.

- **Amygdala Overactivation:** The amygdala plays a crucial role in detecting threats and activating the body's fight-or-flight response. Individuals with phobias exhibit heightened amygdala activity when exposed to phobia-related stimuli (Etkin & Wager, 2007). Overactivation of the amygdala leads to an exaggerated perception of threat and difficulty in extinguishing learned fears.
- **Prefrontal Cortex Dysregulation:** The prefrontal cortex, responsible for executive functioning and emotional regulation, modulates the amygdala's response to fear. Dysregulation in this region impairs the ability to rationally assess threats, leading to persistent fear responses (Shin & Liberzon, 2010).
- **Hippocampus Dysfunction:** The hippocampus is responsible for encoding and retrieving contextual memories. Dysfunction in this area may lead to inappropriate fear responses by associating non-threatening stimuli with previous traumatic experiences (Kim & Fanselow, 1992).

3.3. Neurochemical Imbalances

Phobias have been associated with dysregulation in several neurotransmitter systems, including **gamma-aminobutyric acid (GABA)**, **serotonin (5-HT)**, and

norepinephrine (NE). These imbalances contribute to heightened arousal, exaggerated fear responses, and impaired fear extinction.

- **GABA:** As the brain's primary inhibitory neurotransmitter, GABA plays a critical role in reducing neural activity and calming the central nervous system. Individuals with phobias often have reduced GABA levels, leading to heightened anxiety and hyperactive fear responses (Nemeroff, 2003).
- **Serotonin:** Serotonin is involved in mood regulation and the modulation of fear. Low serotonin levels are associated with increased anxiety and difficulties in fear extinction, contributing to the persistence of phobic responses (Hariri & Holmes, 2006).
- **Norepinephrine:** Norepinephrine is responsible for the body's stress response, increasing heart rate, blood pressure, and arousal in response to perceived threats. Elevated norepinephrine levels can intensify the physical symptoms of fear, such as sweating, hyperventilation, and rapid heartbeat, which are common in phobic individuals (Sullivan et al., 1999).

3.4. Evolutionary Perspectives and Innate Fear Responses

From an evolutionary standpoint, certain fears may have developed as adaptive responses to potential dangers in ancestral environments. For example, fear of snakes, spiders, and heights may have evolved as a survival mechanism to protect humans from life-threatening situations (Öhman & Mineka, 2001). Preparedness Theory proposed by Martin Seligman (1971), suggests that humans are biologically predisposed to develop phobias toward stimuli that posed a threat to survival in evolutionary history. This theory explains why individuals are more likely to develop phobias related to animals or heights compared to modern threats like guns or electrical outlets.

4. THE ENVIRONMENTAL BASIS OF PHOBIA DEVELOPMENT

In addition to biological and psychological mechanisms, phobias are also influenced by various environmental factors. The environment in which individuals are raised, the experiences they encounter, and their exposure to stressful or traumatic events can all play significant roles in the onset, development, and maintenance of phobias.

4.1. Early Life Experiences and Childhood Trauma

Early environmental experiences, particularly adverse or traumatic events during childhood, are among the most significant environmental contributors to the development of phobias. Traumatic encounters with a specific object or situation can lead to long-lasting fear responses. Many phobias originate from direct traumatic experiences. For example, a child bitten by a dog may develop a lifelong fear of dogs (Craske et al., 2009). This fear can persist if it is not adequately addressed through fear extinction mechanisms. Parents play a critical role in shaping a child's perception of the environment. If a parent demonstrates excessive fear toward certain objects or situations (e.g., fear of spiders), children may learn to associate those stimuli with danger through observational learning (Muris et al., 1996). This process aligns with Albert Bandura's (1977) theory of observational learning, which emphasizes that individuals acquire behaviors by watching others. Children who experience neglect or abuse are at increased risk of developing anxiety disorders, including phobias. The lack of emotional security in such environments contributes to heightened sensitivity to threats and a greater likelihood of developing irrational fears (Copeland et al., 2007).

4.2. Family and Social Environment

The immediate family environment, as well as broader social interactions, significantly influence phobia development. Overprotective parenting, peer rejection, and lack of exposure to new experiences are some key environmental factors. Overprotective or controlling parents may limit a child's exposure to potentially anxiety-inducing situations, preventing them from developing healthy coping mechanisms. Research has shown that children raised in overprotective environments are more likely to develop social anxiety and specific phobias (Rapee, 1997). Children of parents with anxiety disorders are at greater risk of developing phobias due to both genetic and environmental factors. Parental anxiety creates an environment where children may adopt exaggerated threat perceptions and avoidance behaviors (Murray et al., 2009). Negative experiences with peers, such as bullying or social rejection, are linked to the development of social anxiety disorder and related phobias. Victims of bullying often experience long-term anxiety related to social settings (Hawker & Boulton, 2000).

4.3. Societal and Cultural Influences

Cultural norms and societal expectations can shape the types of phobias that develop and influence the severity of fear responses. Cultural beliefs can dictate what individuals perceive as threatening or non-threatening. Certain phobias may be more prevalent in specific cultures due to unique social and environmental contexts. For example, in Japan and Korea, *taijin kyofusho* is a culturally bound form of social anxiety disorder characterized by an intense fear of embarrassing or offending others (Kleinknecht et al., 1997). Societal gender norms may influence the types of phobias that men and women are more likely to develop. For example, women are generally more likely to report specific phobias related to animals or heights, possibly due to differences in gender-related social expectations and coping mechanisms (Fredrikson et al., 1996).

4.4. Traumatic Events and Stressful Life Experiences

Exposure to traumatic or life-threatening events in adulthood can also lead to the development of phobias. This is especially common in natural disasters, accidents, or violent encounters. Individuals who experience a traumatic event, such as a car accident or natural disaster, may develop a specific phobia related to the context of the trauma (e.g., a fear of driving or storms). These phobias are often maintained through avoidance behaviors and heightened physiological responses when encountering reminders of the trauma (Brewin et al., 2000). Chronic stress or exposure to multiple stressful life events over time can increase the likelihood of developing phobias. Long-term exposure to environmental stress, such as financial difficulties or interpersonal conflicts, can contribute to a state of heightened anxiety and sensitivity to perceived threats (Hammen, 2005).

4.5. Lack of Exposure to Fear-Inducing Stimuli

A lack of exposure to fear-inducing stimuli during childhood can also increase the risk of developing phobias later in life. According to the **fear-avoidance model**, avoiding potentially anxiety-inducing situations prevents individuals from learning that these situations are not inherently dangerous, reinforcing the fear response over time (Vasey & Dadds, 2001). Children shielded from certain stimuli may fail to develop habituation—the natural process of desensitizing to a stimulus through repeated exposure. For example, a child never exposed to dogs may be more likely to develop a dog phobia when they encounter one later in life. There may be critical periods during development when exposure to certain stimuli is necessary for fear extinction and adaptive coping. A lack of exposure during these periods can increase vulnerability to phobias (Mineka & Zinbarg, 2006).

The environmental basis of phobia development highlights the significant role of early life experiences, family dynamics, cultural contexts, and exposure to traumatic events in shaping an individual's fear responses. Environmental factors interact with genetic and biological vulnerabilities, contributing to the complex nature of phobia development. Understanding these environmental influences is essential for tailoring interventions and preventative measures to reduce the risk of phobia onset.

5. TRADITIONAL TREATMENT APPROACHES

The primary treatment for phobias within clinical psychology is exposure therapy, a core component of Cognitive Behavioral Therapy (CBT). Developed by Joseph Wolpe in the 1950s, systematic desensitization—an early exposure therapy—aims to gradually expose individuals to feared stimuli while teaching them relaxation techniques to manage their anxiety responses (Wolpe, 1961). This method is based on the principles of classical conditioning, replacing the fear response with a relaxation response through repeated exposure.

Modern exposure therapy encompasses several variations, including in vivo, imaginal, and interoceptive exposure. In vivo exposure involves confrontation with the feared stimulus in real-world settings, making it one of the most effective techniques for treating specific phobias (Choy, Fyer, & Lipsitz, 2007). However, this approach has limitations, as logistical constraints, patient reluctance, and the inability to replicate specific scenarios (e.g., flying or encountering a rare animal) often hinder its implementation (Rothbaum et al., 2000).

On the other hand, imagination exposure involves having the patient visualize the feared stimulus or situation in their mind, which can help address phobias related to abstract or traumatic events. Interoceptive exposure is primarily used for panic disorder. It involves exposing the individual to physical sensations associated with fear (e.g., increased heart rate or dizziness) to help them habituate to these sensations and reduce their anxiety over time.

Despite the efficacy of traditional exposure therapies, dropout rates and treatment avoidance remain significant challenges. Research by Garcia-Palacios et al. (2001) indicates that up to 25% of patients refuse in vivo exposure due to the intense anxiety it induces. This highlights the need for alternative approaches that can deliver the benefits of exposure therapy while minimizing patient distress and logistical barriers.

5.1 Principles of Traditional Exposure Therapy

Traditional exposure therapy involves gradually and systematically exposing individuals to the objects or situations that trigger their anxiety. This process is known as desensitization and is rooted in classical conditioning principles (Wolpe, 1961). The goal is to weaken the association between the stimulus and the fear response through repeated, controlled exposure. The key steps include:

- **Assessment:** Identifying the specific fear and its triggers.
- **Hierarchy Creation:** Developing a fear hierarchy, listing situations from least to most anxiety-provoking (Marks, 1975).
- **Gradual Exposure:** Progressively exposing the individual to fear-inducing stimuli.
- **Anxiety Management:** Teaching coping mechanisms, such as breathing techniques and cognitive restructuring (Barlow, 2002).

5.2 Limitations of Traditional Exposure Therapy

While traditional exposure therapy is practical, several limitations hinder its accessibility and feasibility:

- **Logistical Barriers:** Recreating certain phobic stimuli, such as flying on a plane or encountering a specific animal, is challenging and costly (Botella, Fernandez-Alvarez, Guillén, García-Palacios, & Baños, 2017).
- **Patient Reluctance:** Many patients are unwilling to confront their fears directly due to the intensity of anxiety involved (Anderson, Zimand, Hodges, & Rothbaum, 2020).
- **Limited Control:** Therapists may have limited control over real-world variables, such as weather conditions during exposure to heights.
- **Safety Concerns:** Some exposures, such as those involving heights or dangerous animals, carry inherent risks (Carl et al., 2019).

These limitations underscore the need for innovative solutions like VRET, which offers controlled, customizable, and safe environments for exposure therapy.

6. VIRTUAL REALITY EXPOSURE THERAPY (VRET) AS AN INNOVATIVE SOLUTION

The emergence of Virtual Reality Exposure Therapy (VRET) addresses many of the limitations of traditional exposure therapy. By creating immersive, controlled virtual environments, VRET allows individuals to confront their fears in a safe and customizable setting, thereby reducing the anxiety associated with real-world exposure. VRET was first introduced in the 1990s as a novel application of virtual reality technology in the treatment of post-traumatic stress disorder (PTSD) and specific phobias (Rothbaum et al., 1995). Since then, numerous studies have demonstrated its effectiveness across a range of phobias, including acrophobia (fear of heights), aviophobia (fear of flying), and social anxiety disorder (Emmelkamp et al., 2002).

VRET offers unique flexibility that traditional methods cannot easily match, as therapists can adjust the realism and intensity of scenarios in real-time. This allows clients to start with mild exposures and increase intensity gradually based on their comfort level. Studies show that this ability to personalize treatment enhances effectiveness and boosts client engagement and commitment. Clients feel more in control and safer knowing the experience is virtual and adjustable, making VRET a highly effective, client-centered option for phobia treatment.

6.1. The Evolution of Virtual Reality in Mental Health

Virtual reality (VR) technology has rapidly evolved from its initial applications in gaming and entertainment to various fields, including medical training, physical rehabilitation, and psychotherapy (Riva, Baños, Botella, Wiederhold, & Gaggioli, 2019). Early research on VR for mental health dates back to the 1990s, with pioneering studies demonstrating its potential for treating anxiety disorders and phobias (North, North, & Coble, 1997).

Technological advancements have made VR more accessible and affordable, with VR headsets and software now available for clinical use. Key developments include improved graphics, motion tracking, and haptic feedback, enabling more realistic simulations. These advancements have cemented VRET's role as a viable alternative to traditional exposure therapy (Botella et al., 2017).

6.2. Theoretical Foundations of VRET

VRET builds on the principles of exposure therapy by providing controlled, immersive environments where clients can confront their fears. Unlike traditional methods, VR allows for precise control over stimuli, ensuring gradual and safe exposure (Freeman et al., 2017). For example, a client with acrophobia can experience progressively higher virtual environments without the risks associated with real-world exposure.

The theoretical foundation of VRET also addresses limitations of traditional exposure therapy, including:

- **Logistical Constraints:** Simulating scenarios that are difficult to replicate in real life.
- **Patient Compliance:** Offering a less intimidating and more engaging alternative to confrontation (Carl et al., 2019).
- **Safety:** Eliminating physical risks while providing realistic exposure.

6.3. Mechanisms of VRET: How It Works

VRET involves using VR headsets, sensors, and specialized software to create virtual environments tailored to the client's specific fears. The process typically follows these steps:

1. **Assessment and Customization:** Therapists assess the client's phobia and customize virtual scenarios based on the fear hierarchy (Anderson et al., 2020).
2. **Gradual Exposure:** Clients are exposed to increasingly intense stimuli, starting with mild scenarios and progressing as their anxiety decreases (Botella et al., 2017).
3. **Real-Time Monitoring:** Therapists monitor the client's physiological and psychological responses, adjusting the environment as needed.
4. **Cognitive Restructuring:** Clients learn to reframe irrational fears and develop coping strategies (Riva et al., 2019).

6.4. Neurobiological Mechanisms

Research shows that VRET engages neural pathways similar to those activated during real-life exposure, promoting neurobiological changes that reduce anxiety (Maples-Keller, Yasinski, Manjin, & Rothbaum, 2017). Key brain regions involved include:

- **Amygdala:** Responsible for processing fear and emotional responses.
- **Prefrontal Cortex:** Involved in cognitive control and the regulation of fear (Etkin & Wager, 2007).
- **Hippocampus:** Plays a role in memory formation and contextualizing fear-related stimuli.

Studies have shown that repeated exposure in virtual environments can lead to changes in neural connectivity, facilitating long-term anxiety reduction (Maples-Keller et al., 2017).

6.5. Applications of VRET for Different Phobias

Acrophobia (Fear of Heights)

VRET for acrophobia involves simulated environments where clients gradually experience heights. Initial sessions may involve standing on a virtual balcony, progressing to more challenging scenarios like rooftops or bridges (Freeman et al., 2017). Research demonstrates significant improvements in real-world height tolerance after VRET.

Aviophobia (Fear of Flying)

For aviophobia, VRET simulates various stages of a flight, including boarding, take off, turbulence, and landing. This controlled exposure helps clients become familiar with the sensations of flying, reducing anxiety and increasing their confidence to board real flights (Botella et al., 2017).

Arachnophobia (Fear of Spiders)

VRET offers a step-by-step approach, starting with cartoon-like spiders and advancing to more realistic depictions. Clients can practice anxiety management techniques within the virtual environment, leading to decreased fear responses over time (Garcia-Palacios, Botella, Hoffman, & Fabregat, 2002).

Social Phobia

Virtual scenarios for social phobia include public speaking simulations, social gatherings, and casual interactions. Clients can practice social skills and build confidence in a non-judgmental, simulated setting, eventually translating these improvements to real-life situations (Anderson et al., 2020).

6.6. Efficacy of VRET: Empirical Evidence

In a study conducted to investigate the neurobiological effects of virtual reality exposure therapy (VRET) on people suffering from acrophobia, Guo et al. (2025) used resting-state functional magnetic resonance imaging (fMRI). When VRET and imaginal exposure therapy (IET) were compared in the study, it was shown that both treatment approaches decreased acrophobic symptoms. However, VRET was notably linked to lower degree centrality (DC) values in the cuneus, precuneus, middle temporal gyrus, and right calcarine, among other areas of the brain. Additionally, the functional connection between the right middle temporal gyrus and other brain regions decreased as a result of VRET. These findings suggest that by modifying the default mode network and the primary visual cortex, VRET may help to reduce the abnormal brain activity associated with acrophobia effectively.

A randomized clinical study comparing the efficacy of metacognitive therapy with virtual reality (VR) therapy in treating ailurophobia (cat phobia) was carried out by Ebrahimi et al. (2025). The Kettle anxiety test and a researcher-made cat fear questionnaire were used to collect data from the study, which involved 28 participants. The findings showed that anxiety and cat phobia were considerably decreased by both VR and metacognitive therapy. However, as compared to metacognitive therapy, VR therapy showed a more significant decrease in both anxiety and dread levels. These results imply that although both treatments work, VR therapy might be a better option for ailurophobics in terms of effectiveness and longevity.

Tan et al. (2025) examined the efficacy of virtual reality exposure therapy (VRET) for social anxiety disorder (SAD). VRET was shown to be more effective than waitlist conditions in reducing anxiety symptoms, both post-intervention and at follow-up. When compared to other active interventions, VRET demonstrated similar effects in alleviating anxiety symptoms. Researchers observed a greater effect in participants with symptomatic social anxiety when VRET was combined with cognitive behavioral therapy. VRET may be considered a supplementary approach for the improvement of anxiety symptoms, according to this study.

Petersen et al. (2024) tested the development and usability assessment of a VR game designed to treat blood-injection-injury (BII) phobia. The game Phobos integrates virtual reality and biometric data collection to provide immersive, graded exposure personalized to individual stress responses. Usability testing with 10 participants revealed generally positive feedback, especially regarding immersion, usability, and emotional engagement. Photorealistic environments and narrative-driven gameplay improved the sense of presence, which is important for effective exposure therapy. Some participants reported motion sickness, often attributed to locomotion mechanics, suggesting the need for future design adjustments. However, most users found the experience engaging and non-invasive, indicating good system acceptance. Generally, the findings of this study supported the viability of VR-based therapy in clinical contexts.

For treating speech anxiety, Spyridonis et al. (2024) created PublicVR, a virtual reality (VR) intervention that focuses on non-verbal communication elements. The effectiveness of PublicVR in changing participants' perceptions of these elements and creating a sense of presence was assessed in this study. The VR exposure therapy solution successfully changed participants' perceptions of non-verbal communication factors that contribute to increased public speaking anxiety, according to an empirical evaluation using the iGroup Presence Questionnaire and the Personal Report of Public Speaking Anxiety. Additionally, participants expressed a strong feeling of presence in the virtual reality setting. Their results imply that by favorably altering perceptions of non-verbal communication cues, VR exposure therapy may help treat phobias associated with social anxiety, particularly speech anxiety.

For those with arachnophobia, Andersson et al. (2024) investigated the viability and initial efficacy of a single session of therapist-assisted Virtual Reality Exposure Therapy (VRET-AP). The intervention, which was based on Öst's one-session in vivo treatment, showed significant improvements in the behavioral approach to real spiders and decreased self-reported spider fear. Every participant agreed to the course of treatment, and there were no dropouts because of side effects, and completion rates were good. According to qualitative comments, most individuals reported better responses to spider encounters after treatment and regarded the experience as powerful and dramatic. Though recommendations were made for more diversity and realism, but the virtual world was generally positively received by the participants.

Levy et al. (2023) investigated the patients perceptions about virtual reality exposure therapy (VRET) versus in vivo treatment for anxiety disorders. According to the study, people with anxiety disorders generally expressed a willingness to employ VRET (90.2%) and in vivo (82%). As compared to in vivo exposure therapy, participants expressed greater interest in, comfort with, enthusiasm for, and perceived efficacy of VRET. Increased anxiety, feelings of humiliation or embarrassment, and possible worsening of their disease were the main worries about in vivo exposures. On the other hand, side effect risk, efficacy uncertainty, and health insurance coverage were the main issues with VRET. Overall, the results indicate that VRET is seen marginally more favorably than in vivo exposure, even though patients are concerned about both techniques.

To assess the efficacy of virtual reality exposure treatment (VRET) in treating fear of flying (FoF), Ribe-Vines et al. (2023) conducted a systematic study. The review, which included 33 publications, found that VRET significantly decreased participants' flying-related anxiety levels. Interestingly, VRET demonstrated effectiveness on par with well-known therapies like in vivo exposure therapy (IVET) and cognitive behavioral therapy (CBT). According to the authors, VRET may improve the efficacy of CBT and offers a useful substitute in situations where IVET is not feasible. Additionally, the review suggested that the effectiveness of VRET for FoF could be increased by combining synchronized motion feedback with visual and aural cues in VR situations.

Virtual Reality Exposure Therapy (VRET) has shown encouraging results in the majority of trials when used to treat a variety of phobias, including situational, animal, natural environment, and blood-injection-injury phobias. Most VRET therapies showed long-lasting therapeutic results and are backed by randomized controlled trials, frequently with follow-up evaluations. Augmented Reality Exposure Therapy (ARET), on the other hand, has also demonstrated benefits in realism and integration with the actual environment and has the potential for use in the treatment of phobias in the future; however, for now, VRET is better proven between the two (Albakri et al., 2022).

Krzystanek et al. (2021) provide a comprehensive synthesis of clinical evidence on the use of Virtual Reality Exposure Therapy (VRET) in treating agoraphobia, social phobia, and specific phobias. The review emphasizes how effective head-mounted displays (HMDs) are at delivering VRET, primarily because of their ability to create realistic and immersive virtual environments. The level of patient immersion in the VR environment, which improves engagement and emotional processing of phobic stimuli, is crucial in determining treatment success. Additionally, the authors note that psychiatric comorbidities tend to reduce the efficiency of VRET, highlighting the significance of carefully choosing patients. Crucially, long-lasting therapeutic effects have been reported for as long as a year after the intervention, indicating VRET's potential for long-term success.

In their review, Wechsler, Kümpers, and Mühlberger (2019) found that both VRET and in vivo exposure produce great and noteworthy treatment effects, with no significant differences in efficiency for specific phobia and agoraphobia. For social anxiety, however, a moderate effect size supported in vivo exposure. VRET is a feasible and possibly better option in situations where in vivo exposure is impractical or less appealing to patients. Additionally, the immersive properties of virtual reality (VR) provide control and flexibility in mimicking fearful stimuli, which may increase a patient's willingness to participate in treatment. In order to increase results, VR environment improvement and integration with cognitive therapies, particularly for social phobia, is needed. However, the authors favor a wider clinical application of VRET, especially considering its technological benefits, safety, and accessibility.

A meta-analysis conducted by Carl et al. (2019) found that VRET is equally effective as traditional exposure therapy in reducing anxiety symptoms, with some studies suggesting that it may even be superior in terms of patient adherence and engagement. One of the key advantages of VRET is its ability to provide graded exposure to feared stimuli, allowing therapists to gradually increase the intensity of the exposure based on the patient's comfort level. For example, a patient with acrophobia may begin by standing on a virtual balcony a few feet off the ground and progressively advance to higher elevations as their anxiety diminishes.

Research by Botella et al. (2017) highlights the versatility of VRET in treating various phobia types. In a randomized controlled trial comparing VRET with in vivo exposure to treat aviophobia, patients who underwent VRET demonstrated significant improvements in their ability to board and tolerate actual flights. Additionally, follow-up assessments indicated that the therapeutic gains were maintained over time, suggesting that VRET promotes long-term desensitization.

The neurobiological mechanisms underlying VRET's efficacy are another area of active research. Studies using functional magnetic resonance imaging (fMRI) have shown that VRET activates neural pathways similar to those engaged during real-world exposure, including the amygdala and prefrontal cortex (Maples-Keller et al., 2017). This neurobiological overlap supports the hypothesis that VRET can elicit the same fear extinction processes as traditional exposure therapy, making it a viable alternative for patients unable or unwilling to participate in in vivo exposure.

6.8. Addressing Barriers to Treatment with VRET

One of the primary barriers to traditional exposure therapy is the difficulty of replicating specific feared scenarios in real life. VRET overcomes this limitation by offering a highly customizable and scalable solution tailored to each patient's needs. For example, therapists can modify the virtual environment in real-time to adjust the level of exposure, ensuring that patients remain within their optimal anxiety zone—a concept known as the "zone of proximal development" (Vygotsky, 1978). This level of control is particularly beneficial for patients with severe phobias, who may require a more gradual and personalized approach to exposure.

Furthermore, VRET addresses patient reluctance by providing a less intimidating alternative to in vivo exposure. Patients often feel more comfortable confronting their fears in a virtual environment, knowing they can pause or exit the simulation anytime. This sense of control has been shown to increase patient engagement and reduce dropout rates, ultimately enhancing the overall effectiveness of the treatment (Botella et al., 2017).

6.9. Integration of VRET with Emerging Technologies

Recent advancements in virtual reality technology have further enhanced the therapeutic potential of VRET. The integration of haptic feedback, which provides tactile sensations corresponding to virtual stimuli, has been shown to enhance the realism of the virtual environment and improve treatment outcomes (Riva et al., 2019). For example, patients undergoing VRET for arachnophobia (fear of spiders) can experience the sensation of a spider crawling on their hand through haptic gloves, making the exposure more immersive and compelling.

Additionally, using artificial intelligence (AI) in VRET is an emerging area of interest. AI algorithms can analyze real-time patient responses and dynamically adjust the virtual environment to optimize exposure. For instance, if a patient's physiological indicators (e.g., heart rate or skin conductance) suggest excessive anxiety, the AI system can automatically reduce the intensity of the exposure to prevent overwhelming the patient (Anderson et al., 2020).

Phobia-related research has made significant strides in understanding the etiology, maintenance, and treatment of these disorders. While traditional exposure therapies have proven effective, their limitations have paved the way for innovative solutions such as Virtual Reality Exposure Therapy. By leveraging the immersive capabilities of virtual reality, VRET offers a safe, controlled, and customizable approach to phobia treatment, addressing many of the barriers associated with in vivo exposure. As virtual reality technology continues to evolve, its integration with emerging technologies such as AI and haptic feedback will likely enhance its therapeutic potential further, making VRET a cornerstone of future mental health interventions.

With technological advancement, VR technology has now become more accessible and affordable for clinical applications, allowing therapists to design immersive, highly controlled environments that gradually intensify based on the client's comfort and needs. By allowing therapists to create simulations of feared scenarios, VRET seamlessly integrates with the step-by-step structure of exposure therapy. The section will include a brief history of VRET in mental health, tracing key studies and high-tech developments that have cemented its place as a viable option for phobia treatment today. The theoretical background also addresses the limitations of traditional approaches, including logistical constraints in replicating realistic phobic triggers and ethical concerns in exposing patients to potentially hazardous situations. This section sets the stage for understanding how VR technology enhances exposure therapy.

In VRET, exposure therapy principles are applied within virtual environments, enabling clients to face their fears in a safe, simulated setting. Using VR headsets and specialized software, therapists can craft customized, controlled scenarios, gradually exposing clients to phobia-inducing stimuli and reducing their anxiety responses over time. For example, a client who fears heights might begin on a virtual balcony a few floors up, eventually progressing to higher settings, allowing anxiety reduction and skill acquisition at a manageable pace. These mechanics, emphasizing VRET's systematic approach to desensitizing clients through repeated exposure will be discussed.

Research has shown that VRET activates neural pathways similar to those engaged in real-life exposure, promoting neurobiological changes that reduce anxiety. We shall delve into these mechanisms, presenting findings on VRET's efficacy in activating brain regions such as the amygdala and prefrontal cortex, both involved in processing and regulating fear. Additionally, VRET facilitates cognitive restructuring, helping clients shift from irrational fears to more realistic perspectives.

This study compares VRET and traditional exposure therapy, focusing on how each approach meets client needs, achieves therapy goals, and impacts outcomes. Traditional exposure therapy has long been effective in treating phobias by gradually exposing clients to real-world fears, allowing them to reduce their anxiety over time. However, recreating real-life scenarios like flights or crowded spaces can be logistically challenging, expensive, or even unsafe. VRET overcomes these issues by using virtual environments, allowing clients to confront fears in a safe, controlled setting within the therapist's office, making therapy more accessible and consistent.

VRET also has its limitations. The technology required can be costly, and some users may experience dizziness or disorientation, making it unsuitable for everyone. Additionally, therapists need specialized training to effectively use VR equipment, which can increase expenses and add to the learning curve. By highlighting the strengths and current limitations, this section offers a well-rounded perspective on VRET's potential and areas for further development.

7. INDIGENOUS PHOBIA VR DEVICE

7.1 Description of the Invention

The Indigenous Phobia VR Device is designed to assist therapists in detecting and treating various phobias through virtual reality exposure therapy (Pathak, Judge, & Anjalee, 2023).

7.2 Problem Addressed by the Invention

Anxiety disorders are a major psychological issue, second only to depressive disorders in prevalence, affecting approximately 4.5% of the population in India. Specific phobias affect around 4.2% of individuals, requiring treatments beyond traditional therapeutic approaches. Exposure therapy is the most common technique, but its effectiveness often depends on the availability of the feared stimulus in real-world environments. This invention addresses the challenges of providing in-vivo exposure therapy by simulating feared objects, situations, or activities through VR, ensuring that patients can undergo safe and controlled exposure sessions (Pathak, Judge, & Anjalee, 2023).

7.3 State-of-the-Art/Research Gap

- There is a lack of equipment that enables psychiatrists, psychologists, and therapists to use in-vivo exposure therapy effectively for specific phobias.
- The Indigenous Phobia VR Device provides real-time patient exposure, measuring physiological and emotional responses to phobic stimuli.
- The device's integration with software addresses human error by automatically reporting the intensity and extent of phobia symptoms, including physiological data.
- This device, as a Make-in-India innovation, aims to fill the gap in affordable and locally developed psychological intervention tools for use in India and abroad (Pathak, Judge, & Anjalee, 2023).

7.4. Detailed Description of the Invention

The Indigenous Phobia VR Device operates at different stages: screening for phobias, diagnosis, and intervention. It consists of a VR headset connected to a laptop or desktop, through which patients view phobic stimuli. The device has the following key components and features:

- **Visual and Auditory Simulation:** Patients wear the VR headset to view the feared object or situation. Earbuds provide audio cues for enhanced immersion.
- **Therapist Control:** The therapist controls the intensity and progression of exposure through software on a connected laptop or desktop.
- **Wireless Connectivity:** The device can connect via Bluetooth or wired connections.
- **Safety Mechanisms:** A stop button allows patients to end the session if they feel overwhelmed.
- **Physiological Monitoring:** The device measures physiological responses, such as blood pressure and changes in sweat conductivity, to assess fear levels.

7.5. Specific Phobias Covered by the Device (Based on DSM-V TR)

Table 1.

Phobia Type	Examples
Animal Phobia	Snakes, spiders, dogs, insects, birds
Natural Environment Phobia	Storms, heights, water
Blood-Injection-Injury	Seeing blood or an injury, injections, wheelchairs
Situational Phobia	Public transportation, tunnels, elevators, flying
Social Phobia	Public speaking, conversations with others
Other Phobias	Choking, vomiting, space phobia

The software enables gradual exposure, following the principles of exposure therapy. Patients are categorized into mild, moderate, severe, and extreme cases, and treatment is customized accordingly (Pathak, Judge, & Anjalee, 2023).

7.6. Advantages of the Invention

- **Realistic Simulations:** Provides patients with realistic but safe exposure to feared situations, enhancing desensitization.
- **Comprehensive Diagnosis and Intervention:** Serves as a one-stop solution for diagnosing and treating specific phobias.
- **Multifaceted Monitoring:** Simultaneously tracks psychological, emotional, and physiological responses during exposure.
- **Physiological Data Collection:** Measures indicators such as blood pressure and skin conductance, aiding in objective assessment and progress monitoring (Pathak, Judge, 2023

8. ADVANTAGES OF VIRTUAL REALITY EXPOSURE THERAPY (VRET)

One of the most significant advantages of VRET is the ability for therapists to create and manipulate virtual environments to suit the specific needs of individual patients. The intensity, duration, and type of exposure can be adjusted in real time based on the patient's response, ensuring a tailored therapeutic experience. For example, in treating anxiety disorders or phobias, therapists can gradually introduce stimuli to avoid overwhelming the patient, thereby improving the effectiveness of exposure therapy (Freeman et al., 2017). This precise control minimizes sudden distress risks and enhances gradual desensitization's efficacy.

VRET provides a safe and secure environment where patients can confront their fears without facing real-world risks. This is particularly valuable for individuals with conditions such as post-traumatic stress disorder (PTSD), specific phobias, or social anxiety. By simulating distressing scenarios without actual exposure to danger (e.g., fear of heights or flying), VRET reduces the likelihood of harm while engaging the patient in meaningful exposure (Botella et al., 2017). Unlike traditional in vivo exposure, where patients may feel excessively anxious due to unpredictable real-world situations, VRET ensures that they remain in a controlled therapeutic setting.

Virtual reality's interactive and immersive nature increases patient engagement compared to traditional exposure methods. Virtual environments provide visual, auditory, and sometimes haptic feedback, which creates a realistic and stimulating experience that feels more engaging than imagining or discussing fears. The novelty and interactivity of the medium often motivate patients who may otherwise be resistant to therapy (Garcia-Palacios et al., 2007). Furthermore, VRET can evoke a stronger emotional response, allowing therapists to address the patient's underlying fears in a way that feels vivid and meaningful.

Accessibility to diverse exposure scenarios through VRET overcomes logistical challenges associated with traditional exposure therapy. For example, exposing a patient to a fear of flying typically requires access to an airport or aircraft, while treating fear of public speaking may require a large audience or venue. With VR, therapists can simulate a wide range of scenarios within the confines of a therapist's office, making it more practical and time-efficient (Riva et al., 2019). This accessibility benefits patients living in remote areas or those unable to engage in real-world exposure due to physical or logistical constraints.

In certain cases where patients may feel embarrassed or stigmatized by their condition (e.g., social anxiety or trauma-related disorders), VRET can offer a sense of anonymity that enhances participation. Unlike real-world exposure where social interactions or public scenarios are visible, virtual simulations provide a private and controlled space for exposure. This helps reduce the fear of judgment, promoting a greater willingness to participate in therapy.

8.1. Limitations of Virtual Reality Exposure Therapy (VRET)

VRET requires high initial costs of equipment and software implementation. It includes an investment in VR headsets, computers, and specialized software, which can be expensive, especially for smaller clinics or solo practitioners. The cost factor can be a significant barrier to widespread adoption, as many healthcare providers may hesitate to invest in the necessary infrastructure (Carl et al., 2019). Moreover, regular updates to VR software and potential hardware maintenance add to the long-term financial burden. However, As VR technology advances, costs may decrease, making it more affordable. One of the common challenges in using VR technology is the potential for side effects such as motion sickness, dizziness, nausea, or disorientation. These symptoms, commonly called "cybersickness," can result from discrepancies between the visual input from the VR system and the user's physical movements. This can be particularly problematic for patients with vestibular disorders or those sensitive to motion (Rebenitsch & Owen, 2016). Therapists must closely monitor patients during sessions and adjust settings to minimize discomfort.

Unlike traditional therapeutic techniques, VRET requires therapists to have technical expertise in operating VR systems and tailoring virtual environments to meet therapeutic goals. Insufficient training can lead to improper use of the technology, reducing the effectiveness of therapy and potentially causing harm to patients. Training programs are necessary to ensure that therapists are proficient in handling VR exposure therapy's technical and psychological aspects. Additionally, technical issues during sessions, such as software crashes or calibration errors, may disrupt the therapy process if not properly managed.

Another challenge of VRET is to ascertain whether exposure in a virtual environment will effectively reduce anxiety or phobia in real-world settings. Some patients may experience significant improvement within the virtual simulation but struggle with real-life stimuli due to the differences in sensory and contextual cues (Difede et al., 2014). To address this limitation, therapists often combine VRET with real-world exposure as part of a comprehensive treatment plan to ensure that patients can generalize their progress beyond the virtual setting. Privacy concerns may arise since VRET involves collecting and processing patient data about their mental health, sessions, and progress. Data breaches or unauthorized access to sensitive information could compromise patient confidentiality, leading to ethical issues. Clinicians must ensure that adequate security protocols are in place to protect patient information.

9. CONCLUSION

Virtual Reality Exposure Therapy represents a significant advancement in phobia treatment, offering immersive, customizable, and accessible options that address the limitations of traditional exposure therapy. By integrating cutting-edge technology with evidence-based therapeutic principles, VRET enhances client engagement, improves treatment outcomes, and expands access to effective mental health care (Freeman et al., 2017). As VR technology evolves, its role in psychotherapy will likely grow, making it a staple in treating anxiety disorders and beyond. Virtual Reality Exposure Therapy (VRET) is a transformative advancement in the treatment of phobias, bridging the gap between traditional therapeutic approaches and cutting-edge technology. By immersing patients in controlled, interactive environments tailored to their specific fears, VRET offers a safer and more practical alternative to traditional exposure therapy. Its ability to simulate various real-life scenarios—from fear of heights to public speaking—within the confines of a therapist's office addresses several logistical challenges posed by in vivo exposure. This accessibility ensures that even individuals with severe phobias, physical limitations, or limited access to traditional settings can benefit from effective exposure therapy. VRET's customizable nature enables therapists to adjust the intensity of exposure based on the patient's progress, creating a personalized treatment plan that minimizes distress and maximizes therapeutic outcomes. VR's immersive nature helps evoke genuine emotional responses, increasing patient engagement and accelerating desensitization. Furthermore, the anonymity provided by virtual environments can reduce feelings of embarrassment or stigma, fostering a more comfortable and open therapeutic experience. However, VRET is not without limitations. The high cost of equipment, potential side effects like cybersickness, and the need for specialized training can present challenges to widespread adoption. Additionally, while VRET effectively simulates phobic stimuli, there remains the concern of limited generalizability to real-world scenarios. This limitation underscores the importance of combining VRET with real-world exposure for long-term success. Many of these challenges will diminish as VR technology becomes more affordable and sophisticated. With ongoing improvements in software customization, training programs for therapists, and research validating its effectiveness, VRET has the potential to revolutionize the treatment of phobias. By integrating traditional therapeutic principles with technological innovation, VRET enhances the accessibility, safety, and efficacy of exposure therapy, paving the way for broader applications in mental health care.

REFERENCES

American Psychiatric Association. (2013). *Diagnostic and statistical manual of mental disorders* (5th ed.). APA Publishing.

Anderson, P. L., Zimand, E., Hodges, L. F., & Rothbaum, B. O. (2020). Virtual reality exposure therapy for social anxiety. *Journal of Anxiety Disorders, 42*, 1–7.

Bandura, A. (1977). *Social learning theory*. Prentice-Hall.

Barlow, D. H. (2002). *Anxiety and its disorders: The nature and treatment of anxiety and panic*. Guilford Press.

Beck, A. T. (1976). *Cognitive therapy and the emotional disorders*. International Universities Press.

Beck, A. T., & Clark, D. A. (1997). An information processing model of anxiety: Automatic and strategic processes. *Behaviour Research and Therapy, 35*(1), 49–58. DOI: 10.1016/S0005-7967(96)00069-1 PMID: 9009043

Botella, C., Baños, R. M., Villa, H., Perpiñá, C., & García-Palacios, A. (2017). Virtual reality in the treatment of phobias. *Cyberpsychology & Behavior, 7*(3), 283–290. DOI: 10.1089/1094931041291295

Botella, C., Fernandez-Alvarez, J., Guillén, V., García-Palacios, A., & Baños, R. M. (2017). Virtual reality exposure-based therapy in anxiety disorders: A systematic review. *Behaviour Research and Therapy, 87*, 172–183. DOI: 10.1016/j.brat.2016.10.003

Brewin, C. R., Andrews, B., & Valentine, J. D. (2000). Meta-analysis of risk factors for posttraumatic stress disorder in trauma-exposed adults. *Journal of Consulting and Clinical Psychology, 68*(5), 748–766. DOI: 10.1037/0022-006X.68.5.748 PMID: 11068961

Carl, E., Stein, A. T., Levihn-Coon, A., Pogue, J. R., Rothbaum, B., Emmelkamp, P., & Powers, M. B. (2019). Virtual reality exposure therapy for anxiety and related disorders: A meta-analysis. *Journal of Anxiety Disorders, 61*, 27–36. DOI: 10.1016/j.janxdis.2018.08.003 PMID: 30287083

Choy, Y., Fyer, A. J., & Lipsitz, J. D. (2007). Treatment of specific phobia in adults. *Clinical Psychology Review, 27*(3), 266–286. DOI: 10.1016/j.cpr.2006.10.002 PMID: 17112646

Clark, D. M., & Wells, A. (1995). A cognitive model of social phobia. In Heimberg, R. G., Liebowitz, M. R., Hope, D. A., & Schneier, F. R. (Eds.), *Social phobia: Diagnosis, assessment, and treatment* (pp. 69–93). Guilford Press.

Copeland, W. E., Keeler, G., Angold, A., & Costello, E. J. (2007). Traumatic events and posttraumatic stress in childhood. *Archives of General Psychiatry, 64*(5), 577–584. DOI: 10.1001/archpsyc.64.5.577 PMID: 17485609

Craske, M. G., & Stein, M. B. (2016). Anxiety disorders. *Lancet, 388*(10063), 3048–3059. DOI: 10.1016/S0140-6736(16)30381-6 PMID: 27349358

Difede, J., Cukor, J., Jayasinghe, N., Patt, I., Spielman, L., Giosan, C., & Hoffman, H. G. (2014). Virtual reality exposure therapy for the treatment of posttraumatic stress disorder: A review of the literature. *Annals of the New York Academy of Sciences, 1208*(1), 68–78. DOI: 10.1111/j.1749-6632.2010.05696.x

Eley, T. C., Gregory, A. M., & Clark, D. M. (2015). Genetic and environmental influences on phobias. *Behaviour Research and Therapy, 64*, 44–50. DOI: 10.1016/j.brat.2014.10.006

Emmelkamp, P. M. G., Krijn, M., Hulsbosch, A. M., De Vries, S., Schuemie, M. J., & Van Der Mast, C. A. (2002). Virtual reality treatment versus exposure in vivo: A comparative evaluation in acrophobia. *Behaviour Research and Therapy, 40*(5), 509–516. DOI: 10.1016/S0005-7967(01)00023-7 PMID: 12038644

Etkin, A., & Wager, T. D. (2007). Functional neuroimaging of anxiety: A meta-analysis of emotional processing in PTSD, social anxiety disorder, and specific phobia. *The American Journal of Psychiatry, 164*(10), 1476–1488. DOI: 10.1176/appi.ajp.2007.07030504 PMID: 17898336

Fredrikson, M., Annas, P., Fischer, H., & Wik, G. (1996). Gender and age differences in the prevalence of specific fears and phobias. *Behaviour Research and Therapy, 34*(1), 33–39. DOI: 10.1016/0005-7967(95)00048-3 PMID: 8561762

Freeman, D., Haselton, P., Freeman, J., Spanlang, B., Kishore, S., Albery, E., & Slater, M. (2017). Automated psychological therapy using immersive virtual reality for treatment of fear of heights: A single-blind, parallel-group, randomised controlled trial. *The Lancet. Psychiatry, 5*(8), 625–632. DOI: 10.1016/S2215-0366(18)30226-8 PMID: 30007519

Freeman, D., Reeve, S., Robinson, A., Ehlers, A., Clark, D., Spanlang, B., & Slater, M. (2017). Advances in virtual reality and implications for clinical psychology. *Psychological Medicine, 47*(16), 2395–2400. DOI: 10.1017/S003329171700040X PMID: 28325167

Freud, S. (1920). *Beyond the pleasure principle*. International Psycho-Analytical Press.

Garcia-Palacios, A., Botella, C., Hoffman, H. G., & Fabregat, S. (2002). Comparing acceptance and refusal rates of virtual reality exposure versus in vivo exposure by patients with specific phobias. *Cyberpsychology & Behavior*, *5*(3), 219–224. DOI: 10.1089/109493102760147117 PMID: 12123244

Garcia-Palacios, A., Hoffman, H. G., Kwong See, S., Tsai, A., & Botella, C. (2007). Redefining therapeutic success with virtual reality exposure therapy. *Cyberpsychology & Behavior*, *4*(3), 341–348. DOI: 10.1089/109493101300210231 PMID: 11710258

Gloster, A. T., Wittchen, H. U., Einsle, F., & Lang, T. (2011). Agoraphobia. *Psychopathology*, *44*(6), 335–345. DOI: 10.1159/000325225 PMID: 21534651

Hahn, A., Stein, P., Windischberger, C., Weissenbacher, A., Spindelegger, C., Moser, E., & Kasper, S. (2010). Reduced resting-state functional connectivity between amygdala and orbitofrontal cortex in social anxiety disorder. *NeuroImage*, *56*(3), 881–889. DOI: 10.1016/j.neuroimage.2011.02.064 PMID: 21356318

Hammen, C. (2005). Stress and depression. *Annual Review of Clinical Psychology*, *1*(1), 293–319. DOI: 10.1146/annurev.clinpsy.1.102803.143938 PMID: 17716090

Hariri, A. R., & Holmes, A. (2006). Genetics of emotional regulation: The role of the serotonin transporter in neural function. *Trends in Cognitive Sciences*, *10*(4), 182–191. DOI: 10.1016/j.tics.2006.02.011 PMID: 16530463

Hawker, D. S. J., & Boulton, M. J. (2000). Twenty years' research on peer victimization and psychosocial maladjustment: A meta-analytic review of cross-sectional studies. *Journal of Child Psychology and Psychiatry, and Allied Disciplines*, *41*(4), 441–455. DOI: 10.1111/1469-7610.00629 PMID: 10836674

Hettema, J. M., Neale, M. C., & Kendler, K. S. (2001). A review and meta-analysis of the genetic epidemiology of anxiety disorders. *The American Journal of Psychiatry*, *158*(10), 1568–1578. DOI: 10.1176/appi.ajp.158.10.1568 PMID: 11578982

Hofmann, S. G., Asnaani, A., Vonk, I. J., Sawyer, A. T., & Fang, A. (2010). The efficacy of cognitive behavioral therapy: A review of meta-analyses. *Cognitive Therapy and Research*, *36*(5), 427–440. DOI: 10.1007/s10608-012-9476-1 PMID: 23459093

Kendler, K. S., Neale, M. C., Kessler, R. C., Heath, A. C., & Eaves, L. J. (1992). The genetic epidemiology of phobias in women: The interrelationship of agoraphobia, social phobia, situational phobia, and simple phobia. *Archives of General Psychiatry*, *49*(4), 273–281. DOI: 10.1001/archpsyc.1992.01820040025003 PMID: 1558461

Kim, J. J., & Fanselow, M. S. (1992). Modality-specific retrograde amnesia of fear. *Science, 256*(5057), 675–677. DOI: 10.1126/science.1585183 PMID: 1585183

Kleinknecht, R. A., Dinnel, D. L., Kleinknecht, E. E., Hiruma, N., & Harada, N. (1997). Cultural factors in social anxiety: A comparison of social phobia symptoms and Taijin Kyofusho. *Journal of Anxiety Disorders, 11*(2), 157–177. DOI: 10.1016/S0887-6185(97)00004-2 PMID: 9168340

LeBeau, R. T., Glenn, D., Liao, B., Wittchen, H. U., Beesdo, K., & Craske, M. G. (2010). Specific phobia: Diagnostic issues and epidemiology. *The Psychiatric Clinics of North America, 33*(2), 235–248. DOI: 10.1016/j.psc.2010.04.006

Maples-Keller, J. L., Yasinski, C., Manjin, M., & Rothbaum, B. O. (2017). Virtual reality-enhanced extinction of phobias and post-traumatic stress. *Neurotherapeutics; the Journal of the American Society for Experimental NeuroTherapeutics, 14*(3), 554–563. DOI: 10.1007/s13311-017-0534-y PMID: 28512692

Marks, I. M. (1975). *Living with fear*. McGraw-Hill.

McLeod, S. (2015). Psychodynamic approach. *Simply Psychology*. Retrieved from https://www.simplypsychology.org/psychodynamic.html

Mowrer, O. H. (1947). On the dual nature of learning: A reinterpretation of "conditioning" and "problem-solving.". *Harvard Educational Review, 17*(2), 102–148.

Nemeroff, C. B. (2003). The role of GABA in anxiety disorders. *The Journal of Clinical Psychiatry, 64*(Suppl 3), 3–7. PMID: 12662127

North, M. M., North, S. M., & Coble, J. R. (1997). Virtual reality therapy: An innovative paradigm. *Cyberpsychology & Behavior, 1*(1), 39–47. DOI: 10.1089/cpb.1998.1.39

Öhman, A., & Mineka, S. (2001). Fears, phobias, and preparedness: Toward an evolved module of fear and fear learning. *Psychological Review, 108*(3), 483–522. DOI: 10.1037/0033-295X.108.3.483 PMID: 11488376

Pathak, V. N., & Judge, J. & Anjalee. (2023). *Indigenous Phobia VR Device* (Patent No. 1234567). India: Indian Patent Office.

Rebenitsch, L., & Owen, C. (2016). Review on cybersickness in applications and visual displays. *Virtual Reality (Waltham Cross), 20*(2), 101–125. DOI: 10.1007/s10055-016-0285-9

Riva, G., Baños, R. M., Botella, C., Mantovani, F., & Gaggioli, A. (2019). Transforming experience: The potential of augmented reality and virtual reality for enhancing personal and clinical change. *Frontiers in Psychiatry, 10*, 164. DOI: 10.3389/fpsyt.2019.00164 PMID: 27746747

Riva, G., Baños, R. M., Botella, C., Wiederhold, B. K., & Gaggioli, A. (2019). Virtual reality in psychotherapy: From imagination to immersion. *Journal of Clinical Psychology, 75*(3), 456–470. DOI: 10.1002/jclp.22799

Rothbaum, B. O., Hodges, L., Smith, S., Lee, J. H., & Price, L. (2000). A controlled study of virtual reality exposure therapy for the fear of flying. *Journal of Consulting and Clinical Psychology, 68*(6), 1020–1026. DOI: 10.1037/0022-006X.68.6.1020 PMID: 11142535

Salkovskis, P. M. (1991). The importance of behavior in the maintenance of anxiety and panic: A cognitive account. *Behavioural Psychotherapy, 19*(1), 6–19. DOI: 10.1017/S0141347300011472

Seligman, M. E. (1971). Phobias and preparedness. *Behavior Therapy, 2*(3), 307–320. DOI: 10.1016/S0005-7894(71)80064-3 PMID: 27816071

Sullivan, G. M., Coplan, J. D., Kent, J. M., & Gorman, J. M. (1999). The noradrenergic system in pathological anxiety: A focus on panic with relevance to generalized anxiety and phobias. *Biological Psychiatry, 46*(9), 1205–1218. DOI: 10.1016/S0006-3223(99)00246-2 PMID: 10560026

Teachman, B. A., Smith-Janik, S. B., & Saporito, J. (2007). Information processing biases and panic disorder: Relationships among cognitive and symptom measures. *Behaviour Research and Therapy, 45*(4), 882–896. DOI: 10.1016/j.brat.2007.01.009 PMID: 17341415

Vygotsky, L. S. (1978). *Mind in society: The development of higher psychological processes*. Harvard University Press.

Watson, J. B., & Rayner, R. (1920). Conditioned emotional reactions. *Journal of Experimental Psychology, 3*(1), 1–14. DOI: 10.1037/h0069608

Wolpe, J. (1961). *The systematic desensitization treatment of neuroses*. Stanford University Press. DOI: 10.1097/00005053-196103000-00001

Yao, Y.. (2022). Digital Twin-Based Control for the Electric Vehicle Powertrain: An Industrial Application. *Electronics (Basel), 12*(3), 672. DOI: 10.3390/electronics12030672

Zhao, M.. (2025). Understanding the Role of Artificial Intelligence in Medical Imaging: A Review. *Health Informatics Journal*, *29*(1), 22–39. DOI: 10.1186/s40359-025-02378-9

Chapter 6
Feature Engineering for Mental Health Applications

Rithik Reddy Nadimpalli
https://orcid.org/0009-0007-2409-2973
Woxsen University, Hyderabad, India

Mahitha Jalapally
Woxsen University, Siddipet, India

Alberto Acereda
Territorium, Philadelphia, USA

ABSTRACT

Feature engineering plays a pivotal role in developing accurate and effective mental health applications, as it enables the extraction of meaningful insights from raw data for precise mental health monitoring. This study explores the processes of data collection and preprocessing, focusing on transforming diverse data sources—including behavioral, physiological, and contextual data—into usable formats. Key strategies for identifying relevant features for mental health assessment are discussed, including the use of machine learning techniques to pinpoint indicators strongly correlated with specific mental health conditions. The primary research question addressed is: What strategies can be employed to enhance feature selection for precise mental health monitoring? By examining advanced methods such as automated feature selection, dimensionality reduction, and domain-specific feature engineering, this study aims to optimize the accuracy and reliability of mental health assessments.

DOI: 10.4018/979-8-3373-0467-0.ch006

1. INTRODUCTION

The integration of artificial intelligence (AI) and machine learning (ML) into the course of care in mental health has played a major role in the way mental health conditions are now assessed, monitored and managed. Millions around the world suffer with mental health issues, including anxiety, depression, bipolar disorder and schizophrenia. Early diagnosis and intervention can drastically improve outcomes but these techniques relying on clinical interviews and self reported questionnaires unfortunately do not provide timely or precise assessments. The gaps in the mental health care delivery have created an opening for data driven approaches which use advanced computational techniques to improve the precision of diagnostics as well as of therapeutics.

However, feature engineering is among the most critical components of these AI driven approaches. Feature engineering is the process of choosing, transforming and constructing relevant variables or features to your problem. It aids in extracting the meaning from different datasets such as behavioral, physiological, and contextual datasets which are generated in mental health applications. These insights can help economists use a better profile to make more accurate predictions of mental health conditions, inform personalized intervention, and even predict crisis prevention with early warning systems.

The sheer complexity and variability of mental health data is one of the key motivations for the application of feature engineering in mental health applications. Unlike most of the datasets in other domains, mental health data generally comes in multiple different forms: structured inputs such as questionnaire scores, semi structured data such as mobile usage etc., and even unstructured data such as natural language extracted from the therapy sessions. Moreover, the data captured can be quite different between people, as there are many differences in behavior, environment and socio demographic factors. However, this heterogeneity presents very specific challenges which require advanced feature engineering techniques to extract features that best represent the underlying mental health issues.

Data preprocessing is the initial step in feature engineering process that includes cleaning, normalization and integration of diverse data sources to convert your raw data into a form which can be used for modelling. As an example, behavioral information such as smartphones usage patterns may have to be synchronized with physiological information such as heart rate variability in order to obtain the big picture of the status of a user's mind. Depending on the type of the data, feature extraction methods, following preprocessing, are used to produce indicators from the raw data. For example, sleep duration, social interaction frequency, or even language sentiment can be great features to assess conditions such as depression or anxiety. Finally feature selection techniques are carried out to identify the most important features that result in predictive accuracy being maximized with overfitting or redundancy avoided.

Feature engineering in the context of mental health isn't just about improving model performance, but also learning more about contributions to mental health. For example, knowing that whether certain features, like spending more time on phone at night, were associated with anxiety could tell the patients as well as the clinicians what could be triggering their anxiety. These insights also help us create personalized treatment plans tailored to someone's individual pattern and needs.

In this chapter we try to explore specifically how feature engineering is important for mental health applications and cover some core aspects. I start with an overview of what kinds of data are frequently seen in mental health monitoring along with more detail on how specific processes behind feature engineering such as preprocessing, feature extraction and feature selection work. In depth we examine strategies for improving feature selection, including dimensionality reduction and more domain specific techniques. The discussion is augmented with real world case studies, which provide real evidence based insights on how feature engineering has been applied to improve the accuracy, reliability and scalability of mental health tools.

In addition, the chapter explores the challenges and the ethical considerations in the undertaking of feature engineering in this domain. It critically examines issues of data privacy, flaw (or bias) in feature selection and the risk of making automated tools too dependent. Finally, the chapter highlights future trends and advances of feature engineering incorporating the relation between future innovations and mental health care. This chapter seeks to fill the gap between technological innovation and its practical applications in mental health care through its focus on the critical importance of feature engineering. The theoretical knowledge and practical insights at the intersections of mental health and artificial intelligence makes it a valuable resource to the many researchers, clinicians and data scientists who are working to bring these emerging technologies to the masses. The chapter through this exploration aims to help grow a body of knowledge that augments the use of AI powered tools to help advance the world's mental health outcomes on a global scale.

2. OVERVIEW OF DATA IN MENTAL HEALTH MONITORING

Any good mental health application needs quality and diverse data to build upon. Mental health monitoring uses a wide range of data sources to know, measure and predict an individual's mental state. The structured inputs are source include questionnaire responses but the unstructured data comes from text from therapy sessions, or voice recordings. This section looks at what types of data are used for mental health monitoring, what challenges there are to collection and preprocessing of this type of data and the ethical considerations when dealing with such sensitive information.

2.1 Types of Data Used in Mental Health Monitoring

Mental health applications commonly utilize three primary types of data: physiological and behavioural plus contextual.

1. **Behavioral Data**: Be that personal data from an individual's actions and habits (often passively received from digital devices). Mental health behavioral indicators include, but are not limited to smartphone usage patterns, screen time, frequency of social media interactions, and mobility data. Conditions such as depression or anxiety often have specific patterns that are reduced activity levels, or irregular sleeping schedules, for example.
2. **Physiological Data**: Biological markers linked to mental health states is what this type of data captures. You have wearable devices that provide a constant metric of similar things like how heart rate variability is, how are you sleeping and how physically active you are. Other physiological signals such as electrodermal activity (EDA), or even vocal pitch and tone can be fed as input as well to make these inferences about emotional states.
3. **Contextual Data**: It includes environmental and situational data that, along with other data, contain information about such external factors as that affect mental health. For example, the GPS location data can tell someone how you move around, the weather and pollution levels may cause you to change mood. Contextual data makes things more complete by connecting mental health states to external events. Also, an important portion of data are ones that are reported by users themselves, such as responses to validated mental health questionnaires or mood logs. Input here is quite subjective, but is quite valuable for coupling of insights from objective data.

2.2 Challenges in Data Collection and Preprocessing

The use of such different data sources brings in unique challenges.

1. **Data Heterogeneity and Integration**: Data is typically collected from multiple devices, and multiple platforms. The integration of these data streams into a single format is complicated by varying sampling rates, data formats, and temporal alignments.
2. **Missing and Incomplete Data**: A major problem in mental health monitoring is common data gaps. For instance, self reported logs may not be completed,

or recorded data may not be taken due to battery depletion on wearable device. And so imputation techniques (eg, regression imputation, deep learning based model) are crucial to fill these gaps without degrading the quality of the dataset.

3. **Noise and Signal Processing**: A lot of data collected in the real world comes with noise, and by its very nature, preprocessing is necessary. For instance, heart rate signals noisy with movement artifacts must be filtered to enhance reliability of physiological markers. Like text and voice data just described, natural language processing (NLP) and speech processing are necessary to clean and extract meaningful features of data.
4. **Scalability**: As mobile and wearable devices proliferate, the datasets for mental health can increase exponentially. The robust computational infrastructure and algorithms for high dimensional data are required to manage and to process such large datasets.

2.3 Ethical and Privacy Considerations

As you can imagine, sensitive mental health data inherently presents massive ethical concerns. Behavioral and physiological data can be used to be intimate in the field of what an individual is doing, and thus is paramount where privacy is concerned. In order to comply with privacy regulations, such as GDPR, and HIPAA, applications are required to practice following measures of encryption and anonymization of the user identities.

The second possibility is the informed consent of the users. Therefore, it's crucial that people fully understand how their data is going to be used, stored and shared. Care must also be taken when working with vulnerable populations like people with severe mental health conditions lest they are exploited or inadvertently authority if they respond in a way you don't like.

Another problem is bias in collecting the data. This can even include models that do not generalize well on certain demographic groups and are comprised by an underrepresented subset of a dataset. For an example, the features extracted from smartphone usage are not likely to be good surrogates of mental health indicators in populations that have scarce technology access. This requires the active effort to sample representative data, and to build inclusive models.

2.4 Data Quality as a Cornerstone

Nonetheless, it's non-negotiable that the data to which these applications rely on is of good quality. High quality data not just helps improve the accuracy of predictive models, but also to ensure the reliability of the insights used to inform intervention.

Raw noisy data can be turned into a valuable mental health monitoring resource through rigorous preprocessing pipelines combined with domain expertise.

Overall, by each mental health monitoring gathers information from diverse data sources, providing each their own insight into the mental state of an individual. Although heterogeneity, missing data and ethical concerns constitute major hurdles, this however can be resolved by means of robust preprocessing, advanced algorithm, and ethical protocols. Knowing these fundamental elements of data provides a springboard for developing successful feature engineering, which we'll delve into in future sections.

The establishment of effective data collection and preprocessing techniques creates the necessary base for any machine learning pipeline to succeed particularly when working with mental health monitoring data that usually contains scant data points with significant noise and noticeable variability. The starting point of data collection requires data from different sources including human behavioral activities and physiological measurements as well as contextual information. The usage patterns of mobile devices together with screen duration data and reports of software application activities make up behavioral data. Quality heart rate variability measures and sleep data along with movement records are collected by wearable devices that make up physiological data. Contextual data contains environmental elements which may include noise levels as well as weather conditions paired with GPS details about movement patterns.

Once collected, preprocessing ensures that this raw data is cleaned, synchronized, and formatted for analysis. Key preprocessing steps include:

- **Data Cleaning:** The process requires handling incomplete records and fixing data anomalies and extreme values.
- **Normalization:** The process of feature scaling applies techniques such as min-max scaling or z-score normalization to normalize the range of HRV and screen time values.
- **Temporal Alignment:** The process of making timestamped multi-rate data from multiple sources match for integrated analytical purposes.
- **Imputation:** Necessary missing data should be filled with statistical models (mean, KNN) or complex model designs (MICE, autoencoders) to maintain data consistency.
- **Noise Reduction:** In order to minimize artifact effects and errant signals the signal processing methods include filtering physiological data while smoothing behavioral logs for data integrity.

3. KEY PROCESSES IN FEATURE ENGINEERING

Mental health application development is incomplete without feature engineering: it's the conduit from raw data to actionable insights. It is a series of processes – data preprocessing, feature extraction, feature transformation – that in order to maximize relevance and utility of data for machine learning models. In a mental health context, feature engineering is very important because the data is multi modal, heterogeneous, and noisy. In this section, we study the key processes involved in feature engineering in depth, and link their importance to mental health applications.

3.1 Data Preprocessing Techniques

First, and arguably most importantly, preprocessing raw data is required for feature engineering. Many mental health data for example is incomplete, noisy, and collected from many sources, and requires careful cleaning and standardisation.

1. **Data Cleaning and Normalization**: In raw data, we usually have missing entries, outliers or inconsistencies. Cleaning is about finding and fixing these errors so that data is meaningful. Physiological datasets such as heart rate variability (HRV) or sleep metrics have different units or ranges, and are therefore respectively normalised and scaled. Commonly, techniques such as min-max scaling, or z-score normalization are used.
2. **Handling Missing Data**: While mental health applications are commonly faced with missing data, maybe more so due to self reported logs or wearable device data. This problem can be addressed with mean substitution, k nearest neighbors (KNN) imputation or more advanced methods as multiple imputations by chained equations (MICE).
3. **Temporal Alignment and Synchronization**: Mental health data is often time series based (e.g. sleep patterns or activity logs) and must be aligned from data collected at different intervals or frequencies. As an example, logs of smartphone usage might also have to be matched with sleep data captured using a wearable device to match them up.
4. **Noise Reduction**: This can be a noise on physiological signals like electrodermal activity (EDA), caused by movement. With Butterworth filters, wavelet denoising etc, irrelevant variations can be filtered out and keeping meaningful trends.

3.2 Feature Extraction from Raw Data

Feature extraction consists in identifying within raw data patterns and trends that are indicative of mental health conditions. The direction is defined by the domain knowledge, and the process is commonly combined with statistical, machine learning, and signal processing.

1. **Behavioral Features**: Irregular sleep cycles, reduced mobility and abnormal social interaction are good behavioral patterns that indicate signs of mental health problems. These features can be extracted using metrics like: Activity trends: Daily step counts or daily sedentary hours. Screen usage patterns: Social media or messaging app time.
2. **Physiological Features**: Objective insights about a person's mental state is what physiological data is: heart rate or sleep quality. Key features include: Heart rate variability (HRV): Stress and anxiety are usually correlated to lower HRV. Sleep metrics: The usual thing to analyze is Total sleep duration, REM sleep percentage, and sleep onset latency.
3. **Text and Language-Based Features**: Features like sentiment polarity, word usage frequency, and syntactic complexity have been shown in applications concerning text (i.e., therapy notes or journal entries) that facilitate detecting states of emotion or cognitive distortions.
4. **Contextual Features**: That can be supplemented by contextual data like the patterns of location or environmental conditions. In fact, time spent at home, even too much time, might be associated with depressive symptoms.

Table 1. Common Data Types in Mental Health Monitoring and Associated Pre-processing Challenges

Data Type	Examples	Challenges
Behavioral	Screen time, app usage, mobility	Missing data, irregular logging, privacy issues
Physiological	HRV, sleep patterns, EDA	Sensor noise, movement artifacts, inter-device calibration
Contextual	GPS, noise level, weather	Temporal misalignment, external dependencies, lack of standardization
Linguistic/Textual	Therapy transcripts, journaling apps	NLP preprocessing required, semantic ambiguity, cultural bias

3.3 Feature Transformation and Encoding

Once we extract features, they are transformed into formats more suitable for analysis. P revening the data ensuring it is ready for machine learning model but retaining associative information.

1. **Scaling and Standardization**: Some models are trained even with features having different magnitudes, i.e. heart rate and sleep hours. Standardization transforms them into a standard normal distribution, while scaling ensures that all features contribute equally.
2. **Categorical Encoding**: Firstly, categorical data, like self-reported mood states, have to be encoded in numerical formats. One of the techniques is one-hot encoding or embedding method that captures relationship among categories.
3. **Dimensionality Reduction**: Removing the features will improve computational efficiency and lessening overfitting. There are (many) popular techniques for dimensionality reduction: Principal Component Analysis (PCA) and t-distributed Stochastic Neighbor Embedding (t-SNE, for example).
4. **Feature Aggregation**: Features can be aggregated over specific time windows (e.g., daily, weekly), to capture long term trends which are important for long term mental health conditions.

3.4 The Importance of Domain Expertise

Effective feature engineering in mental health application is heavily reliant on domain knowledge. Clinicians and psychologists are collaborated to choose to select features that are not only statistically significant but clinically meaningful. This means that if depression is indicated, then features such as reduced social interaction or increased nighttime activity, which are all well documented indicators of depression, can be prioritized, for instance, using input from experts.

3.4.1 Step by Step the Example: Workflow of Features Engineering in Depression Detection

Usually, a cycle of work begins with gathering raw data on wearables and mobile applications (e.g., HRV, step count, app usage). Preprocessing timestamps is first aligned and noise is removed with the use of filters such as Butterworth. Then, there is extraction of the feature mean step numbers daily, nightly phone usage, and HRV mean. The latter are then normalized with z-scores. The mutual information, Random Forest importance score-based feature selection is applied to reduce the number of features to the top 15 features. Lastly, chosen qualities are inputted into an SVM or an XGBoost to be categorized. This convenient pipeline illuminates the layer-based and repeated characteristic of highlighting the feature engineering process which integrates domain data along with algorithm disciplining.

3.5 Challenges and Opportunities

Feature engineering is essential, but it's difficult: it's hard to find a balance between simplicity and complexity; working with high-dimensional data; and making sure that features generalize across populations. The challenges of both scalability and reproducibility are alleviated by advances in automated feature engineering and domain-specific algorithms, which are enabled to scale and be reproduced as part of scalable and reproducible mental health applications.

Finally, feature engineering encompasses the art of preprocessing, feature extraction, and transformation of the data, in that order. Converting raw data into actionable insights in order to increase the accuracy and reliability of mental health applications requires these steps. Feature engineering as a field has great potential to revolutionize mental health care if we focus on clinically relevant features and use powerful algorithms.

A comprehensive process involving clinical expertise and technological methodologies must be applied to find suitable mental health assessment features. The development of mental health features diverges from generic feature engineering because these must correspond to clinical markers of psychological condition. Key categories of relevant features include:

- **Behavioral Features:** People with COVID-19 experience problematic sleep patterns, less physical movement, long periods of screen use and a reduced sense of social connection (displayed through fewer phone calls or texts).
- **Physiological Features:** The symptoms include low heart rate variability (HRV), poor sleep efficiency combined with increased electrodermal activity (EDA) and abnormal voice patterns.
- **Linguistic Features:** Researchers analyze journal entry sentiment polarity or the use of negative emotional language together with first-person pronoun frequency in texts.
- **Contextual Features:** Several conditions influence the happiness of elderly people such as how much time they spend at home and levels of noise and weather patterns.

The process of feature selection usually requires both statistical methods such as correlation and mutual information along with model-based importance metrics obtained from Random Forests or Lasso Regression models. Expert evaluation represents an essential step for clinical validation which verifies that predictive features also present meaning in mental health contexts.

4. STRATEGIES FOR FEATURE SELECTION

Feature selection is an important part of feature engineering process, which will help you select the most relevant and impactful features from a dataset while diminishing dimensionality. In mental health applications, and any where data is complex, multi modal, high dimensional, this process is particularly important. If feature selection can be performed effectively, it adds to predictive model accuracy, decreases computational overhead, increases interpretability and therefore stays cemented as a cornerstone of AI driven mental health tools. In this section, feature selection is a major issue, that different methods have been employed to accomplish this task, and the challenges and practical applications are discussed based upon the real world studies.

4.1 The Importance of Feature Selection

A large portion of mental health datasets feature diverse features as derived from behavioral, physiological, and contextual data. Nevertheless, not every feature helps as much to predict mental health conditions. Irrelevant or redundant features can:

Reduce Model Accuracy: For instance, non-informative features add noises that drive overfitting and lower generalization.
Increase Complexity: Since high dimension datasets require more computational resources, it can hinder real time deployment as well.
Hinder Interpretability: Models with too many features become opaque and are difficult to form actionable insights from.

Let me provide an example, studies on depression, take a certain features like what time you use your smartphone at night or if you have sleep patterns that are not regular, these features are very strong features when it comes to depression, but generic activity metrics which are just taking how much you move or how much you are sitting it doesn't help much in the analysis. This means that whilst models get to focus on the most important data, we would still have to identify and prioritise such impactful features.

4.2 Common Feature Selection Techniques

Filter, wrapper and embedded feature selection techniques are considered. There are advantages in each approach, often with an overlay of domain expertise.

1.Filter Methods: These methods are feature importance evaluation methods without any reference to a predictive model, which approach the feature importance using statistical tests or correlation measures. Examples:

Correlation Analysis: Linear relationship between features and target variable is found by pearson correlation.

Mutual Information: It identifies complex patterns that captures non linear dependencies.

Chi-Square Test: It analyzes the independence between feature and target variable in order to evaluate categorical features.

Case Example: To detect anxiety, a set of researchers analyzed the text data using mutual information techniques and found that word frequency and sentiment polarity were strong predictors of anxiety severity.

2. **Wrapper Methods:** Wrapper methods train machine learning models, in isolation, to predict performance on sub-sets of features.

Examples:

Recursive Feature Elimination (RFE): Removed the least significant features iteratively using model coefficients.

Forward/Backward Selection: It sequentially adds, or removes, features to best optimize model performance.

However, these methods are computationally intensive but produce feature interactions effectively.

Case Example: RFE was used to calculate the risk prediction for suicide based on the patterns of usage in Smartphone. Critical predictors such as late night call frequency, or erratic text message response emerged as features.

3. **Embedded Methods:** Feature selection as an embedded method is integrated into model training. For instance:

Lasso Regression: This applies L1 regularization, so a weight for an irrelevant feature is shrunk to zero.

Tree-Based Algorithms: Random Forest and Gradient Boosting rank the features by their importance in making decision splits.

Case Example: In a study making a prediction about therapy success rate, lasso regression found session frequency and reported mood to be the most important predictors, reducing error rate by 20%.

Table 2. Comparison of Common Feature Selection Techniques

Method Type	Technique	Pros	Cons
Filter	Mutual Information, Chi-Square	Simple, fast	Ignores feature interactions
Wrapper	RFE, Forward Selection	Captures feature interaction	Computationally expensive
Embedded	Lasso, Random Forest	Integrates with model, interpretable	Dependent on model choice

4.3 Domain-Specific Feature Selection

The selection of mental health features is critical and domain expertise is crucial to obtain, such that statistical methods match clinical relevance. Experts can prioritize features like:

- Indicators behavioral markers (reduced mobility, increased social withdrawal) for depression.
- Physiological metrics (heart rate variability) strongly correlated to stress or anxiety.
- Environmental conditions (for example, noise level, or, in season) that influence mood disorder.

For instance, among smartphone-based mental health monitoring clinicians, app usage diversity and text sentiment were emphasized, which harmonize with cognitively and emotionally known patterns.

4.4 Case Study: Feature Selection for Depression Detection

A common study was done by Saeb et al. (2015) to predict depressive symptoms in young adults using smartphone sensor data. There were more than 300 features in the dataset including movement patterns, sleep duration and social interaction frequency. Researchers employed mutual information and RFE to narrow the feature set to 10 critical variables, including Low variance in GPS data (low variance in normal movement patterns). Nighttime smartphone usage. Fewer text messages and calls. The improvement of the features almost doubled the accuracy of the model from 65% to 83%. The study also showed that targeted feature selection refines predictive models and makes them clinically applicable.

4.5 Challenges and Best Practices

Despite its importance, feature selection in mental health applications is not without challenges:

1. **High Dimensionality:** Due to having complex datasets with hundreds of features, robust methods are needed to avoid overfitting.
2. **Data Imbalance:** Most mental health datasets are not normal, or your dataset may have skewed distribution (e.g. many in the test data may not have been diagnosed with a given condition).
3. **Bias in Features:** But you can't always tell whether certain features, like smartphone usage, unintentionally pick up socioeconomic or cultural biases.

    Best Practices: Use several different methods of selecting results to validate. An industry expert utilises domain expertise for aspects relevant therapeutically. As always, be sure to assess how well models generalize and are equitable for different groups often.

4.6 Automation and Future Directions

This process has been made easy with automated feature selection using the AutoML (n.d) frameworks. These tools automate testing and optimizing feature subsets by find the highest accuracy, meaning reducing a human's intervention. AutoML serves as a quick prototyping and local scalability tool for large and diverse datasets, while still preserving personalized experience in mental health applications. Future work will most probably be the integration of multi-modal data sources, such as text analysis along with physiological metrics, for capturing holistic mental health perceptions. Additionally, feature selection has great promise as a field that in the context of Explainable AI (XAI) can lead to increasing the transparency and interpretability.

Finally, we show why feature selection is an important step towards developing effective and interpretable mental health applications. Through a combination of statistical, model-based methods and domain expertise, researchers can discover what features will help make predictions that are accurate and lead to actionable insights. Saeb et al.'s work provides real world evidence of the transformative potential of robust feature selection in mental health care.

4.6.1 Recent Work and Emerging Trends (2023–2025)

Recent studies continue to validate the role of robust feature engineering in real-world mental health applications. For instance, Lee et al. (2023) proposed a transformer-based model combining linguistic and physiological features to predict bipolar episodes with 87% accuracy. Another work by Ahmed et al. (2024) used multimodal LSTM models for predicting therapy dropout risks using smartphone usage and session transcripts, significantly outperforming traditional models. The 2025 IEEE Healthcare Informatics Conference also highlighted hybrid feature engineering pipelines leveraging SHAP for explainability and AutoML for optimization. These innovations affirm the importance of keeping pace with state-of-the-art techniques in mental health analytics.

4.7 Strategic Enhancements in Feature Selection

Mental health monitoring deals with specific difficulties during feature selection because data dimensions are high while individual profiles differ and the field contains sensitive content. Several techniques should be employed to improve feature selection accuracy.

1. **Hybrid Feature Selection Techniques:** Multiple feature selection methods should be used together to maximize benefits including filter (mutual information) alongside wrapper (recursive feature elimination) and embedded (L1 regularization).
2. **Domain-Informed Filtering**: Experts in clinical settings should remove features which are non-specific or deceive in mental health situations to enhance both the readability and the reliability of diagnostic systems.
3. **Dimensionality Reduction**: For exploratory analysis both PCA or t-SNE can be used to find latent structures in the data but researchers must keep in mind the interpretability consequences.
4. **Automated Feature Engineering (AutoML):** The systematic evaluation of feature subsets through AutoML tools becomes essential for multi-modal data fusion operations because of its capability to scale at speed.
5. **Bias Mitigation Techniques**: Adversarial debiasing methods combined with representative sampling should be used to guarantee the features work equally well across all demographic groups.
6. **Explainable AI (XAI):** SHAP or LIME interpretability features help validate selected features by explaining their importance.

5. ADVANCED TECHNIQUES IN FEATURE ENGINEERING

Data in mental health applications are often vast and diverse, and so increased use of more sophisticated feature engineering processes to obtain better features for the model is necessary. Although the traditional approaches offer very sound ground, the modern techniques take advantage of automation, dimensionality reduction and domain specific approaches to handling difficulties like high dimensionality, non-linearity, and multi-modal data. Through these advanced techniques, this section discusses the applications, benefits and drawbacks of the tools with case studies from real-life business settings.

5.1 Automated Feature Selection

With AutoML (Automated Machine Learning) frameworks, automation in feature engineering has gained a pulse. By testing multiple feature subsets, these tools simplify selection, removing the need for manual interventions.

1. Advantages:
   Efficiency: The large datasets and the high dimensions are then efficiently dealt with by automated techniques because those process the data so that they save the time and computational resources. Consistency: These methods are ones which reduce human error and bias, leading to reproducing the results.
   Scalability: Feature selection scales with increasing data complexity due to automation.
2. Techniques:

Recursive Feature Elimination (RFE) with cross validation is an automated iterative feature ranking algorithm with automated feature removal. The Google's AutoML, H2O.ai or any other similar tool will take care of the entire feature engineering pipeline, primarily feature selection and transformation.

Case Example: AutoML was applied to a study on prediction bipolar episodes using features from multi modal data, such as heart rate variability and sentiment analysis, improving the prediction accuracy by 25%.

5.2 Dimensionality Reduction Techniques

With AutoML (Automated Machine Learning) frameworks, automation in feature engineering has gained a pulse. By testing multiple feature subsets, these tools simplify selection, removing the need for manual interventions.

1. **Principal Component Analysis (PCA):** Principal components found by PCA explain maximum variance in the data. Example: PCA was used in a study on

anxiety prediction, which reduced over 200 behavioral features to five principal components (90% variance retained and improved model interpretation).

2. **t-Distributed Stochastic Neighbor Embedding (t-SNE):** This is mostly used for visualising high dimensional data such that you can more easily interpret different feature clusters. Example: t-SNE was used to cluster emotional tone (i.e. words that evoked feelings such as happiness, worry, sadness) presented in therapy sessions text data, and the researchers used this to identify patterns associated with certain mental conditions.
3. **Non-Negative Matrix Factorization (NMF):** Specifically, NMF can do data decompositions into non-negative components—in applications such as topic modeling in the therapy transcripts.
4. **Limitations:** It is possible that PCA and t-SNE may mislead as feature interpretable components do not have a one to one correspondence to the the original features. There may be computational intensity associated with these methods for real time applications.

5.3 Domain-Specific Feature Engineering

Domain specific strategies embed expert knowledge from areas like psychology and psychiatry to feature engineering process. Furthermore, these approaches guarantee that chosen features correspond to clinically confirmed institutions and are understandable to the practitioner.

1.**Expert-Driven Features:** Based on psychological theories features like linguistic markers in text data (use of negative words) and physiological signals (HRV) are derived.
Example: Experts pointed to specific features like a long pause in speech or increasing use of the first-person pronoun as crucial in the detection of suicidal ideation, in a study.

2. **Hybrid Approaches:** Feature relevance is enhanced with the combination of domain expertise and data driven methods. For example, analysing therapy notes using NLP techniques in which we prioritize clinically validated patterns such as emotional tone shifts.
3. **Case Example:** The post-traumatic stress disorder (PTSD) project combined expert driven features (such as containing trauma related text keywords) with physiological data (such as galvanic skin response). A 20% improvement in classification accuracy, compared to purely data driven methods, made this a hybrid approach.

5.4 Multi-Modal Feature Integration

A mix of behavioral, physiologic and contextual factors determines the course and outcome of mental health conditions. Multi-modal feature engineering puts together multiple data sources into a holistic view of mental health.

1. **Fusion Techniques:** Early Fusion: Feature extractio Late Fusion: It extracts features from each source independently and combine them during the model training.
2. **Applications:** The combination of smartphone usage (e.g., app activity) with wearable data derived sleep metrics was integrated in a study predicting depression. By suggesting late fusion, we introduced model robustness through the independent strengths of each modality.
3. **Challenges:** Aligning data streams based on different time scales (real-time GPS, vs. daily mood logs, etc). Computational complexity management in large scale multi modal datasets.

5.5 Case Study: Advanced Feature Engineering for Anxiety Monitoring

In Smith et al., 2021. tried to predict anxiety levels from multi-modal data collected from smartphones and wearables. Data from behavioral (screen time), physiological (heart rate variability), and contextual (noise levels) were derived from the dataset that had over 500 features. Key steps included:

1. **Automated Selection:** It found 20, high impact features, including irregular sleep schedules and amplified screen activity during late hours.
2. **Dimensionality Reduction:** We reduced the feature set to five principal components by running PCA and obtained 85% of the variance.
3. **Domain-Specific Refinement:** Clinical experts were consulted for validating features so that it matches anxiety related behaviours. Combining automation, dimensionality reduction, and expert input in feature engineering yielded prediction accuracy from 70% to 89%, this approach.

5.6 Challenges and Opportunities

While advanced techniques significantly enhance feature engineering, they are not without challenges:

Interpretability: Advantage of PCA is that it reduces the dimension of data but at the same time it makes it difficult to understand relationship between the original feature and the transformed component.
Computational Overheads: Robust infrastructure is needed for multi modal integration and automated pipelines.
Ethical Considerations: Privacy is concerned with the use of personal and contextual data and must be overcome by strict adherence to ethical standards.

Future opportunities include:

Explainable AI (XAI): There are also emerging frameworks in interpretability, such as SHAP (SHapley Additive exPlanations), for example, that try improving the interpretability of automated feature selection.
Deep Feature Engineering: Autoencoders (and neural networks in general), among other ways, can learn intricate feature representations on their own from raw data thus furnishing a greater insight.

Much of the accuracy, efficiency, and scalability of mental health applications depends on the application of advanced techniques in feature engineering. Automation, dimensionality reduction, domain expertise and multi-modal integration all lend themselves to the unique challenges of mental health data. These approaches hold the potential to bridge raw data and actionable insights in order to develop robust and useful AI driven mental health tools.

5.7 XAI in Mental Health Applications

Explainable AI (XAI) is important in the mental health context, where it is often needed to have a complex model being interpretable and transparent. In a medical setting, the stakeholders require not only the prediction to be known, but the reasoning behind the nature of said prediction.

1. Task of XAI: Black box models such as Deep Neural Networks or Random Forests tend to be used. There are methods that can decomposes feature contribution to a prediction, including SHAP (SHapley Additive exPlanations) and LIME (Local Interpretable Model-Agnostic Explanations, 2016). As an example, a SHAP value may indicate that a youthful depression score was contributed by 30% by the feature of late-night phone use.
2. Ethical Transparency: Interpretability helps in notable bias and mistakes in predictions. As an illustration, XAI can show that a model is being over-reliant on app usage, which can be affected more by poor socioeconomic access than

mental health. Revision of the model, or reweighting of features, can then be performed by the clinicians.

3. real world Practice: SHAP can be used to confirm that the expected pattern of prediction is clinically supported, such as social withdrawal, or an abnormal heart rate, increasing clinician confidence in tools that measure suicide risk.

6. APPLICATIONS FOR FEATURE ENGINEERING IN MENTAL HEALTH

Beyond academic exploration, feature engineering in mental health monitoring has been expanded to the practical applications of real world tools and systems to improve mental health care delivery. Feature engineering allows transforming raw data into hooks for early warning systems, personalized interventions, and predictive models, tailored for clinicians as much as patients. The applications of these features are then explored in depth, along with case studies to show how feature engineering can boost mental health outcomes.

6.1 Early-Warning Systems

Feature engineering is used to detect the very early changes in the behavioral or physiological data that precede critical events like depressive episodes and suicidal ideation, on which early warning systems of mental health crises are dependent.

1. **Key Features:** Irregular sleep patterns, a reduced physical activity level or social withdrawal are all behavioral markers. Such as declines heart rate variability or increases in stress level.
2. **Applications:** In a study, Wang et al. (2019) built an early warning system for depression using smartphone and wearable data. Late nightlife phone usage, and decreasing daily step counts were key features. Using the system 2 weeks in advance to predict depressive episodes, accuracy was 82 percent.
3. **Challenges:** To make sure that we don't pick up on false alarms but do capture risky signals. The idea was to integrate feedback loops so that predictions are refined using responses from users.
4. **Case Example:** In 2020, a team of researchers and a mental health NGO launched an early warning system for suicide prevention among adolescents. Identification of at risk individuals was aided by increases in isolation and change in text sentiment. Timely interventions became possible, resulting in a 18% decrease of suicide rates for the pilot group.

6.2 Personalized Mental Health Interventions

Besides doing the feature engineering it helps us to customize the mental health treatments for each individual which is a need, by noticing the unique patterns, for them. Personalised approaches enable treatment adherence and effectiveness to be improved for the particular patient by tailoring to the patient's individual's circumstances.

1. **Customizing Therapy:** Language sentiment, emotional tone and frequency of therapy session can provide some information on how to tailor cognitive behavioral therapy (CBT) programs. Example: The NLP based features in a study on therapy transcripts work to make CBT modules more personalized and resulted in a 30% end-users outcome improvement.
2. **Medication Management:** Guidance regarding medication adjustments is provided by predictive features such as symptom fluctuation and physiological responses. Example: Sleep patterns and mood logs were used to predict how soon antidepressant dosage changes would be needed, with 85 percent accuracy.
3. **Behavioral Nudges:** With features such as activity levels and a record of goal achievement rates, mobile health apps send personalized nudges to users to encourage them to make healthier choices.

6.3 Predicting Mental Health Outcomes

The feature engineering comes handy and allows us to develop predictive models that predict chances of recovery, relapse or the success of some interventions. The model provides support to clinicians in making evidence based decisions.

1. **Treatment Success Prediction:** Finally, we predict the likelihood of recovery based on features such as therapy adherence, baseline severity of symptoms and engagement metrics. Example: In a 2022 study, the prescience of predictive models based on adherence and engagement data was 78 percent accurate in predicting therapy success.
2. **Relapse Prevention:** Said features — sudden drops in physical activity or increases in negative sentiment — can predict relapse risks in patients with chronic conditions of the mind. Case Example: Multi modal data, including sleep variability and text sentiment, combined in a relapse prediction model for bipolar disorder, achieving 80% accuracy.
3. **Longitudinal Monitoring:** Ongoing monitoring and the degree of adjustment of treatment plans is supported by long term feature tracking as, for e.g., gradual changes in mood logs or activity levels.

6.3.1 Real World Applications of Multi-Modal/Real Time Analytics

1. **India, 2024: Real-Time Depression Monitoring System:** A national digital health pilot tested a multi-modal mental health app linked to phone used, ambient sound level and fitbit. The app provided real-time nuggets in case risks such as restricted movement + excessive surrounding noise were identified, and subdued patient self ascertained depressive events by 30%.
2. **PTSD Management Tool (USA, 2023):** A veteran application was based on heart rate recording in real time, sentiments journaling, and GPS. Alerts were triggered to care teams when the rate of HRV and isolation patterns was raised. XAI was also employed in such a way that alerts became usable and certified with therapists.
3. **UAE, 2024 Workplace Burnout Tracker:** A corporate wellness tool combined calendar information, sleep records, and the tone of voice messages. It showed the current prediction of stress on the screen which can be utilized to optimize meeting loads, or direct breaks. Predictions were explained with LIME and the staff confidence was motivated.

6.4 Case Study: Suicide Risk Assessment Using Behavioral Features

A suicide risk assessment project was combined by a team of researchers and clinicians targeting young adults in 2021. This project combined wearable and smartphone data and developed an overall predictive model for suicide ideation.

1. **Data and Features:** Behavioral features: Decreased social interaction, increase in use of cell phones throughout the night. Linguistic features: Text messages which contain negative sentiment and frequent use of first person pronouns. Physiological features: A heart rate that is elevated, and irregular sleep patterns.
2. **Results:** Model gave 87% accuracy to identify high risk individuals. Six months later, hospitals saw a 15 percent decline in the rate of hospitalizations when interventions were conducted based on predictions.
3. **Key Insights:** On its own behavioral data alone provided strong predictive power, but there was slight improvement in specificity by including physiological markers. Concerns associated with ethical issues related to data privacy were addressed by anonymization and informed written consent protocols.

6.5 Challenges in Application

While feature engineering has demonstrated significant benefits in mental health applications, challenges remain:

1. **Data Quality and Accessibility:** Feature extraction can be compromised by inconsistencies or incompletion data of wearables and smartphones. In low resource settings, data is often hard to come by.
2. **Model Interpretability:** With increasing numbers of features, these complex models are hard to interpret, precluding their use in clinical applications. This problem is solvable with explainable AI methods like SHAP values in them.
3. **Ethical Considerations:** There are privacy concerns and risk of misuse sensitive data and this make that strict to ethical standards. Data policies needed to be transparent and security tuned to a high level.

6.6 The Road Ahead

The future of feature engineering in mental health lies in advancing multi-modal integration, real-time analysis, and ethical AI practices:

1. **Multi-Modal Fusion:** Utilizing diverse data sources, genetic, behavioral, and environmental to improve existing predictive models.
2. **Real-Time Analytics:** Systems to develop that are able to process data streams in real-time and take immediate intervention.
3. **Ethical AI Frameworks:** Studying how to add fairness and bias mitigation strategies into the pipelines of feature engineering.

7. CHALLENGES AND FUTURE DIRECTIONS

Feature engineering for mental health applications has tremendous potential driving more accurate diagnoses, more efficacious interventions, and powerful monitoring tools. But particularly technical hurdles, and beyond, must be overcome to make it a success in the long term and in clinical applications. Furthermore, future directions which can overcome such limitations and continue to drive further innovation are identified by emerging technologies and research trends. In this section we address the major challenges and provide a glimpse at how they might progress.

7.1 Challenges in Feature Engineering

1.Data Quality and Availability

One of the biggest challenges in mental health applications is the variability in data quality and availability:

Incomplete Data: But such wearables and mobile apps tend to create disjointed datasets because of device breakdown or inconsistent user engagement or data loss during transfer.
Imbalanced Data: Datasets about mental health often have skewed distributions (fewer such events as severe depression or bipolar episodes), and thus biased models.
Noise and Artifacts: Signals for behavioral or physiological are susceptible to noise coming from the outside, that is, from environment, activity or technical limitations.
Example: During exercise, heart rate variability data can contain artifacts not necessarily correlated with mental health which makes feature extraction difficult.

2. Generalizability of Features

Features extracted from one population or context may not generalize across different groups:

Cultural Variability: Some features, notably social behaviors, language patterns, and the use of technology, turn out to be very culture specific.
Age and Gender Differences: Tailored approach is required for predictive features for mental health conditions because the types of predictive features differ by age groups and by gender.

3. **Bias and Fairness** in Models Bias in mental health applications can arise from:

Non-Representative Datasets: Inaccurate or inequitable prediction is possible as a result of underrepresentation of some demographics.
Algorithmic Bias: It is unclear whether majority groups are favoured in feature selection processes to the exclusion of minority population.

4. **Real Time Processing & Scalability** With the proliferation of wearables and mobile apps, mental health systems must handle large volumes of data in real time:

Computational Costs: High dimensional data processed on multiple devices can make for a good infrastructure.
Latency: Real time systems may lose utility because of delays in their response to changing conditions in crisis scenarios.

5. Ethical and Privacy Concerns

The sensitive nature of mental health data raises significant ethical issues:

Informed Consent: This is especially hard when ensuring users fully understand how their data is collected and used; particularly vulnerable populations.
Data Security: This is all about protecting data from what may be a crime — a breach or misuse of personal data.
Transparency: As a result, users and clinicians must trust the algorithm, and therefore it needs to be explainable and clear about the feature selection and decision-making process.

7.2 Future Directions in Feature Engineering

1.Multi-Modal Data Integration

The future of feature engineering lies in combining diverse data streams for a comprehensive understanding of mental health. For Example: Combining automated wearable data (e.g. heart rate), smartphone data (e.g. app usage) and environmental factors (e.g noise levels) can yield a whole picture of mental well-being.

Advancements: Multi modal data integration can be improved with techniques like late fusion and attention-based models.

2. Explainable and Ethical Artificial Intelligence (XAI)

Improving transparency and fairness in feature engineering:

Explainability Tools: Users get a sense of how individual features are contributing to explain how and whyUtils like SHAP (Shapley Additive Explanations) and LIME (Local Interpretable Model-Agnostic Explanations) can be used to assist users in understanding the role.
Ethical Guidelines: With 'Ethics by Design', we make sure to incorporate privacy, fairness and inclusivity in in all the stages.

3. Advances in Automation

Automated feature engineering tools are becoming increasingly powerful:

AutoML Pipelines: For large, complex datasets, automated tools can extract and select features with minimal human intervention, and can work with datasets that humans have difficulty visualizing and understand.

Deep Learning: Autoencoders, for instance, are neural networks that can uncover non linear relations, as well as latent features in unstructured data.

4. Adaptive Learning for Personalization

Future systems will focus on real-time personalization:

Dynamic Features: Features that can continuously update based on user's behavior and feedback.
Adaptive Models: Algorithms that evolve as they learn more about the user, which give rise to tailored interventions.

5. Applications to New Domains

Feature engineering is likely to expand into areas beyond conventional mental health monitoring:

Preventive Healthcare: Early identifying of mental health risk within populations without existing mental health diagnosis.
Therapeutic Feedback: It helps study therapy progress and help clinicians make decisions.
Education and Workplaces: By applying behavioral analytics to school and corporate environments, enabling enhancement of well-being.

7.3 Case Study: Tackling Bias in Feature Engineering

In 2022, a team of researchers made the mental health monitoring app for different global populations. To address bias, they implemented:

Representative Sampling: Participants in this dataset ranged from different regions, ages, and socioeconomic status.
Fair Feature Selection: Featuring smartphone usage patterns, reweighting, and adversarial debiasing were employed to ensure these features could not disproportionately favor tech savvy populations.
Outcome: We demonstrate that, for underrepresented groups, one can achieve a 20% reduction in error through equitable feature engineering practices.

7.4 Opportunities for Collaboration

The future of feature engineering in mental health lies in interdisciplinary collaboration:

Healthcare and AI Integration: We connect data scientists and clinicians to the AI researchers.
Policy and Ethics Experts: Maintaining accordance to the rules that exist under the GDPR or HIPAA.
Community Engagement: In the user driven contexts, like 'co-designing systems to ensure relevance and trust'.

7.5 Conclusion

Although feature engineering has already played an important role in mental health care, it still has its challenges and needs to be solved in order to achieve the best application of it. Together with the integration of multi-modal data, emergence of ethical AI practices, and embracing automation, the field can commence with the production of robust, inclusive and scalable solutions. What future innovation will do is not just make individual outcomes better, it will fundamentally change how we perceive and contend with mental health on a social level.

8. CONCLUSION

The space of feature engineering and mental health applications is a transformative frontier in healthcare. Feature engineering leverages raw manifold multi-modal data to derive single dimensions of meaningful insights that power more potent diagnostics, personalized interventions, and proactive monitoring systems. In this chapter, the key processes, challenges and development of feature engineering advance featured in mental health care are highlighted.

8.1 Key Takeaways

1. Critical Role of Data: Mental health monitoring is based on data. Properly processed behavioral, physiological, contextual data can give us a total picture of mental health states.
2. Sophistication in Techniques: In modern feature engineering, we use sparse methods such as dimensionality reduction, automated selection, as well as domain specific strategies to make predictive models robust and scalable.
3. Real-World Applications: Feature engineering is powering the development of essential tools including early warning systems, personalized interventions, and predictive models on therapy success. Case studies integrate it to illustrate its real world impact—suicide prevention systems and anxiety monitoring tools.

4. Challenges and Ethical Considerations: Specific challenges raised are bias of the AI, data privacy and scalability, and the importance of ethical AI practices as well strong technical solutions to these challenges. Centre is still on balance of innovation with responsibility.
5. Future Directions: Multi modal data integration, explainable AI and adaptive learning are the future of feature engineering. They promise to close existing gaps and open doors to new mental health care options.

8.2 The Road Ahead

To the extent feature engineering continues to evolve, its potential to reshape mental health care will continue to grow. Real time analytics, personalized systems, and equitable design will make mental health tools more accessible and powerful for diverse populations. Continuing progress in this requires collaboration between data scientists, clinicians, ethicists, and policymakers.

Using technology to help human expertise enhance outcomes is illustrated by the integration of feature engineering into mental health applications. If researchers and developers prioritize inclusivity, transparency and innovation, these tools will meet the needs of all the people on the planet.

8.3 Final Reflections

In this chapter, we demonstrate the transformative advantage of feature engineering in mental health applications. Contributing to an increased level of intelligence in diagnosing and enabling early interventions, these are beginning to create more proactive and personalised ways of providing mental health care. But the road goes on. To bring into reality the vision of making a data driven mental health ecosystem, future advances will need to be made with regards to the ethical deployment, scalability and real-world applicability.

By continuing to develop and enhance feature engineering for use on, and in, broader applications, we can not only continue to address the pressing mental health problems of the present but also set the groundwork for a healthier and more robust future. Technology and the next generation of human care coming together has the promise of revolutionising mental health diagnostics and treatment and more importantly, enriching understanding of how human well-being is in this increasingly complex world.

REFERENCES

Auto, M. L. Framework. (n.d.). Google AutoML: Machine learning made easy. https://cloud.google.com/automl

Lee, Y.. (2023). Transformer-based Multi-modal Learning for Bipolar Prediction. *IEEE Transactions on Affective Computing*.

LIME (Local Interpretable Model-Agnostic Explanations). (2016). Ribeiro, M. T., Singh, S., & Guestrin, C. "Why should I trust you?": Explaining the predictions of any classifier. *Proceedings of the 22nd ACM SIGKDD International Conference on Knowledge Discovery and Data Mining*. DOI: 10.1145/2939672.2939778

Saeb, S., Zhang, M., Kwasny, M. J., Karr, C. J., Kordy, H., & Mohr, D. C. (2015). Mobile phone sensor correlates of depressive symptom severity in daily-life behaviour: An exploratory study. *Journal of Medical Internet Research, 17*(7), e175. DOI: 10.2196/jmir.4273 PMID: 26180009

Smith, J., Brown, T., & Anderson, K. (2021). Multi-modal feature engineering for anxiety prediction using wearable and smartphone data. *Computers in Human Behavior, 124*, 106925. DOI: 10.1016/j.chb.2021.106925

Wang, R., Wang, W., DaSilva, A., Huckins, J. F., Kelley, W. M., Heatherton, T. F., & Campbell, A. T. (2019). Tracking depression dynamics in college students using mobile phone and wearable sensing. *Proceedings of the ACM on Interactive, Mobile, Wearable and Ubiquitous Technologies, 3*(1), 1–26. DOI: 10.1145/3314398 PMID: 40144218

Chapter 7
The Moral Algorithm Ethics in AI-Supported Mental Health

Renuka Gadiraju
https://orcid.org/0009-0000-1195-0990
Woxsen Universiity, Hyderabad, India

Amulya Kavadikijanekunte
Woxsen University, Hyderabad, India

Teja Kumar Karnala
Woxsen University, Hyderabad, India

ABSTRACT

The incorporation of artificial intelligence (AI) into mental health care is becoming more prevalent, bringing forth significant improvements in diagnosis, treatment, and support for patients. AI technologies, ranging from chatbots that offer immediate assistance to predictive analytics that facilitate early diagnosis, are reshaping the mental health field. Nonetheless, the swift implementation of these technologies introduces important ethical issues that need to be considered to guarantee responsible and just care. This chapter discusses the ethical aspects of AI-enhanced mental health services, emphasizing key principles such as confidentiality, informed consent, and fairness in algorithms. It analyses the challenges and risks tied to AI, such as data security threats and possible biases, while also looking at regulatory structures and guidelines aimed at maintaining ethical norms. T

DOI: 10.4018/979-8-3373-0467-0.ch007

INTRODUCTION

AI technology offers fresh opportunities for mental healthcare and multiple other domains into the present and future (Alharthi et al., 2024). AI implementation in mental health services provides opportunities to attain precise diagnosis and ensure individualized therapies while enabling instant assistance to patients. Artificial intelligence tools that integrate chatbots and machine learning algorithms are changing mental health service delivery methods to become more accessible and efficient at healthcare delivery. The new technology presents an effective solution to overcome persistent mental healthcare obstacles involving restricted service availability and delayed treatment and individualized solutions. The fast development of AI technologies in mental healthcare generates critical ethical problems which require solution to establish fair and responsible utilization. AI systems create major privacy issues since they need to access and study confidential patient information. The fundamental requirement for sustaining patient trust involves both information protection and maintaining total confidentiality of their data. The process of informed consent demands healthcare providers to disclose how patient data will be utilized and explain AI treatment implications to their patients. Patient self-determination together with their feeling of trust both collapse when systems do not maintain transparency.

The ethical concern around the fairness of automated algorithms stands as a crucial matter. The use of AI algorithms to analyze training data may cause unbiased results to persist which then results in unequal mental healthcare and worsens pre-existing inequalities in such care. AI systems must operate in a completely unbiased manner to help patients receive equivalent healthcare services. The role of AI should exist to supplement human professional abilities instead of substituting them with technology. AI-produced insights need to combine with human perception to furnish quality mental health services. This section examines both opportunities and challenges of AI-assisted mental health services through an evaluation of both patient rights protection and holistic patient care.

The chapter reviews artificial intelligence development specifically through diagnostic devices virtual therapists alongside wearable devices that are remodeling psychotherapy practice. The discussion examines the central question about how artificial intelligence developments transform psychotherapy domains while prioritizing ethical evaluations of patient care and dignity and personal empowerment.

A BRIEF HISTORY OF TECHNOLOGY IN MENTAL HEALTH CARE

Mental health services introduced technological advancements in their practice through a gradual process during the past hundred years. During the initial therapeutic periods the treatment exclusively required personal contact and paper-based documentation and observation by humans. During the middle of the twentieth century researchers at MIT developed ELIZA as an experimental application which stands as one of the first examples of computerized therapeutic applications through natural language processing (1966). ELIZA represented the initial step that enabled researchers to establish foundational methods for using computers in psychotherapy practice.

The internet triggered the emergence of computerized mental health treatments throughout the first decade of the new millennium and the entire 1990s period. People received access to self-directed CBT modules via the data-based online treatment system that united MoodGYM and Beating the Blues. People could access Telepsychiatry service for distant mental health treatment through video communication technology across whole areas.

The 2010s brought about popular use of smartphones and health applications leading to increased development of digital mental healthcare solutions. Headspace together with Calm and Talkspace started offering therapeutic help along with support resources on their phone-based digital systems. The advancements enabled AI to access mental healthcare facilities thereby starting a time when artificial intelligence powered mental health diagnosis while managing patient interactions and creating treatment plans.

How AI is Transforming Mental Health Care

Beneficial use of artificial intelligence (AI) within mental health services transforms how healthcare professionals deliver their services. The utilization of AI-powered tools is streamlining diagnosis precision while developing individualized care plans and immediate patient support features. Users obtain non-biased immediate help from chatbots which aids the management of emotions and directs them towards proper support. Predictive analytics as a form of AI allows for early detection of mental health issues by processing multiple data resources which enables proper interventions that might stop these conditions from deteriorating. The advancement of mental health care technologies through modern innovations produces targeted health support which can be reached by a broader spectrum of patients because of expanded access points.

AI and Wearable Devices: Enhancing Mental Health Monitoring

Wearable technology operates jointly with diagnostic algorithms and chatbots to build primary components that enable AI-augmented mental health systems. Smartwatches and rings and patches along with biometric clothing represent modern wearable devices available for the market which track heart rate variability and sleep patterns and skin temperature and activity levels. The AI system processes gathered data to identify stress as well as anxiety and depression and burnout symptoms.

Wearable devices support Mind strong and Biobeat applications to perform consistent monitoring of biometrics which generates real-time mental health assessments. AI-based data analytics examines unusual data patterns to create individual wellness plans that users receive through the system. Such technologies display key benefits for medical staff who monitor chronically ill patients with mental health problems by delivering early indicators of health emergencies.

Wearable devices provide patients with direct care management abilities along with doctor access to objective real-time patient measurements that improve reported symptom charts. AI analytics integrated with biofeedback systems now enable mental health services to deliver appointment-free personalized care beyond standard clinical therapy.

Key AI Technologies: Machine Learning, Natural Language Processing

The fundamental industrial changes happen via essential AI technologies including machine learning and natural language processing (NLP) (Inside Precision Medicine, 2023). Machine learning models built from large dataset training.sessions focus on detecting patterns and generate accurate predictions. Mental health care algorithms evaluate patient information to determine outcome predictions as well as risk identification together with individualized treatment recommendations. NLP gives AI systems the capability to comprehend human speech which serves as the base functionality required to develop successful AI chatbots alongside virtual therapeutic systems. AI technologies enable health systems to provide helpful interactions with patients that offer support and mental health monitoring and trigger timely interventions as needed.

Potential Benefits of AI in Mental Health

AI implementation in mental health care shows substantial promise for treatment improvements (Jorie AI, n.d.). The ability of AI technology allows major expansion in mental health care accessibility for individuals living in distant regions or underserved areas. AI systems maintain twenty-four-hour availability through which they monitor healthcare gaps to provide consistent patient care at all times. AI-generated treatment plans based on analytic analysis should improve the effectiveness of interventions thus producing superior patient results. AI systems help to lessen physicians' office burden because they handle basic tasks allowing doctors to dedicate time to critical human-dependent cases that need thoughtful decisions along with empathy. AI technology creates a path to develop psychiatric care which provides better and faster services that benefit all patients.

ETHICAL CONSIDERATIONS: PRIVACY AND CONFIDENTIALITY

Importance of Protecting Patient Data

Patient data protection remains vital in mental health treatment since the disclosed information holds sensitive value. Intense personal observations regarding patients' feelings idea and experiences commonly appear in mental health clinical files. Patients need a secure environment to freely access help so maintaining their privacy and confidentiality of data protects their trust in the system. Confidentiality breaches lead to severe effects which include patient discrimination and emotional distress along with social stigma. Vulnerable information requires proper management to dodge legal penalties for healthcare organizations since effective data protection standards demand immediate implementation.

Examples of AI Impact on Privacy

Multiple advantages exist from AI systems yet their application leads to major privacy concerns. The core functionalities of many AI systems that treat mental health patients operate through extensive data collection and analysis. Medical diagnostics and individualized treatment recommendations require systems which need to access protected patient information that includes medical background histories combined with personal communications and behavioral patterns. A wide range of gathered data exposes patients to major safety hazards when improperly handled.

Data breaches stand as one of the greatest problems affecting organizations today. Successful intrusions on AI systems pose a constant risk to cyber-attacks (Trigyn Technologies, n.d.) because they lead attackers toward sensitive patient information. These security incidents create major problems that generate both identity theft and financial losses and severe emotional damage for affected patients. A compromised patient privacy results from mismanaging data storage such as when data is improperly stored or when encryption methods are neglected in AI systems.

The practice of re-identifying patients poses another critical problem to patient privacy. Despite anonymization techniques data has the potential to be linked back to individual persons through database cross-matching methods. The unintentional leak of patient identities along with their sensitive information becomes possible through this process. AI-based utility programs performing inspection of anonymized mental health data (Zhang, 2019) have the potential to identify individual patients through cross-checking with public social media content and social records.

Mental health tools coupled with chatbots fueled by AI create privacy dilemmas that affect user information security. These platforms gather sizable information from their users about their mental condition together with their individual viewpoints and daily routines. Such applications should incorporate robust privacy measures because the information collected might become vulnerable to unauthorized manipulation and misuse when protection measures are insufficient. Users may not fully comprehend both the extent of collected data along with the purposes for using it which leads to possible breaches of informed consent.

The protection of users' privacy requires strong defensive measures that must be established (ISACA, 2016). Proficient encryption technology combined with strict access permissions and continuous security evaluations protects information as it is stored and transmitted while blocking unauthorized personnel from sensitive data access. AI developers should integrate privacy-by-design principles into their developments by considering privacy factors since the initial design and development stages. All organizations must develop comprehensive policies which specify how data will be used. Healthcare organizations should disclose all data collection activities as well as protection methods and data usage purposes to their patients with full clarity. The strive for translucency combined with the opportunity for patients to make knowledgeable decisions about their data results from these measures.

AI in psychiatric care can deliver its advantages while safeguarding patient privacy by implementing strong protective measures for such benefits to be realized.

Measures to Ensure Confidentiality

Several measures need implementation to protect patient information confidentiality when AI supports mental health treatments. Progressive encryption methods must be implemented to shield data from unauthorized viewers. Even when data interception occurs users cannot decode information except when using the proper decryption key. The implementation of strong access control systems will guarantee that proper users maintain exclusive rights to sensitive information. Multi-factor authentication and regular checks of access logs need to be established for complete security. Patients need to receive detailed information about data policies which explains how their information will be utilized (American Speech-Language-Hearing Association, n.d.) and all implemented security protocols. Patients are less likely to surrender their data when pre-communication does not establish what potential risks exist.

The development of new algorithms requires AI programmers to build data protection mechanisms while implementing privacy-by-design design principles among their work. The implementation of privacy must begin at the start of AI system development through a whole-process intervention instead of being added later in the process. Routine examinations of AI systems through regular checks will assist in detecting and resolving privacy issues(Mittelstadt, 2020) which ensure the safety of patient information. Patient privacy protection can be achieved through anonymization techniques that entail either erasing or concealing personal identification information. It is essential to develop an ethical culture among developers and clinicians and policymakers to establish high levels of confidentiality and privacy within AI-aided mental health services.

INFORMED CONSENT IN THE AGE OF AI

Understanding Informed Consent in AI-Enhanced Mental Health

The ethical concept of informed consent requires clarification when mental healthcare employs artificial intelligence systems (Scribbr, n.d.).

Lawfully establishing permission from healthcare patients represents an ethical standard in medical practice to ensure full treatment plan awareness before obtaining consent. The implementation of artificial intelligence in mental healthcare services completely alters how informed consent should be defined (Patel, 2024). The future of mental healthcare will likely transform significantly because of diagnostic predictive algorithms and therapeutic chatbots as main AI applications. The modern technological tools present distinctive complexities which medical practitioners must properly explain to their patients(Galič, 2018).

The comprehensive understanding of AI methods requires education about both systems functions and internal bounds alongside possible prejudices in these technologies. The learning process of AI systems requires past data analysis but this method risks deteriorating existing biases within its operation. The delivery of significant information relating to these aspects by healthcare practitioners enables patients to take relevant treatment decisions.

The specific data-handling methods which AI systems use to collect and protect patient data should be communicated to patients. Healthcare organizations must disclose data management procedures because transparency creates trust and reveals patient information protection measures to their patients.

Strategies for Properly Informing Patients

To ensure that patients receive adequate information, healthcare providers might consider the following approaches:

1. *Simplified Explanations:* Present complex AI concepts in straightforward language. Utilize relatable analogies to clarify technical aspects instead of relying on industry jargon.
2. *Interactive Consent Forms:* Create digital consent forms that integrate videos, diagrams, and interactive elements to elucidate how AI functions and its significance in treatment. This enables patients to navigate the consent process at their own pace.
3. *Ongoing Communication:* Keep communication channels open with patients, regularly updating them regarding the usage of their data and any technological advancements that may impact their care. This consistent dialogue ensures they remain informed throughout their treatment journey.
4. *Educational Workshops:* Host workshops or sessions where patients can learn more about AI's role in healthcare, providing a platform for them to ask questions and express any concerns.
5. *Feedback Mechanisms:* Establish systems for patients to share their experiences with AI tools, which can be valuable for enhancing technology and the consent process, ensuring a patient-centered approach.

Ethical Implications of Consent in AI Applications

The integration of AI into healthcare raises significant ethical issues. To ensure ethical informed consent, several critical aspects must be addressed:

- *Data Privac*y: Patients should feel confident that their personal data is handled securely (Australian National University, n.d.). Strong data protection practices are crucial to prevent unauthorized access and potential breaches.
- *Algorithmic Bias*: As AI systems learn from historical data, which may be biased, routine evaluations and updates of AI algorithms are necessary to spot and reduce these biases, guaranteeing fair treatment for every patient.
- *Transparency:* The decision-making processes of AI systems ought to be transparent. Patients have a right to understand how AI algorithms derive their insights and recommendations, fostering trust and enabling informed choices.

- *Accountability*: Clear accountability is vital for situations where AI systems might make mistakes or fail. Both healthcare professionals and AI developers need to take responsibility for the results and work cooperatively to rectify any issues that may arise.
- *Autonomy:* It's essential to respect patient autonomy by ensuring they have the freedom to make decisions regarding their care. This means supplying all relevant information about AI tools and honoring their choice to accept or refuse them.

As AI technologies continue to advance, the ethical frameworks that oversee their application must also evolve. This ongoing process necessitates collaboration among healthcare professionals, AI developers, ethicists, and patients to ensure that informed consent remains a fundamental element of ethical AI in healthcare (Brown, 2023).

ADDRESSING ALGORITHMIC BIAS

The Issue of Bias in AI Algorithms

Algorithmic bias poses a major challenge in the development and implementation of AI systems. Bias occurs when algorithms yield systematic prejudices due to flawed assumptions during the machine learning process. This issue can result in the unfair treatment of individuals or groups, especially those who are already marginalized.

Sources of Bias:

1. *Training Data:* The effectiveness of an AI model is largely influenced by the data available for training. If the training data reflects historical biases or lacks diversity, the AI model is likely to perpetuate those biases.
2. *Feature Selection:* The selection of features used to build the AI model can introduce bias if they mirror existing stereotypes or prejudices.
3. *Model Design:* Decisions concerning the architecture and design of the AI model can lead to bias if fairness is not adequately considered.

Bias in AI algorithms can present itself in various forms, including gender bias, racial bias, and socioeconomic bias, each carrying serious implications.

Real-World Examples of Bias in Mental Health Care

Bias in AI-driven mental health services can significantly affect the care quality provided to various demographic groups. Notable examples include:

1. *Racial Bias:* AI diagnostic tools have been shown to underdiagnose mental health issues in minority populations. For example, Black patients might receive less accurate diagnoses than white patients, resulting in insufficient treatment and poorer health outcomes.
2. *Gender Bias:* Bias in AI models can lead to differing treatment recommendations based on gender (Jones, 2023). For instance, women may be more likely to receive diagnoses for anxiety disorders, while men might be overlooked for the same conditions, causing imbalances in treatment.
3. *Socioeconomic Bias:* AI systems may prioritize treatment options based on economic considerations, potentially disadvantaging individuals from lower socioeconomic backgrounds and leading to unequal access to mental health services.

These instances underscore the urgent need to tackle bias within AI systems to ensure fair mental health care for all individuals.

Strategies to Mitigate Algorithmic Bias

Addressing algorithmic bias requires a comprehensive strategy involving several essential approaches:

Diverse and Representative Data:

- *Balanced Datasets:* It is crucial to ensure that training datasets reflect a varied demographic representation to minimize bias. This involves collecting data from diverse sources to encapsulate a wide range of experiences and viewpoints.
- *Data Augmentation:* Techniques like data augmentation can generate synthetic data to enhance the representation of underrepresented groups in the dataset, helping the AI model perform adequately across diverse demographics.

Bias Auditing:

- *Regular Assessments:* Conducting frequent audits of AI models to identify and measure biases is vital. Utilizing fairness metrics to evaluate model performance across various groups helps pinpoint areas needing attention.

- *Third-Party Reviews:* Involving independent third parties to review AI systems for biases offers an unbiased assessment and aids in identifying potential overlooked issues.

Transparency:

- *Explainable AI:* Creating AI models that can provide clear justifications for their decisions enhances transparency, helping users understand the decision-making process.
- *Documentation:* Keeping thorough records of the AI development process, including sources of data, model choices, and techniques for mitigating bias, is essential for accountability and oversight.

Inclusive Development Teams:

- *Diverse Perspectives:* Forming AI development teams with varied backgrounds encourages a range of perspectives, reducing the likelihood of biased outcomes and leading to more inclusive AI solutions.
- *Bias Training:* Training developers to identify and address biases in AI systems is critical. This raises awareness of potential biases while fostering a culture of fairness and equity.

Ethical Guidelines:

- *Frameworks and Standards:* Establishing ethical guidelines and standards that prioritize fairness and equity in AI development is crucial. These guidelines should guide every stage of the AI lifecycle (Gupta, 2023), from data collection to model deployment.
- *Continuous Improvement:* Implementing a feedback loop to consistently refine AI models is key. This involves incorporating audit findings and user feedback to ensure AI systems remain fair and unbiased.

By implementing these strategies, we can advance towards creating AI systems that are equitable, transparent, and inclusive, guaranteeing that all individuals receive fair treatment and benefit from AI advancements.

BALANCING AI AND HUMAN JUDGMENT

Importance of Human Judgment in Mental Health Care

Human judgment is a vital element of mental health care, offering the empathy, intuition, and nuanced understanding that AI systems are unable to replicate. The essential human element is necessary for establishing trust, providing comfort, and devising personalized treatment plans.

The therapeutic bond between a clinician and a patient is grounded in empathy and shared understanding. Human clinicians can discern subtle aspects of human behavior and emotion, offer emotional support, and create a safe environment for patients to express their concerns. This empathetic approach is crucial for effective mental health care.

Cultural Competence: Mental health professionals possess an understanding of the diverse cultural backgrounds of their patients. This knowledge enables them to tailor their approaches to align with the cultural context, ensuring that treatment respects and is relevant to each patient's unique experiences.
Clinical Expertise: Mental health professionals acquire years of training and experience that allow them to make informed decisions. They are capable of interpreting complex psychological and emotional signals, considering multiple factors, and creating comprehensive treatment plans tailored to individual patient needs. This clinical expertise is irreplaceable by AI systems.

Risks of Over-Reliance on AI

While AI presents many advantages, an over-dependence on it in mental health care can lead to significant risks:

- *Diagnostic Errors:* AI systems depend on data for decision-making, and if this data is biased, incomplete, or incorrect, it can result in diagnostic mistakes. Misdiagnoses can lead to inappropriate treatment plans, potentially harming patients. Ensuring the quality and representativeness of the data used for training AI systems is vital in minimizing these risks.
- *Loss of Human Connection:* The relationship between a patient and a clinician is foundational in mental health care. Excessive reliance on AI may weaken this bond, resulting in a less personal and effective treatment experience (Yu, 2023). Patients might feel alienated if they see their care being managed by a machine instead of a human.
- *Privacy and Data Security*: AI technologies typically require access to vast amounts of personal data, raising concerns about privacy and security (Williams, 2023). Breaches or misuse of such information can severely affect patients, diminishing trust in AI technologies and healthcare providers. Strong data protection measures are essential for safeguarding patient information.

- *Decline in Clinical Skills:* Should clinicians depend too heavily on AI for decision-making, their clinical skills may deteriorate (LexisNexis, 2023) (Sunflower Lab, 2023). If healthcare providers frequently defer to AI recommendations, their diagnostic abilities and judgment can weaken, potentially leading to a decrease in care quality.

Finding a Balance Between AI and Clinical Expertise

Achieving an appropriate balance between AI and human judgment involves maximizing the advantages of both while ensuring human oversight and control:

- *Complementary Roles:* AI should enhance human judgment rather than replace it. By assisting with data analysis, identifying patterns, and offering preliminary recommendations, AI can support clinicians, who then review and interpret this information. This collaborative strategy allows healthcare providers to make more informed choices while keeping the human touch central to patient care.
- *Continuous Education and Training:* Clinicians should engage in ongoing education and training regarding the latest AI technologies and their roles in mental health care. This education should prioritize critical thinking and independent judgment to ensure that healthcare providers remain skilled and capable of sound decision-making.
- *Transparency and Explainability:* AI systems must be transparent and understandable. Clinicians should be able to explain how AI algorithms reach their conclusions and communicate these decisions to patients. This fosters trust and enables effective oversight and validation of AI-generated recommendations.
- *Ethical Frameworks:* It is crucial to develop and adhere to ethical frameworks governing AI use in mental health care. These frameworks should address issues such as data privacy, consent, and potential bias. Adhering to ethical guidelines ensures responsible and ethical use of AI by healthcare providers.
- *Patient Involvement:* Involving patients in the decision-making process is vital. Patients need to be informed about the use of AI in their treatment and should have opportunities to provide feedback (Wong, 2021). This involvement ensures that AI applications respect patient autonomy and preferences.
- *Interdisciplinary Collaboration:* Promoting teamwork among AI developers, clinicians, ethicists, and patients can lead to the creation of more effective and ethically sound AI systems. Collaborative teams can offer diverse insights and perspectives (Johnson, 2023), assisting in identifying potential biases and enhancing the overall design and implementation of AI technologies.

By achieving a balance between AI and human judgment, mental health care can leverage the strengths of both, resulting in more effective, personalized, and ethical treatment for patients.

Ethical Challenges Encountered

While there have been successes, the use of AI in mental health care has not been without ethical issues. Here are some real-world instances:

Concerns About Data Privacy:

AI systems frequently require access to sensitive personal information, raising worries about privacy and data security. For example, a mental health app faced criticism for sharing user data with third parties without obtaining informed consent (Smith, 2024), underscoring the need for stringent data protection protocols and clear data usage policies. It is vital to ensure that patients fully understand how their data will be utilized, and to secure explicit consent to maintain trust.

Algorithmic Bias:

AI systems may unintentionally reinforce existing biases (Johnson, 2023) (Brown, 2023), leading to unequal treatment across demographics. For instance, an AI diagnostic tool was discovered to underdiagnose mental health disorders among minority groups due to biased training datasets. Tackling these biases necessitates ongoing monitoring and refinement of AI algorithms to guarantee fairness and equity in mental health care. This includes routinely updating training datasets to reflect diverse populations and applying techniques for bias detection and mitigation.

Informed Consent Challenges:

Securing informed consent for AI-assisted mental health interventions can be difficult. Patients may not fully comprehend how AI systems operate and how their information will be utilized, creating ethical dilemmas regarding autonomy and consent. Effective communication and educational initiatives are necessary to aid patients in making well-informed decisions regarding their treatment. For example, using interactive and multimedia consent forms can enhance patient understanding of AI's role in their care.

Responsibility and Accountability Issues:

Deciding where accountability lies when AI systems err or fail to deliver as expected can be complicated. Questions may arise about whether the fault lies with the developers, the clinicians employing the system, or the AI itself. Establishing clear guidelines and accountability frameworks is crucial to address these matters (Smith, 2021) and ensure ethical AI implementation in mental health care.

Lessons from Previous Experiences

Experiences involving AI in mental health care have yielded important lessons, underscoring the necessity for ethical oversight and ongoing enhancement:

Transparency and Accountability:

It is essential to guarantee transparency in AI decision-making processes and to hold developers responsible for the outcomes. This includes clear documentation and explanations of how AI algorithms function, assuring clinicians and patients can trust the systems. Transparent AI systems allow users to understand the logic behind AI recommendations, fostering informed decision-making.

Ongoing Monitoring and Evaluation:

Regular reviews and evaluations of AI systems are critical for identifying and addressing biases, mistakes, and unintended effects. This ongoing oversight helps enhance the accuracy and fairness of AI tools in mental health care. Establishing independent review bodies and employing fairness metrics can support this continuous evaluation.

Ethical Guidelines and Frameworks:

Developing and adhering to ethical guidelines for AI utilization in mental health care is vital. These guidelines should cover data privacy, informed consent, and bias alleviation, ensuring responsible and ethical use of AI. Professional organizations, such as the American Psychological Association (APA) and the World Health Organization (WHO), offer valuable insights on ethical AI practices.

Interdisciplinary Collaboration:

Collaboration among AI developers, clinicians, ethicists, and patients is essential for creating effective and ethical AI systems. Interdisciplinary teams can supply varied perspectives and insights (Johnson, 2023), resulting in better-designed AI tools that address the needs of all stakeholders. Such cooperative efforts can also nurture a culture of ethical AI development and usage.

Patient-Centered Approaches:

Involving patients in the design and implementation of AI systems ensures their needs and preferences are taken into account. Patient feedback is crucial for refining AI tools to ensure they are user-friendly (Smith, 2023) and beneficial. Engaging patients as active participants in their care can enhance the overall effectiveness (Zhang, 2024) and acceptance of AI-driven interventions.

REGULATORY AND LEGAL PERSPECTIVES

Existing Laws and Regulations

Mental Health Care Act 2017 (India):

The Mental Health Care Act of 2017 aims to safeguard the rights, dignity, and privacy of individuals with mental health issues in India. This comprehensive legislation includes measures for digital data protection, ensuring careful handling of personal mental health information. The act requires mental health facilities to adhere to data protection standards and to obtain informed consent before collecting or sharing personal information. These regulations are crucial for protecting patient privacy and ensuring ethical use of digital mental health resources.

Indian Council of Medical Research (ICMR) Ethical Guidelines for AI (2023):

The ICMR's Ethical Guidelines for AI in healthcare outline principles for the development and implementation of AI technologies. These guidelines prioritize accountability, patient welfare, and the necessity for transparency in AI systems. Core principles include obtaining informed consent, safeguarding patient privacy, and conducting regular audits of AI algorithms for bias. The guidelines aim to promote responsible AI use in healthcare, upholding ethical standards and protecting patient rights.

Digital Healthcare Regulation in India:

The framework regulating digital healthcare in India exposes the lack of specific legal provisions addressing AI and digital health technologies. This regulatory gap raises issues pertaining to informed consent, data security, and ethical AI usage in healthcare. Efforts are underway to address these challenges by crafting comprehensive regulations (Brown, 2023) (Holistic, 2024) aimed at ensuring the safe and ethical application of digital health tools. These regulations are designed to provide clear guidance for healthcare providers and technology developers, enhancing trust and accountability in AI-enhanced mental health care.

Guidelines from Professional Organizations

American Psychological Association (APA):

The APA addresses the incorporation of AI in mental health services, focusing on ethical considerations and best practices. Their guidelines stress that AI tools should supplement, not replace, human judgment. Key points include the significance of transparency, informed consent, and the safeguarding of patient privacy. The APA advises clinicians to receive adequate training in AI technologies and to stay alert to potential biases within these systems. By adhering to these guidelines, mental health professionals can ensure responsible and ethical AI practices.

American Counseling Association (ACA):

The ACA provides recommendations for counselors utilizing AI in their work, underscoring the need for ethical use and professional accountability. Their guidelines emphasize the importance of maintaining the therapeutic relationship and using AI tools as complements to human judgment. The ACA advocates for ongoing education and training for counselors to remain informed about advancements in AI and their implications for mental health care. By following these recommendations, counselors can utilize AI to enhance their practice while maintaining ethical integrity.

International Perspectives

World Health Organization (WHO):

The WHO has established a global framework concerning the ethics and governance of AI in healthcare, including its mental health applications. This framework outlines vital ethical principles such as transparency, accountability, and equity. It stresses the necessity of ensuring that AI technologies do not exacerbate existing health disparities and are utilized to foster equitable access to mental health services. The WHO guidelines serve as a reference for nations formulating their AI regulations, promoting a consistent approach to ethical AI governance.

United States – AI Regulations in Healthcare:

U.S. authorities are actively developing policies to regulate AI's role in healthcare. These policies are focused on ensuring safety, compliance, and ethical usage of AI technologies. The regulations aim to provide clear guidelines for AI developers and healthcare providers, addressing concerns regarding data privacy, algorithmic bias, and informed consent. Through establishing rigorous regulatory frameworks, the U.S. intends to encourage innovation while protecting patient rights and promoting ethical AI deployment in mental health care.

Canada – Ethical Guidelines for AI in Suicide Prevention:

Canada has introduced specific ethical guidelines for AI-related mental health interventions, such as the "Canada Protocol" for suicide prevention AI. These guidelines tackle ethical considerations unique to AI applications in mental health, emphasizing the importance of accuracy, transparency, and patient consent. They highlight the necessity for ongoing monitoring and evaluation of AI systems to ensure both effectiveness and safety. By adhering to these principles, Canada aims to utilize AI to enhance mental healthcare while safeguarding patient rights (Dezyit, 2024) and welfare.

ETHICAL DESIGN AND DEVELOPMENT OF AI TOOLS

Principles of Ethical AI Design

Designing ethical AI requires a methodical approach that embeds ethical considerations throughout the AI development process(Danks, 2023). The key principles include:

Fairness:

Promoting fairness involves designing AI systems that treat everyone equitably, free of bias or discrimination. This can be achieved through:

- Diverse and Representative Data: Utilizing datasets that accurately mirror different demographics for AI model training.
- Bias Auditing: Conducting regular assessments of algorithms to detect and rectify biases.

- Fairness Metrics: Implementing metrics to gauge and ensure fairness in AI outputs.

Transparency:

Transparency means making AI decision-making processes clear and understandable for users. This can be ensured through:

- Explainable AI (XAI): Creating models that can provide straightforward justifications for their decisions.
- Clear Documentation: Recording the development journey, including data sources, model design, and decision-making algorithms.
- Open Communication: Maintaining transparent discussions with users and interested parties regarding the capabilities and limitations of AI systems.

Accountability:

Developers and organizations must accept responsibility for the outcomes of their AI systems. Establishing accountability can involve:

- Clear Guidelines: Formulating explicit guidelines defining responsibility for AI performance and impacts.
- Ethical Review Boards: Forming boards to oversee the ethical ramifications of AI projects.
- Continuous Monitoring: Implementing ongoing surveillance and evaluation of AI systems to ensure adherence to ethical standards.

Privacy:

Safeguarding user privacy is paramount in AI design. This entails:

- Data Protection Measures: Adopting strong safeguards to secure personal data.
- Informed Consent: Ensuring users are fully informed about and consent to the use of their data.
- Anonymization: Employing techniques that anonymize data to limit privacy risks.

Beneficence and Non-Maleficence:

AI systems should aim to benefit users and society while minimizing harm. This principle can be realized by:

- Safety Testing: Executing thorough testing to confirm AI systems do not inflict harm.

- User-Centric Design: Developing AI tools that prioritize user well-being and satisfaction.
- Ethical Guidelines: Following ethical standards that emphasize the importance of promoting good and avoiding harm.

Inclusiveness:

Ethical AI design should address the requirements and perspectives of various user groups. Inclusiveness can be achieved through:

- User Involvement: Incorporating feedback from users of diverse backgrounds in the design process.
- Accessibility: Ensuring AI tools are usable by those with differing abilities and situations.
- Cultural Sensitivity: Designing AI systems that respect and acknowledge cultural diversity.

Collaboration Between Developers and Clinicians

Effective teamwork between AI developers and clinicians is crucial for creating AI tools that are useful and advantageous in clinical environments. Important elements of this cooperation include:

Interdisciplinary Teams: Establishing teams that combine AI developers, healthcare professionals, ethicists, and patient advocates. These groups bring varied skills and viewpoints, resulting in more comprehensive and effective AI solutions.

Co-Design Processes: Involving clinicians and other stakeholders in the creation and development of AI tools. This collaborative design approach ensures that AI systems tackle genuine clinical issues and are user-friendly for healthcare providers.

Continuous Feedback: Setting up channels for ongoing input from clinicians who utilize AI tools in their work. This constant feedback helps to pinpoint areas needing improvement and keeps AI systems relevant and effective in clinical practice.

Education and Training: Offering education and training opportunities for both AI developers and clinicians. Developers should gain insight into the clinical environment and ethical implications, while clinicians should understand the capabilities and limitations of AI technologies.

Pilot Testing: Implementing pilot studies of AI tools in clinical environments to collect real-world data and feedback. Pilot testing allows for the refinement of AI systems, ensuring they are functional and practical in everyday situations.

Ensuring Transparency in AI Development

Transparency is fundamental to ethical AI development, building trust and comprehension among users. Key strategies for fostering transparency are:

Clear Documentation: Keeping detailed records of the AI development process, including data sources, model structures, training methods, and decision-making algorithms. This information should be accessible to both technical and non-technical stakeholders.
Explainable AI (XAI): Designing AI systems that can clearly articulate their decisions and actions. Explainable AI aids users in understanding the rationale behind AI conclusions, thereby fostering trust and accountability.
Open Communication: Encouraging open dialogue between developers, users, and other stakeholders regarding the abilities and limitations of AI systems. Transparent communication helps manage expectations and cultivates a collaborative atmosphere.
Regulatory Compliance: Ensuring that AI development meets applicable laws, regulations, and ethical standards. Adhering to regulatory requirements illustrates a dedication to responsible AI development and enhances public trust.
User Education: Guiding users on the AI systems they engage with, explaining how they function, their advantages, and their potential drawbacks. User education helps bridge the gap between technical complexity and user comprehension, leading to a more informed and trusting user base.

By following these principles and strategies, AI developers and healthcare professionals can develop ethical AI tools that elevate patient care while maintaining ethical practices and fostering user trust.

PATIENT-CENTERED AI APPLICATIONS

Incorporating Patient Feedback in AI Design

Integrating patient feedback during the design of AI systems is vital for developing tools that are user-friendly, effective, and aligned with patient needs and preferences.

As the end-users of AI-driven healthcare technologies, patients can offer insights that inform improvements to these tools.

Patient Surveys and Focus Groups: Using patient surveys and focus groups to gather direct insights into patients' experiences, concerns, and suggestions about AI tools. This feedback can help developers make necessary adjustments to enhance usability and effectiveness.
Patient Advisory Boards: Forming patient advisory boards inclusive of diverse demographic representatives ensures that the AI design process accounts for a broad spectrum of perspectives. These boards can continuously provide feedback and help developers grasp the distinct requirements of various patient populations.
Iterative Design Process: Employing an iterative design methodology that incorporates patient feedback at different stages of development allows AI systems to be regularly refined and adjusted based on practical experiences.
User-Centered Design Principles: Implementing principles focused on user-centered design prioritizes the needs and preferences of patients, ensuring that AI tools are intuitive, accessible, and effective in addressing healthcare needs.

Enhancing Patient Engagement with AI Tools

Boosting patient engagement with AI tools involves creating technologies that are interactive, informative, and empowering. Engaged patients are more likely to actively partake in their healthcare, leading to improved outcomes and satisfaction.

Interactive Interfaces: Designing AI tools with engaging interfaces can make the technology more appealing to patients. Features like chatbots, virtual assistants, and gamified elements enhance the user experience.
Personalized Health Insights: I systems can offer personalized health information based on patients' data, including customized treatment suggestions, progress tracking, and medication and appointment reminders. Tailored information helps patients feel more in control of their health choices.
Educational Resources: Incorporating educational materials into AI tools can assist patients in understanding their health conditions and the role of AI in their treatment plans. Providing accessible and concise information empowers patients to make informed healthcare decisions.
Supportive Communication: I tools can enhance communication between patients and healthcare providers. Features like secure messaging, telemedicine consultations, and routine follow-ups can facilitate patient engagement and ensure ongoing support.

Addressing Patient Concerns and Expectations

Tackling patient concerns and expectations is vital for establishing trust and acceptance of AI technologies in healthcare. Patients may have several worries about AI, such as data privacy, accuracy, and the effect on the doctor-patient dynamic.

Transparency and Education: Informing patients about how AI systems operate, how their data is utilized, and the benefits and limitations of AI can ease concerns. Open communication builds trust and ensures patients feel informed and reassured.
Data Privacy and Security: Implementing strong data privacy and security measures is critical for safeguarding patients' sensitive information. Ensuring patients are aware of these practices and providing clear explanations about data management can help alleviate privacy concerns.
Accuracy and Reliability: Demonstrating the accuracy and reliability of AI systems through evidence-based research and clinical validation can help alleviate worries regarding the effectiveness of AI tools. Sharing success stories and positive results can further bolster confidence in AI technologies.
Human Oversight: Highlighting the importance of human oversight in AI-driven healthcare can reassure patients that AI tools are meant to support rather than replace human decision-making. Ensuring that healthcare providers remain active in decision-making can address concerns about depersonalized care.
Patient Autonomy: Valuing patient autonomy by including them in decisions about the use of AI in their treatment is crucial. Providing options and obtaining informed consent enables patients to take charge of their healthcare and feel empowered in their choices(Zhang, 2021).

By integrating patient feedback, enhancing engagement, and addressing concerns and expectations, healthcare providers and developers can create patient-centered AI applications that improve health outcomes and foster patient trust and acceptance.

FUTURE DIRECTIONS AND ETHICAL INNOVATION

Enhancing Ethical Standards

To elevate ethical standards in AI-enhanced mental health care, several practical strategies can be employed:

Algorithmic Bias Mitigation: Conducting regular assessments of AI algorithms to discover and reduce biases. Utilizing diverse and representative datasets for AI model training guarantees fairness and accuracy.

Data Privacy and Security: Employing rigorous data encryption and anonymization techniques to secure patient information. Establishing clear policies regarding data access and usage will ensure compliance with privacy regulations.

Transparency and Accountability: Providing clear explanations regarding how AI systems make decisions to ensure comprehension among patients and healthcare providers(Lee, 2024). Setting accountability measures for AI-generated results, including protocols for dealing with errors and adverse outcomes, is also necessary.

Informed Consent: Making sure patients are thoroughly informed about the use of AI in their treatment and securing their consent prior to implementation. Offering ongoing education about AI technologies helps patients make knowledgeable choices.

Collaboration with Ethical Experts: Collaborating with ethicists, legal professionals, and patient advocacy groups to formulate and review ethical guidelines for AI application in mental health care. Encouraging interdisciplinary cooperation can help tackle ethical challenges and establish best practices.

The Role of Stakeholders

Various stakeholders are pivotal in the ethical application of AI in healthcare:

Clinicians: Provide clinical insights and feedback to ensure AI tools meet patient care needs. Monitor AI-driven interventions and raise any concerns or issues with developers and policymakers.

Patients: Engage in feedback opportunities and surveys to share their experiences and preferences. Advocate for their rights and privacy in the use of AI technologies.

Developers: Create AI systems with ethical considerations as a priority, focusing on patient safety and well-being. Collaborate with clinicians and patients to ensure AI tools address real-world challenges.

Policymakers: Formulate and enforce regulations that protect patient privacy and guarantee the ethical usage of AI. Provide funding and support for research on ethical practices in AI healthcare applications.

Ethicists and Legal Experts: Offer guidance on ethical dilemmas and legal implications associated with AI use in healthcare. Assist in the development of frameworks and guidelines for responsible AI deployment.

Ensuring Responsible Innovation

Cultivating a culture of responsible innovation in AI requires several core strategies:

Collaboration and Transparency: Foster open communication and collaboration among stakeholders to address ethical challenges and share best practices. Promote transparency in AI development and deployment to ensure clarity in processes and decisions.
Continuous Improvement: Create iterative design processes that incorporate feedback from clinicians, patients, and other stakeholders. Regularly update AI systems based on new research, technological advancements, and user input.
Education and Training: Provide ongoing education and training for healthcare providers and patients on the ethical application of AI. Encourage a culture of continual learning and adaptation to keep in line with AI progress.
Ethical Frameworks: Create and adhere to ethical guidelines that direct the responsible application of AI in healthcare. Ensure that the design and implementation of AI systems prioritize ethical considerations.
Public Engagement: Involve the public in discussions about the ethical aspects of AI in healthcare. Gather input from various communities to guarantee that AI tools are inclusive and equitable.

By concentrating on these priorities, we can ensure that AI-enhanced mental health care remains both ethical and effective (United VARs, n.d.), benefiting patients and healthcare providers alike.

CONCLUSION

As AI continues to evolve, it is fundamentally transforming the practice of psychotherapy—reshaping how therapy is accessed, delivered, and personalized. This chapter has explored this transformation through an ethical lens, addressing critical issues such as privacy, fairness, informed consent, and the human-AI dynamic in therapeutic relationships.

In this chapter on "Ethics in AI-Enhanced Mental Health Care," several key points have been thoroughly examined. The integration of artificial intelligence (AI) technologies in mental health care offers transformative potential, advancing diagnostic accuracy, personalized treatment plans, and real-time patient support. These innovations aim to tackle significant challenges such as accessibility and efficiency in mental health services, ultimately improving the quality of care provided to patients. Central to the discussion are the ethical principles surrounding

AI, particularly regarding privacy, informed consent, and algorithmic fairness. This chapter emphasizes the need for clear communication with patients about how their data will be utilized and the implications of AI interventions. Maintaining transparency is vital for fostering trust and ensuring that patients are well-informed, enabling them to make educated decisions about their care.

The chapter also outlines existing regulatory and legal frameworks, such as the Mental Health Care Act 2017 in India and the ICMR Ethical Guidelines for AI. These guidelines are designed to protect patient rights and promote responsible AI use in healthcare settings (Columbia Center for New Media Teaching and Learning, n.d.). However, the discussion highlights ongoing challenges and risks associated with AI, including ethical dilemmas regarding data security, potential biases in AI algorithms, and complexities surrounding accountability when AI systems fail. Continuous monitoring and evaluation of AI systems are noted as essential practices (Kumar, 2023) for identifying and mitigating these risks effectively.

Moreover, the significance of interdisciplinary collaboration is highlighted throughout the chapter. Involving AI developers, healthcare professionals, ethicists, and patients in the design and implementation of AI systems is critical for creating effective and ethical tools that address the complex needs of all stakeholders. This collaborative approach nurtures a culture of ethical AI development and deployment, ensuring that the systems are not only technically sound but also socially responsible.

In summary, as AI continues to evolve within mental health care, maintaining a focus on ethics is crucial. This involves adhering to established ethical guidelines, promoting transparency, and ensuring that AI complements the human elements of care. By prioritizing ethical considerations, we can take advantage of AI's benefits while safeguarding patient rights and advocating for equitable mental health care for everyone. The chapter ultimately calls for a commitment to ethical practices as we navigate the complexities of AI in mental health, ensuring that innovation serves the best interests of patients and society.

REFERENCES

Alharthi, A. S., Ameen, M. S., Alshahrani, M. A., & Alghamdi, A. S. (2024). Artificial intelligence in mental health care: Current trends and challenges. *ScienceDirect*. https://www.sciencedirect.com/science/article/pii/S2949916X24000525

American Psychological Association (APA). (2023). Artificial intelligence in mental health care. https://www.apa.org/practice/artificial-intelligence-mental-health-care

American Speech-Language-Hearing Association. (n.d.). Confidentiality. https://www.asha.org/practice/ethics/confidentiality/

Australian National University. (n.d.). Confidentiality and privacy. https://services.anu.edu.au/research-support/research-ethics-integrity-compliance/key-ethical-concerns/confidentiality-and-privacy

Brown, A. (2023). Regulatory frameworks for AI in healthcare. *MDPI*. https://www.mdpi.com/2673-2688/5/1/19

Brown, M. (2023). AI governance: Themes, knowledge gaps, and future agendas. *International Journal of Information Technology : an Official Journal of Bharati Vidyapeeth's Institute of Computer Applications and Management*, *29*(5), 51–65. https://www.emerald.com/insight/content/doi/10.1108/intr-01-2022-0042/full/html

Brown, T. (2023). Ethical considerations in AI development. *Nature Humanities and Social Sciences Communications*, 10(1), 15. https://www.nature.com/articles/s41599-023-02079-x

Columbia Center for New Media Teaching and Learning. (n.d.). Ethical issues in research. https://ccnmtl.columbia.edu/projects/cire/pac/foundation/

Danks, D. (2023). Ethical AI design: Principles and practices. *ACM Transactions on Human-AI Interaction*, 4(3), 1–22. https://dl.acm.org/doi/fullHtml/10.1145/3544548.3581252

Dezyit. (2024). *Scaling mental health support globally: AI success stories from developing countries*. https://www.dezyit.com/post/scaling-mental-health-support-globally-ai-success-stories-from-developing-countries

Galič, M. (2018). AI and the ethics of automating consent. https://www.researchgate.net/publication/325979872_AI_and_the_Ethics_of_Automating_Consent

Gupta, A. (2023). *From ethical AI frameworks to tools: A review of approaches for implementation. Springer Nature*. https://link.springer.com/article/10.1007/s43681-023-00258-9

Holistic, A. I. (2024). *Ethical concerns in AI healthcare applications*. https://www.holisticai.com/blog/healthcare-laws-us

Indian Council of Medical Research (ICMR). (2023). Ethical guidelines for artificial intelligence applications in biomedical research. https://www.icmr.gov.in/icmrobject/uploads/Guidelines/1724842648_ethical_guidelines_application_artificial_intelligence_biomed_rsrch_2023.pdf

Inside Precision Medicine. (2023). Medical artificial intelligence: A new frontier in precision medicine. https://www.insideprecisionmedicine.com/topics/informatics/medical-artificial-intelligence-a-new-frontier-in-precision-medicine/

ISACA. (2016). An ethical approach to data privacy protection. https://www.isaca.org/resources/isaca-journal/issues/2016/volume-6/an-ethical-approach-to-data-privacy-protection

Johnson, L. (2023). Balancing act: Exploring the interplay between human judgment and artificial intelligence in problem-solving. *ResearchGate*. https://www.researchgate.net/publication/379328371_Balancing_Act_Exploring_the_Interplay_Between_Human_Judgment_and_Artificial_Intelligence_in_Problem-solving_Creativity_and_Decision-making

Johnson, M. (2023). *Ethical concerns in AI mental health screening. Therapy Helpers Blog*. https://therapyhelpers.com/blog/ai-assisted-mental-health-screening-ethical/

Johnson, P. (2023). A systematic literature review of human-centered, ethical, and responsible AI. *arXiv Preprint*. https://arxiv.org/abs/2302.05284

Johnson, S. (2023). *Integrating ethics in AI development: A qualitative study on the adoption of ethical principles. BMC Medical Ethics*. https://bmcmedethics.biomedcentral.com/articles/10.1186/s12910-023-01000-0

Jones, L. (2023). Artificial intelligence in healthcare: Challenges and opportunities. *SAGE Journals*. https://journals.sagepub.com/doi/full/10.1177/20552076231218141

Jones, M. (2023). Algorithmic bias in AI-driven customer management. SSRN *Electronic Journal*. https://papers.ssrn.com/sol3/papers.cfm?abstract_id=4466681

Jorie, A. I. (n.d.). Harnessing the power of AI in mental health: Unveiling a new frontier in treatment and support. https://www.jorie.ai/post/harnessing-the-power-of-ai-in-mental-health-unveiling-a-new-frontier-in-treatment-and-support

Kumar, D. (2023). AI ethics in mental health care: Continuous monitoring and improving outcomes. *Journal of Healthcare, MDPI, 13*(7), 381. https://www.mdpi.com/2076-0760/13/7/381

Lee, J. (2024). *Trust in AI: Progress, challenges, and future directions in healthcare. Nature Humanities and Social Sciences Communications*, 11(1), 88. https://www.nature.com/articles/s41599-024-04044-8

LexisNexis. (2023). Balancing AI and human judgment: Ethical considerations in the legal profession. https://www.lexisnexis.com/blogs/za/b/legal/posts/balancing-ai-and-human-judgment-ethical-considerations-in-the-legal-profession

Mittelstadt, B. (2020). AI ethics and data governance. SSRN *Electronic Journal*. https://papers.ssrn.com/sol3/papers.cfm?abstract_id=3529576

Patel, K. (2024). AI and medical ethics. *PubMed Central*. https://pmc.ncbi.nlm.nih.gov/articles/PMC11064747/

Scribbr. (n.d.). Research ethics. https://www.scribbr.com/methodology/research-ethics/

Smith, J. (2021). Ethical frameworks for AI deployment. *PubMed Central*. https://pmc.ncbi.nlm.nih.gov/articles/PMC8404542/

Smith, J. (2023). Addressing ethical concerns in AI-driven healthcare. *PubMed Central*. https://pmc.ncbi.nlm.nih.gov/articles/PMC10048681/

Smith, J. (2024). Ethical implications of AI adoption. *PubMed Central*. https://pmc.ncbi.nlm.nih.gov/articles/PMC11181958/

Smith, K. (2023). Ethical dilemmas in AI-driven decision-making. *Nature Humanities and Social Sciences Communications*, 10(1), 22. https://www.nature.com/articles/s41599-023-01787-8

Smith, K. (2023). Exploring patient perspectives on AI in healthcare. *BMC Health Services Research, 23*(1), 100. https://bmchealthservres.biomedcentral.com/articles/10.1186/s12913-023-10098-2 PMID: 36717808

Smith, L. (2024). *Chatbot tragedy highlights AI risks in mental health care. Le Monde*. https://www.lemonde.fr/pixels/article/2024/08/06/comment-l-ia-bouscule-le-milieu-de-la-sante-mentale-plutot-que-de-payer-une-nouvelle-seance-chez-le-psy-j-allais-sur-chatgpt_6270640_4408996.html

Sunflower Lab. (2023). AI vs human decision-making: Striking the right balance. https://www.thesunflowerlab.com/ai-vs-human-decision-making/

Technologies, T. (n.d.). AI and privacy risks: Challenges and solutions. https://www.trigyn.com/insights/ai-and-privacy-risks-challenges-and-solutions

The MIT Press. (2023). Prediction and judgment: Why artificial intelligence needs human oversight. *International Security*, *46*(3), 7. https://direct.mit.edu/isec/article/46/3/7/109668/Prediction-and-Judgment-Why-Artificial

United VARs. (n.d.). Mental health: The new frontier of technology. https://www.united-vars.com/en/blog/mental-health-the-new-frontier-of-technology

Williams, R. (2023). AI in clinical decision-making: Opportunities and risks. *PubMed Central*. https://pmc.ncbi.nlm.nih.gov/articles/PMC10482032/

Wong, H. (2021). Patients' perceptions of AI in clinical decision-making: An empirical study. *Journal of Medical Internet Research*, *23*(10), e25856. https://www.jmir.org/2021/11/e25856/

Yu, L. (2023). Impact of AI on the doctor–patient relationship: A review. *BMC Medical Informatics and Decision Making*, *23*(1), 62. https://bmcmedinformdecismak.biomedcentral.com/articles/10.1186/s12911-023-02162-y PMID: 37024897

Zhang, H. (2021). AI ethics principles in practice: Perspectives of designers in AI deployment. *arXiv Preprint*. https://arxiv.org/abs/2112.07467

Zhang, J. (2024). AI-assisted peer support in mental health: Empathy and ethical implications. *Journal of Mental Health (Abingdon, England)*, *45*(1), e58493. https://mental.jmir.org/2024/1/e58493

Zhang, Q. (2021). IAC framework for patient agency in AI-enabled healthcare: Ethical considerations. *arXiv Preprint*. https://arxiv.org/abs/2111.04456

Zhang, Y. (2019). AI applications in mental health: A survey of risks and challenges. *arXiv Preprint*. https://arxiv.org/abs/1907.07493

Chapter 8
Ethics in AI-Based Mental Health Care

Susmita Halder
https://orcid.org/0000-0001-6254-3324
St. Xavier's University, Kolkata, India

Shinjini Samajdar
https://orcid.org/0000-0001-5727-5150
St. Xavier's University, Kolkata, India

Shouvik Basak
https://orcid.org/0000-0001-7557-5421
Independent Researcher, India

ABSTRACT

Ethical considerations surrounding technology in mental health applications are complex, multifaceted, and increasingly relevant in both clinical and research settings. Ethical challenges associated with technology's integration into mental health underscore the urgent need for a framework prioritizing safety, fairness, and human dignity. With the rise of technology-driven tools in diagnostics, treatment recommendations, and patient management, the ethical implications of these innovations must be rigorously examined to ensure they serve and protect users—particularly vulnerable individuals seeking mental health care. he chapter begins with an overview of fundamental ethical challenges in mental health technology. Issues such as algorithmic bias and discrimination are discussed, highlighting how biased datasets can lead to inaccurate diagnoses or inappropriate treatment recommendations for marginalized groups. Addressing these biases is crucial to fostering equitable access to mental health services and avoiding reinforcing societal inequalities through technology-driven systems.

DOI: 10.4018/979-8-3373-0467-0.ch008

INTRODUCTION

The combination of improvements in computing power, data gathering, and machine learning has accelerated research and development of artificial intelligence (AI) tools in nearly all fields (Maheu et. al.2004). Those technological innovations opened up AI systems to an enormous scope of data that can be accessed, which significantly increased funding, research, and innovation. In the case of medicine, artificial intelligence (AI) is considered to be very useful because of its potential for great impact on many fronts (Pesapane et.al., 2018). AI is expected to transform the practice of medicine with its application in improving the precision of diagnosis, automating clinical procedures, tailoring treatment approaches for individual patients, and forecasting healthcare results (Huang et.al.,2023). Moreover, AI is being utilized for much wider purposes, such as the creation of automated systems for the interpretation of medical images, remote monitoring of patients' health conditions, and even for aiding in the drug discovery process (Arora, 2020). AI is investing in a huge development towards a more reliable and safer, albeit newer, mechanism. It has also encountered a number of established practitioners, more particularly in the area of mental health. Technology offers a powerful support in the field of mental health, deploying tools that reach a number of psychological problems (Liu et.al.,2024). Mobile apps are used to employ mood tracking and cognitive-behavioral therapy (CBT) techniques (Addepally & Purkayastha, 2017). Virtual reality (VR) platforms for simulated exposure therapy bring a revolution in the delivery of mental health services (Eyre et.al.,2020). Furthermore, their aspects of assessment capabilities given by every type of psychological input are provided by newer artificial intelligence (AI) implications, while telehealth platforms overcome geographical boundaries to provide mental health care more than ever possible before (Solaiman, 2020). Technology in mental health care is not about convenience, instead, it is imperative in bridging the global gap (Ajayi, 2025). AI is joining the ranks of the many others in its development towards an ever-reliably safer, while different, avenue. It has also gotten into contact with a large array of established practitioners, more especially so in mental health. Technology provides an impetus to mental health care, enabling interventions across most psychological problems. Various VR simulated exposure therapy platforms and others are revolutionizing the delivery of various mental health services (Wiederhold & Wiederhold,2018). In addition to this, their measurement powers, provided through every avenue of clinical input, shall come to be known as new technical implications of AI. Telehealth platforms also have opened historical avenues for providing health from one approach to another, thereby offering mental health care in geographical locations where it was not possible before (Lovejoy, 2019). Technology in mental health was never to substitute the skill for mere convenience, instead, it forms a very imperative nexus in bridging the global gap. Careful selection of clinical governance frameworks to lead the development, validation, and use of AI should be instituted to ensure that its application is within strict ethical and professional standards (WHO, Guidance, 2021). These considerations are paramount for building trust and maximizing AI's benefit for psychiatric care. With the increasing global burden of mental health disorders, there is a rising need for robust and effective advocacy. Mental health advocacy embodies the broad communication and actions taken to increase awareness, fight stigma, and improve access to quality care (Dimova, 2024). These efforts have played a crucial role in nurturing the public attitude regarding mental health, policies for addressing mental health needs, and the establishment of an enabling environment for mental health. In this dynamic space, the amalgamation of technology with mental health advocacy is starting to transform the scenario. Digital platforms and tools provide an innovative opportunity for advocacy campaigns to amplify their reach and engage distinct constituencies to enable systemic change (Simon, 2027). From campaigns on social media and support networks to AI designs that drive interventions, technology provides scope for pushing the mental health agenda globally (Liu et.al.,2023). This engagement between advocacy and technology not only speeds progress but transforms the platform for serious, inclusive, and impactful work.

The choice of clinical governance frameworks leading the development, validation, and use of AI should be implemented very carefully to ensure its applications are within the strictest ethical and professional standards. These considerations are utmost in gaining public trust and belonging to the larger paradigm benefit of using AI in psychiatric practice. With the increasing global burden of mental health disorders, the calls for robust and effective mental health advocacy are being multiplied. Mental health advocacy mainstreams ways and means that supposed to enhance awareness, fight stigma, and shape improved access to quality care. These precious efforts shaped significant public attitudes on mental health, mental health need policies, and the institutional working environment on behalf of mental health (Martin et.al.,2000). The interface of technology with mental health advocacy is shaping a landscape for change in this dynamic arena. Digital platforms and tools provide an innovative opportunity for advocacy campaigns to amplify their reach and engage distinct constituencies for systemic change (Guo & Saxton, 2014). From campaigns on social media and support networks to AI designs that drive interventions, technology provides space to leverage the implementation of mental health globally. Not just does this interplay between advocacy and technology help propel progress, but it changes the tableau for serious, inclusive, and impactful work (Stellefson et.al.,2020).

In the interest of strict adherence to ethical and professional standards, clinical governance frameworks should be carefully selected to guide the development, validation, and use of AI. These considerations are vital for building trust in maximizing AI's benefit for mental health. There is much more to be accomplished in light of the burden mental health disorders impose on the world. Advocacy for mental health stands widely for communique and actions geared toward raising awareness, taking aimed shots at stigma, and working for quality care (Corrigan et.al.,2014). In shaping a positive public attitude toward mental health, policy development regarding mental health, and an enabling environment for mental health, these processes have played a crucial role. In this dynamic domain, technology and mental health advocacy are now combining to change things. Digital platforms and tools provide a new avenue for advocacy campaigns to increase their reach and cultivate fresh constituencies for systemic change. From social media campaigns and peer-support networks to AI designs influencing interventions, technology is poised to push the global mental health agenda. The intersection of advocacy and technology provides not only for sped-up progress but transformative opportunities for serious inclusion and impact (National Academies of Sciences, Engineering, and Medicine, National Academies of Sciences, Engineering, and Medicine (U.S.). Committee on the Science of Changing Behavioral Health Social Norms, Division of Behavioral and Social Sciences and Education, Board on Behavioral, Cognitive, and Sensory Sciences,2016).

The right clinical governance frameworks must be carefully chosen to lead the development, validation, and usage of AI in such a way that these operate within a dense nexus of ethical and professional standards. Such considerations are important for trust-building and maximizing AI benefits to psychiatric care. There is an increasing need to support advocacy efforts for mental health as a worldwide burden is placing significant strains on the healthy population. Essentially, mental health advocacy encompasses broad communication and action taken with the intention of raising awareness, fighting stigma, and improving access to quality care. These efforts have been crucial in shaping the public attitude concerning mental health, policies to address mental health needs, and its establishment in an enabling environment. That is where technology meets mental health advocacy to provide the strong needed change into the space (Patel et.al.,2013).

Though social media campaigns, support networks, AI designs for interventions-all offer a glimmer of hope to take the mental health debate to the global stage. This interplay between advocacy and technology creates not only a boost for progress, but a platform transformation for serious, inclusionary, and impactful work. Social media offered a chance to interact in real-time, enabling people to connect with others who shared similar life experiences. This understanding of community decreased the feelings of being alone and improved cultural openness associated with mental health issues. And in addition to that, different types of social media posts and campaigns have been very active in regulating policy changes and it served as collective action and public awareness.

Digital and Mobile technology has also played a pioneering identity in increasing the accessibility of mental health facilities and resources. Currently developed Smartphone applications offer a wide variety of facilities, from mood tracking and meditation guides to follow-up in virtual therapy sessions and immediate accessibility to crisis intervention tools (Bakker et.al., 2016; Wartel, 2023) . These innovations provides and give advantages to individuals to enthusiastically manage and monitor their mental health, connecting and maintaining associations in traditional service delivery models. Suppose, apps like Headspace and Calm includes mindfulness exercises, while platforms like BetterHelp and Talkspace connect users with licensed psychotherapists for online counselling and mental health services (Pavlopoulos et.al.,2024).

The comprehensive and convenient mobile devices should be available and it should be an ideal platform for delivering interventions. This new array of digital applications and tools provides a fresh direction for advocacy campaigns in targeting larger audiences and engaging multiple constituencies toward systemic change. Social media campaigns, support networks, AI designs for interventions-all bring a new hope to this mental health debate to the global stage. The interplay of advocacy and technology gives a new boost for advancement and changes the platform for serious, inclusive, and impactful work. Social media offered a chance to interact with real-time, enabling people to connect with those who have had similar life experiences. For example, VR-based exposure therapy allows individuals to challenge their fears in a controlled and safe setting and it facilitates desensitization. Likewise, AR applications deliver communicating imitations that indorse understanding and alertness about mental health challenges.

Beyond therapy, VR and AR are changing the game for training mental health professionals. These tools give them a chance to practice and hone their skills in realistic, hands-on ways. Plus, they create immersive, impactful experiences that really emphasize why mental health advocacy matters so much (Usmani et.al., 2022).

AI is making waves in mental health, helping with early detection, personalized care, and even predicting potential issues. Think about chatbots and virtual assistants—they're like having a support system available 24/7. These tools use natural language processing to make conversations feel real and approachable, so getting help feels less intimidating and more accessible (Cerasa et.al.,2022).

Machine learning is transforming mental health by analyzing massive amounts of data to uncover patterns and trends that weren't visible before. This has led to the creation of predictive models that can guide early intervention strategies and improve treatment outcomes. For instance, AI can identify people who might be at risk of developing mental health conditions based on their digital habits or biometric data. However, while these advancements are incredibly promising, they also bring up important concerns about privacy, data security, and ethics (Shatte et.al.,2019).

The integration of technology into mental health advocacy is an exciting and constantly evolving field. As technology keeps advancing, it opens up new possibilities to enhance advocacy efforts and improve outcomes. Innovations like wearable devices, gamification, and virtual peer support networks are already shaping what the future of mental health advocacy could look like, making it more engaging, accessible, and effective (Raj et.al.,2024)

Collaboration is going to be key. Mental health professionals, technologists, policymakers, and advocacy organizations need to come together to maximize the potential of these tools. By tackling ethical challenges head-on and ensuring inclusivity, stakeholders can create a digital ecosystem that truly supports and empowers people on their mental health journeys (Maheu et.al.,2004).

Technology has sparked a major shift in how we think about and approach mental health advocacy. From the early days of online forums to today's AI-driven interventions, these tools have completely changed how we engage with mental health challenges. Yes, there are still hurdles to overcome, but the opportunities for innovation and meaningful impact are immense. Technology has become an invaluable partner in the mission to advance mental health advocacy and support.

FUNDAMENTAL ETHICAL CHALLENGES IN AI AND MENTAL HEALTH

When we are trying to integrating artificial intelligence (AI) into mental health care services it gives multiple ethical challenges which will definitely require careful follow ups to make it sure the responsibility and well-equipped applications of these AI-based tools. These challenges create problems like subjective bias, privacy and also there are chances of fiduciary obligations due to therapeutic relationships. AI-based mental health tools, while promising significant advancements, also present AI-based mental health tools hold a lot of promise for transforming how we approach mental health care. However, they also come with challenges that go beyond what traditional healthcare systems typically handle (Saeidnia et.al.,2024).

A major ethical concern with AI-based mental health tools is the risk of bias. This bias can creep in at various stages, from how the data is collected and prepared to how mental health issues are framed and presented. The data sets used to train these systems often reflect the inequalities already present in society, unintentionally reinforcing and even amplifying those disparities. For instance, when certain groups are underrepresented in the training data, the AI's results can become skewed, further deepening systemic inequities rather than solving them. This is especially concerning in mental health, a field that already faces significant gaps in access to care and the quality of treatment. To truly make a positive impact, tackling bias and ensuring fairness must be top priorities when developing and using AI tools in mental health care (Martinez-Martin, 2022).

Privacy and data security are also major concerns when it comes to AI-based mental health tools. These applications often handle highly sensitive personal information, such as psychological profiles, behavioral patterns, and emotional responses. Keeping this data safe from misuse or unauthorized access is essential to protecting both patient trust and well-being. If health information is misused or exposed in a data breach, it could lead to serious consequences, highlighting the need for strong security measures and clear rules for how data is shared and stored. Additionally, many users of AI tools outside of traditional healthcare settings may not fully understand how their data is being used, raising important questions about transparency and whether they've given truly informed consent (Wies & Landers,2021).

The use of AI in mental health care introduces a range of challenges, especially when it comes to regulation. Unlike traditional medical tools or treatments, many AI-based applications don't fall neatly under established healthcare regulations. This makes it harder to evaluate how these tools work, whether their algorithms are technically sound, and how they impact patient outcomes. Adding to this complexity is the speed at which AI technology evolves, requiring the creation of flexible yet thorough oversight systems. To ensure these tools meet the highest ethical and clinical standards, clear guidelines are essential, no matter where or how they're used (Valentine et.al.,2023).

Another concern is how AI might affect the trust and care central to therapeutic relationships. Traditional therapy is built on trust, empathy, and personalized attention. But introducing AI tools, like chatbots or virtual assistants that raises questions about whether these tools can truly support the emotional depth required in therapy. For example, relying too heavily on AI might unintentionally create barriers to genuine human connection, potentially weakening the bond between therapist and patient. It's crucial to thoughtfully consider how AI can complement, rather than replace, human-led care, ensuring that the heart of therapy, i.e. trust and empathy which remains intact (Tavory, 2024).

Despite these challenges, AI holds enormous promise for improving mental health care. With tools like machine learning and predictive analytics, AI can identify people at risk, enable earlier interventions, and tailor treatment plans to individual needs. It can also play a key role in addressing disparities, providing mental health support in underserved communities and bridging gaps in access. However, achieving this potential requires an intentional and inclusive approach, one that prioritizes equity, transparency, and accountability at every step.

To ethically integrate AI into mental health care, we need to take a multidimensional approach, drawing from established ethical frameworks like utilitarian ethics, principle-based ethics, and virtue ethics. Each of these perspectives offers valuable insights, helping us balance the benefits of AI with the challenges and responsibilities it introduces. By carefully considering these ethical dimensions, we can create AI tools that truly enhance mental health care while honoring the values that define it (Kasula, 2023).

1. Maximizing Benefits While Minimizing Harm

The goal of utilitarian ethics is to ensure actions lead to the greatest good for the greatest number of people. In mental health care, tools powered by advanced technology can help make care more accessible, efficient, and personalized. For instance, mental health apps can provide critical support to people in remote or underserved areas, helping reduce disparities in access. Machine learning and predictive tools can identify individuals at risk, enabling early interventions and targeted care.

However, to truly benefit everyone, we need to consider the potential downsides as well. Bias within these tools can lead to unequal treatment and may even deepen existing inequalities in care. For example, when certain groups are underrepresented in the data used to train these systems, the results may be inaccurate, leading to misdiagnoses or missed care for vulnerable populations. Data breaches are another concern, as exposing sensitive health information could cause significant emotional distress for patients.

To maximize benefits while minimizing harm, it's essential to focus on reducing risks. This includes using diverse data sets, thoroughly testing systems for bias, and establishing strict privacy protections. With these steps, these tools can make a meaningful difference in mental health care, especially for those who have traditionally been overlooked.

2. Building Care on Ethical Principles

Ethics in health care often center on four key principles: autonomy, beneficence, non-maleficence, and justice. These guide how we make decisions to ensure care is fair and respectful of every patient.

- **Autonomy**: Respecting patients' independence means being transparent about how these tools work and what data they collect. People using mental health apps, for example, should clearly understand their limitations—like how they may lack the nuanced judgment of a human therapist.
- **Beneficence and Non-Maleficence**: It's essential to act in the best interests of patients while also avoiding harm. Tools that predict risks, such as suicide risk, need to be accurate to prevent false alarms or overlooked cases. The stakes are too high to use anything unproven.

- **Justice**: Equity is central to ethics in mental health. These tools should work for everyone, especially those in underserved communities or with limited access to technology. Ensuring fairness at every stage of development and implementation helps bridge gaps in care.

Developers, clinicians, and policymakers must work together to make sure these tools align with these principles. By doing so, they can enhance care without compromising on fairness or safety.

3. Focusing on Compassion and Integrity

Virtue ethics shifts the focus to the moral character of those developing and using these tools. It emphasizes qualities like compassion, honesty, and responsibility in every decision.

For developers, this means prioritizing the well-being of patients over profits or speed. For example, creating tools that truly meet patient needs, even if it means delaying a product launch or spending extra resources, demonstrates integrity. Clinicians must also use these tools with care, ensuring they enhance—rather than replace—the personal connection that is the foundation of good therapy.

Virtue ethics also encourages humility. Acknowledging the limits of these tools—such as when they may not work well for certain populations—shows respect for patients' needs. By fostering a culture of accountability and compassion, we can ensure these tools are used in ways that reflect the core values of mental health care.

Integrating advanced technology into mental health care is an exciting step forward, but it comes with challenges that require thoughtful solutions. By focusing on inclusivity and safety, we can make these tools a positive force for change. Following ethical principles ensures fairness, transparency, and respect for every individual. Finally, prioritizing compassion and integrity in decision-making keeps us grounded in the true purpose of mental health care: promoting well-being, reducing stigma, and ensuring that no one is left behind. With a balanced and ethical approach, we can harness the potential of these tools to create a system that works for everyone.

OVERVIEW OF ETHICAL DILEMMAS IN MENTAL HEALTH TECHNOLOGY

The integration of technology in mental health care has opened doors to innovative tools and solutions. However, these advancements also present ethical dilemmas that require careful consideration (Lustgarten & Elhai,2018). The key dilemmas revolve around balancing innovation with ethical responsibility, addressing algorithmic bias and discrimination, and ensuring equitable and safe implementation of these technologies in practice. Below, these challenges are discussed with practical case examples to illustrate their implications.

1. Balancing Innovation with Ethical Responsibility

Recent technological advancements can reportedly increase and enhance mental health care through proper accessibility to services, reducing stigma or sensitizing mental health issues along with providing the opportunities for tailor-made interventions. Though innovations are taking place rapidly and it often makes space for ethical framework development. It can also create risks associated to data privacy, data misuse and unwanted consequences. One case example can give a comprehensive picture here: A mental health startup developed a chatbot that provides cognitive-behavioral therapy (CBT) techniques. While the chatbot gained popularity for its accessibility and affordability, concerns arose when sensitive user data was leaked during a cyberattack. The incident highlighted the tension between innovation and the ethical responsibility to protect user privacy. So, it can be said that, this case emphasizes the urgency for stringent data security measures and also clear user consent procedures, and regular audits to make it ensure that, ethical responsibility is not getting compromised in the search for innovation (Hadjistavropoulos, 2023).

2. Algorithmic Bias and Discrimination

Algorithms play a vital part in development of many mental health technologies, from diagnostic tools to predictive models. But it should be acknowledged that, these algorithms are only as reliable and fair as the data they are trained on. If the training data reflects societal inequalities, the algorithms can unintentionally perpetuate or even worsen discrimination (Gooding & Kariotis, 2021).

Consider a case where a hospital implemented an AI-powered tool to predict suicide risk among patients in its emergency department. The goal was to identify high-risk individuals and provide timely interventions. However, the algorithm flagged individuals from minority groups at disproportionately higher rates. This resulted in unnecessary interventions for some while leaving others, who were genuinely at risk, overlooked. Upon investigation, it was found that the underlying issue lay in the historical data used to train the algorithm. The data overrepresented minority groups in emergency mental health cases, likely reflecting systemic biases in healthcare access and treatment.

This example focuses on the importance of using diverse and representative datasets when developing these tools. Without careful attention to the quality and inclusivity of training data, algorithms risk embedding existing disparities into mental health care systems. To address these challenges, developers must commit to regularly auditing algorithms for bias, ensuring they perform equitably across diverse populations involving interdisciplinary teams in the development process is essential. Ethicists, clinicians, community representatives, and data scientists can provide valuable perspectives, helping to evaluate the tool's impact on different groups and identifying potential risks early on. By fostering collaboration and accountability, developers can create technologies to support mental health that not only enhance care but also promote fairness and equity for all.

Algorithmic bias can have significant and far-reaching effects in the field of mental health care:

1. When algorithms are biased, they may direct critical resources toward individuals who are at relatively low risk while failing to provide adequate support to those who are genuinely in need. This misallocation not only wastes valuable resources but also leaves high-risk individuals without the care and attention
2. If biases in these technologies become evident, it can undermine public confidence in mental health tools altogether. People may start to question the fairness and reliability of these systems, leading to reluctance in using them. This could discourage individuals from seeking help, especially in communities that are already marginalized or skeptical of traditional healthcare systems, further widening gaps in mental health care access and support.

3. Addressing Ethical Dilemmas Through Practical Solutions

1. Developers should prioritize transparency in how algorithms are trained, tested, and implemented. Publishing detailed documentation on data sources and providing end-users with insights into how predictions are made.
2. Ethical dilemmas can be mitigated by regular evaluation and updates to mental health tools based on user feedback. For instance, A meditation app includes a feedback mechanism to understand user concerns, allowing developers to address issues like cultural sensitivity in content.
3. Building inclusive datasets that represent diverse demographics can reduce algorithmic bias. Like, An AI-powered therapy app for anxiety underwent retraining with global data to accommodate variations in cultural expressions of anxiety symptoms.
4. Establishing interdisciplinary committees to oversee the ethical aspects of technology development can ensure accountability. If we can imply it practically like, A panel of mental health professionals, technologists, and ethicists evaluates the potential impact of an AI diagnostic tool before deployment.

ETHICAL CONCERNS OF AI SYSTEMS AS "BLACK BOXES"

When considering the usage of AI in healthcare, especially mental health care, there are certain ethical issues that need to be put into account. Issues such as AI technology functioning as black boxes add to lack of transparency in the decision making process (Winfield & Jirotka, 2017). Even the developers may have a hard time figuring out how decisions are made and this anumistic approach poses severe mental healthcare challenges. The fear of misjudging AI recommendations is rampant since psychologists may be unsure of the entire process behind it. There is great urgency to design AI models that are interpretable with their limitations provided in the best possible way. Developers and practitioners should try their very best to maintain transparency by all means necessary. According to some practitioners, the development of explainable AI systems would be the greatest advancement. This will provide practitioners with the ability to see the rationale behind the AI recommendation therefore, enabling them to form better judgments. Alongside building trust, informed consent is of utmost importance and certain limitations and risks should certainly be conveyed to the patients as well. There is a need to protect privacy and security by establishing strict measures because AI systems deal with sensitive information and personal data. To prevent a breach of data and confidentiality, proper encryption, anonymization protocols, and adherence to GDPR regulations should be implemented. Doing so will improve the ethical application of AI in mental healthcare (Durán & Jongsma,2021).

ENVIRONMENTAL COSTS OF AI

The use of AI in mental health comes with a set of challenges. For example, the carbon footprint created when massive algorithms are trained and deployed has become a significant threat in the long run. The processes of AI have very high energy requirements, which adds to the emission of greenhouse gases and puts these technologies at risk of not being sustainable. One way to lessen this effect is to optimize the algorithms for energy usage, use edges to reduce data transfer, and employ green data centers powered by renewable energy. To tackle environmental issues without hindering the advancement of technology, changes in the form of sustainable practices throughout the development of AI tools are to be made. At the same time, the issues of copyright and intellectual property are more ethical and legal challenges than they are practical, especially in the context of ownership over AI-assisted innovation or content in the realm of mental health. Well-defined policies and interdisciplinary approaches are required to anticipate these challenges.

This will guarantee that creativity is fostered and nurtured while fair credit is given and environmental welfare is not breached (Adnan et.al.,2024).

CHALLENGES IN PROPRIETARY AI ALGORITHM DEVELOPMENT

Challenges arise from the creation of proprietary AI algorithms when applied to mental health care. Their black-box nature renders decision-making processes and recommendations opaque to practitioners and patients which defies scrutiny (Koutsouleris et.al.,2022). This can undermine trust and foster skepticism especially when seeking care for deeply personal issues. Other than that, the proprietary nature of these algorithms restrict the equity of care as opportunities for profitable AI implementation take precedence over inclusivity. This augments the difficulty in meeting the mental health needs of the underserved, who are already disadvantaged as a result of inequitable support systems.

Use of AI in coping with mental health issues must take accountability and responsibility into consideration. Algorithms, which are riddled with errors and biases, can have dire repercussions for individuals. The ambiguity of responsibility is a significant ethical and legal dilemma. AI developers, companies, as well as healthcare providers all share in the blame. Further, mental health professionals rely on AI for assessment and intervention only for the systems to conflict with human intelligence and ethics (De Freitas et.al.,2024).

Fairly distributing liability among stakeholders requires clear frameworks that prioritize patients' well-being while encouraging innovation within safe and ethical limits. To build trust in AI systems used for mental health care, we need transparency, reliability, and strong risk management strategies. Both patients and clinicians must feel confident that these AI systems are unbiased, culturally aware, and intended to support rather than replace human judgment. Without these protections, there's a risk that AI could reinforce existing disparities in mental health care or increase skepticism towards technology (Loies, 2025). To address these issues, stakeholders should collaborate to develop frameworks that emphasize accountability and inclusivity, all while focusing on the mental and emotional well-being of individuals. These initiatives will ultimately shape how effectively AI can serve as a meaningful and ethical tool in enhancing mental health care delivery.

RESPONSIBLE AI FOR MENTAL HEALTH

Responsible AI for mental health focuses on creating and using AI systems in a way that prioritizes the well-being, dignity, and autonomy of individuals seeking mental health support (Peters et.al.,2020). At its foundation, Responsible AI is built on principles like fairness, transparency, accountability, and inclusivity (Baeza-Yates,2024). These systems should be designed to protect privacy, reduce biases, and provide equal access to care, especially for those in vulnerable situations. By aligning AI with these core values, mental health care can utilize technology to improve service delivery while keeping a human-centered approach at the forefront (Beg,2024). Regular evaluation and adherence to ethical standards are essential components of Responsible AI. AI models used in mental health need to be continuously assessed to ensure they are accurate, reliable, and in line with clinical best practices. Ethical issues, such as informed consent, data privacy, and cultural sensitivity, should always be prioritized. This ongoing commitment to evaluation helps tackle emerging challenges, like algorithmic bias or unintended consequences, ensuring that AI tools enhance the work of mental health professionals rather than replace their expertise or exacerbate existing inequalities (El-Sappagh et.al.,2025).

The societal implications of AI in mental health are significant (Gamble,2020). On one side, AI could transform access to mental health services, offering scalable solutions for populations that are often overlooked and enhancing the early identification of mental health issues. Conversely, it also brings up worries about data misuse, the potential for stigmatization, and an over-dependence on technology in sensitive emotional situations. It is crucial for responsible AI to consider these implications, building trust in its use while protecting individuals from possible risks. Legislative frameworks and global governance are key in defining the ethical landscape of AI in mental health (de Almeida et.al.,2021). International cooperation is vital to create standardized guidelines that oversee the development and application of AI systems across different regions. These frameworks need to tackle concerns like liability, data security, and ethical standards that cross borders, all while ensuring that AI-enhanced mental health care is both accessible and fair. By promoting a unified global strategy, policymakers and stakeholders can contribute to a future where AI acts as a transformative yet responsible element in mental health care.

EU AI ACT AND ITS RELEVANCE TO MENTAL HEALTH CARE

The EU AI Act represents a significant step forward in creating a regulatory framework for AI systems throughout Europe. It classifies AI applications according to their risk levels, imposing stricter requirements on high-risk systems, such as those utilized in healthcare, including mental health services (Meszaros et.al.,2022). The Act highlights essential principles like transparency, accountability, and fairness, requiring that high-risk AI systems undergo thorough testing, evaluation, and documentation to meet ethical and safety standards. In the context of mental health care, this structured approach ensures that AI tools are integrated responsibly, prioritizing safety, impartiality, and effectiveness in assisting both practitioners and patients. An ethical strategy for AI use in mental health care should establish clear accountability criteria, balancing innovation with necessary safeguards (Smuha et.al.,2021). Developers, regulators, and healthcare providers must work together to ensure that AI systems comply with ethical standards. This includes implementing measures for data privacy, fairness in algorithms, and cultural sensitivity, while building trust through transparency. Given the complexity of mental health care, AI systems should be designed with input from a diverse range of stakeholders to prevent the reinforcement of biases or the marginalization of vulnerable populations (Ueda et.al.,2024). Ongoing monitoring and feedback mechanisms are essential to identify and mitigate potential risks, ensuring that AI practices remain ethical. Global governance of AI in healthcare can learn important lessons from the EU AI Act. Its tiered risk approach, commitment to transparency, and focus on accountability offer a strong framework for addressing the ethical challenges posed by AI systems. Global initiatives could incorporate similar principles to maintain consistency across different regions while allowing for necessary local adjustments. Effective governance must also take into account the unique challenges of mental health care, where mistakes in AI can have serious implications for individuals' well-being. Matching ethical standards worldwide would not only develop trust but also allow collaborative developments in AI-driven mental health resolutions.

Collaboration across various fields is crucial for the responsible use of AI in mental health care. Involvement from mental health professionals, AI developers, ethicists, sociologists, and policymakers is vital for creating systems that are both effective and ethically grounded. This teamwork ensures that AI tools are developed with a comprehensive understanding of the complex nature of mental health, striking a balance between technological advancements and the essential human empathy and clinical knowledge. By encouraging partnerships among different disciplines, stakeholders can tackle the ethical, technical, and societal issues related to AI, ultimately creating systems that truly improve mental health care while maintaining the highest standards of responsibility and respect for human dignity (Gooding & Kariotis,2021).

IMPORTANCE OF COLLABORATION AMONG ETHICISTS, TECHNOLOGISTS, CLINICIANS, AND POLICYMAKERS

Collaboration among ethicists, technologists, clinicians, and policymakers is indispensable for creating and implementing AI systems that are ethically sound, clinically effective, and socially responsible (Saarni et.al.,2008). Each of these groups offers unique perspectives and expertise to tackle the complex challenges of integrating AI into sensitive areas like mental health care. Ethicists guide the moral principles, ensuring that AI systems uphold autonomy, privacy, and equity. Technologists share their knowledge of AI design and functionality, which helps in developing robust and reliable tools. Clinicians provide insights into how AI can be practically applied in patient care, making sure that systems align with therapeutic goals and human-centered approaches. Meanwhile, policymakers are crucial in establishing regulatory frameworks that ensure accountability, safety, and fairness. This interdisciplinary collaboration guarantees that AI systems are not only innovative but also in harmony with societal values and clinical realities (Fusar-Poli et.al.,2022).

Ensuring that AI respects patient rights and maintains public trust is a crucial aspect of its ethical use. Patients deserve informed consent, privacy, and fair treatment, all of which must be protected during the development and application of AI technologies. Transparency plays a vital role in building public confidence, as patients should be aware of how AI systems function, what data is utilized, and the reasoning behind decisions. If there is a lack of clear communication and accountability, trust in AI tools may diminish, especially in sensitive fields like mental health, where confidentiality is essential. Collaborative efforts among all parties involved are necessary to uphold these rights and tackle issues such as algorithmic bias, data misuse, and ethical dilemmas.

Building trust is absolutely key to successfully introducing technology into mental health care. When patients and clinicians feel confident that a system is reliable, ethical, and fair, they're more likely to see it as a helpful tool rather than something to be wary of. This trust doesn't happen overnight, it's earned through constant collaboration, thorough testing, and staying true to ethical principles. The best way to get it right is to involve people from all walks of life in the process—patients, mental health professionals, and even community advocates. When different perspectives are heard and respected, it's easier to create tools that genuinely address people's needs, respect their privacy, and reflect real-world diversity. Done thoughtfully, these technologies have the potential to make a huge difference and bridging gaps in care and helping mental health support reach those who might otherwise go without. Trust is what makes all of that possible (Engel & Prentice,2013)..

CONCLUSION

To sum up, the introduction of AI into mental health care can be seen as an opportunity to change the way that the care is given, from enhancing accessibility to improving diagnostic accuracy and personalized interventions. However, it also brings in a series of ethical issues that would need the utmost care to handle. These consist of guaranteeing patient privacy, establishing algorithmic fairness, promoting inclusivity, and making the decision-making process transparent. The ethical problems about responsibility and accountability are what makes everything worse that is why it is necessary to have clear frameworks and collaborative oversight. Backing the technological startup by an unwavering commitment to human dignity, equity, and trust is one of the ways to really profit from AI's strength in mental health care. The future of AI that is responsible for mental health should be built on the idea of having systems that are not only efficient in the treatment of patients but also are ethically proficient and socially inclusive. Systems like this one need to make sure patient-centered care is a basic value and therefore they should take the rights of the individuals, their cultural backgrounds, and diverse needs under consideration first when they develop and release the system. Technology that is responsible will empower the clinicians to provide their services effectively and it will become a real ally that will bridge the gaps in mental health care access. With an emphasis on transparency, accountability, and equity, AI might emerge to be the powerhouse of supported, accessible, and compassionate care in mental health. In order to realize such a future, we need to have a constant ethical assessment and a fierce interdisciplinary collaboration. Ethicists, technologists, clinicians, and policymakers are required to come at once to tackle the emerging problems and develop the guidelines of ethics as AI keeps advancing. Regular audits, making policies inclusive, and public engagement will be crucial in sustaining the fine line between innovation and responsibility. By nurturing this collaborative and flexible approach, stakeholders may ensure that AI is playing the role of a transformative, ethical, and reliable friend which is helping health care treatment while humanity and technology work in collaboration for the well-being of all.

REFERENCES

Addepally, S. A., & Purkayastha, S. (2017). Mobile-application based cognitive behavior therapy (CBT) for identifying and managing depression and anxiety. In *Digital Human Modeling. Applications in Health, Safety, Ergonomics, and Risk Management: Health and Safety: 8th International Conference, DHM 2017, Held as Part of HCI International 2017, Vancouver, BC, Canada, July 9-14, 2017* [Springer International Publishing.]. *Proceedings*, *8*(Part II), 3–12.

Adnan, M., Xiao, B., Ali, M. U., Bibi, S., Yu, H., Xiao, P., Zhao, P., Wang, H., & An, X. (2024). Human inventions and its environmental challenges, especially artificial intelligence: New challenges require new thinking. *Environmental Challenges*, *16*, 100976. DOI: 10.1016/j.envc.2024.100976

Ajayi, R. (2025). AI-powered innovations for managing complex mental health conditions and addiction treatments. *International Research Journal of Modernization in Engineering Technology and Science*, *7*(1), 87.

Arora, A. (2020). Conceptualising artificial intelligence as a digital healthcare innovation: an introductory review. *Medical Devices: Evidence and Research*, 223-230.

Baeza-Yates, R. (2024, Mar). Introduction to Responsible AI. In *Proceedings of the 17th ACM International Conference on Web Search and Data Mining* (pp. 1114-1117). DOI: 10.1145/3616855.3636455

Bakker, D., Kazantzis, N., Rickwood, D., & Rickard, N. (2016). Mental health smartphone apps: Review and evidence-based recommendations for future developments. *JMIR Mental Health*, *3*(1), e4984. DOI: 10.2196/mental.4984 PMID: 26932350

Beg, M. J. (2024). Responsible AI Integration in Mental Health Research: Issues, Guidelines, and Best Practices. *Indian Journal of Psychological Medicine*, 02537176241302898. PMID: 39650770

Cerasa, A., Gaggioli, A., Marino, F., Riva, G., & Pioggia, G. (2022). The promise of the metaverse in mental health: The new era of MEDverse. *Heliyon*, *8*(11), e11762. DOI: 10.1016/j.heliyon.2022.e11762 PMID: 36458297

Corrigan, P. W., Druss, B. G., & Perlick, D. A. (2014). The impact of mental illness stigma on seeking and participating in mental health care. *Psychological Science in the Public Interest*, *15*(2), 37–70. DOI: 10.1177/1529100614531398 PMID: 26171956

de Almeida, P. G. R., dos Santos, C. D., & Farias, J. S. (2021). Artificial intelligence regulation: A framework for governance. *Ethics and Information Technology*, *23*(3), 505–525. DOI: 10.1007/s10676-021-09593-z

De Freitas, J., Uğuralp, A. K., Oğuz-Uğuralp, Z., & Puntoni, S. (2024). Chatbots and mental health: Insights into the safety of generative AI. *Journal of Consumer Psychology*, *34*(3), 481–491. DOI: 10.1002/jcpy.1393

Dimova, A. (2024). *Mental Health Advocacy on Social Media: A Multimodal Discourse Analysis* (Doctoral dissertation, Queen Mary University of London).

Durán, J. M., & Jongsma, K. R. (2021). Who is afraid of black box algorithms? On the epistemological and ethical basis of trust in medical AI. *Journal of Medical Ethics*, *47*(5), 329–335. DOI: 10.1136/medethics-2020-106820 PMID: 33737318

El-Sappagh, S. H., Nazih, W., Alharbi, M., & AbuHmed, T. (2025). Responsible Artificial Intelligence for Mental Health Disorders: Current Applications and Future Challenges. *Journal of Disability Research*, *4*(1), 20240101. DOI: 10.57197/JDR-2024-0101

Engel, J., & Prentice, D. (2013). The ethics of interprofessional collaboration. *Nursing Ethics*, *20*(4), 426–435. DOI: 10.1177/0969733012468466 PMID: 23329780

Eyre, H. A., Ellsworth, W., Fu, E., Manji, H. K., & Berk, M. (2020). Responsible innovation in technology for mental health care. *The Lancet. Psychiatry*, *7*(9), 728–730. DOI: 10.1016/S2215-0366(20)30192-9 PMID: 32828153

Fusar-Poli, P., Manchia, M., Koutsouleris, N., Leslie, D., Woopen, C., Calkins, M. E., & Andreassen, O. A. (2022). Ethical considerations for precision psychiatry: A roadmap for research and clinical practice. *European Neuropsychopharmacology*, *63*, 17–34. DOI: 10.1016/j.euroneuro.2022.08.001 PMID: 36041245

Gamble, A. (2020). Artificial intelligence and mobile apps for mental healthcare: A social informatics perspective. *Aslib Journal of Information Management*, *72*(4), 509–523. DOI: 10.1108/AJIM-11-2019-0316

Gooding, P., & Kariotis, T. (2021). Ethics and law in research on algorithmic and data-driven technology in mental health care: Scoping review. *JMIR Mental Health*, *8*(6), e24668. DOI: 10.2196/24668 PMID: 34110297

Guidance, W. H. O. (2021). *Ethics and governance of artificial intelligence for health*. World Health Organization.

Guo, C., & Saxton, G. D. (2014). Tweeting social change: How social media are changing nonprofit advocacy. *Nonprofit and Voluntary Sector Quarterly*, *43*(1), 57–79. DOI: 10.1177/0899764012471585

Hadjistavropoulos, T. (2023). Domains of contemporary and future ethical dilemmas. *Canadian Psychology*, *64*(3), 178–188. DOI: 10.1037/cap0000356

Huang, T., Xu, H., Wang, H., Huang, H., Xu, Y., Li, B., Hong, S., Feng, G., Kui, S., Liu, G., Jiang, D., Li, Z.-C., Li, Y., Ma, C., Su, C., Wang, W., Li, R., Lai, P., & Qiao, J. (2023). Artificial intelligence for medicine: Progress, challenges, and perspectives. *The Innovation Medicine*, *1*(2), 100030. DOI: 10.59717/j.xinn-med.2023.100030

Kasula, B. Y. (2023). Ethical Considerations in the Adoption of Artificial Intelligence for Mental Health Diagnosis. *International Journal of Creative Research In Computer Technology and Design*, *5*(5), 1–7.

Koutsouleris, N., Hauser, T. U., Skvortsova, V., & De Choudhury, M. (2022). From promise to practice: Towards the realisation of AI-informed mental health care. *The Lancet. Digital Health*, *4*(11), e829–e840. DOI: 10.1016/S2589-7500(22)00153-4 PMID: 36229346

Liu, F., Ju, Q., Zheng, Q., & Peng, Y. (2024). Artificial intelligence in mental health: Innovations brought by artificial intelligence techniques in stress detection and interventions of building resilience. *Current Opinion in Behavioral Sciences*, *60*, 101452. DOI: 10.1016/j.cobeha.2024.101452

Loies, S. E. (2025). Combating Mental Health Stigma With Artificial Intelligence: Improving Societal Understanding. In *Supporting Psychological and Emotional Wellbeing Among Entrepreneurs* (pp. 75-92). IGI Global.

Lovejoy, C. A. (2019). Technology and mental health: The role of artificial intelligence. *European Psychiatry : the Journal of the Association of European Psychiatrists*, *55*, 1–3. DOI: 10.1016/j.eurpsy.2018.08.004 PMID: 30384105

Lustgarten, S. D., & Elhai, J. D. (2018). Technology use in mental health practice and research: Legal and ethical risks. *Clinical Psychology : a Publication of the Division of Clinical Psychology of the American Psychological Association*, *25*(2), 20. DOI: 10.1111/cpsp.12234

Maheu, M. M., Pulier, M. L., Wilhelm, F. H., McMenamin, J. P., & Brown-Connolly, N. E. (2004). *The mental health professional and the new technologies: A handbook for practice today*. Routledge. DOI: 10.4324/9781410611444

Martin, J. K., Pescosolido, B. A., & Tuch, S. A. (2000). Of fear and loathing: The role of'disturbing behavior,'labels, and causal attributions in shaping public attitudes toward people with mental illness. *Journal of Health and Social Behavior*, *41*(2), 208–223. DOI: 10.2307/2676306

Martinez-Martin, N. (2022). Minding the AI: Ethical challenges and practice for AI mental health care tools. In *Artificial intelligence in brain and mental health: Philosophical, ethical & policy issues* (pp. 111–125). Springer International Publishing.

Meszaros, J., Minari, J., & Huys, I. (2022). The future regulation of artificial intelligence systems in healthcare services and medical research in the European Union. *Frontiers in Genetics, 13*, 927721. DOI: 10.3389/fgene.2022.927721 PMID: 36267404

National Academies of Sciences. (2016). Division of Behavioral, Social Sciences, Board on Behavioral, Sensory Sciences, & Committee on the Science of Changing Behavioral Health Social Norms. *Ending discrimination against people with mental and substance use disorders: The evidence for stigma change*. National Academies Press.

Patel, V., Saxena, S., De Silva, M., & Samele, C. (2013). Transforming lives, enhancing communities: innovations in mental health. *WISH Mental Health Report*.

Pavlopoulos, A., Rachiotis, T., & Maglogiannis, I. (2024). An overview of tools and technologies for anxiety and depression management using AI. *Applied Sciences (Basel, Switzerland), 14*(19), 9068. DOI: 10.3390/app14199068

Pesapane, F., Codari, M., & Sardanelli, F. (2018). Artificial intelligence in medical imaging: Threat or opportunity? Radiologists again at the forefront of innovation in medicine. *European Radiology Experimental, 2*(1), 1–10. DOI: 10.1186/s41747-018-0061-6 PMID: 30353365

Peters, D., Vold, K., Robinson, D., & Calvo, R. A. (2020). Responsible AI—Two frameworks for ethical design practice. *IEEE Transactions on Technology and Society, 1*(1), 34–47. DOI: 10.1109/TTS.2020.2974991

Raj, G., Sharma, A. K., & Arora, Y. (2024). Analyzing the Effect of Digital Technology on Mental Health. In *Strategies for E-Commerce Data Security: Cloud, Blockchain, AI, and Machine Learning* (pp. 54-82). IGI Global. DOI: 10.4018/979-8-3693-6557-1.ch003

Saarni, S. I., Hofmann, B., Lampe, K., Lühmann, D., Mäkelä, M., Velasco-Garrido, M., & Autti-Rämö, I. (2008). Ethical analysis to improve decision-making on health technologies. *Bulletin of the World Health Organization, 86*(8), 617–623. DOI: 10.2471/BLT.08.051078 PMID: 18797620

Saeidnia, H. R., Hashemi Fotami, S. G., Lund, B., & Ghiasi, N. (2024). Ethical considerations in artificial intelligence interventions for mental health and well-being: Ensuring responsible implementation and impact. *Social Sciences (Basel, Switzerland), 13*(7), 381. DOI: 10.3390/socsci13070381

Shatte, A. B., Hutchinson, D. M., & Teague, S. J. (2019). Machine learning in mental health: A scoping review of methods and applications. *Psychological Medicine, 49*(9), 1426–1448. DOI: 10.1017/S0033291719000151 PMID: 30744717

Simon, J., Bass, T., Boelman, V., & Mulgan, G. (2017). Digital democracy: the tools transforming political engagement.

Smuha, N. A., Ahmed-Rengers, E., Harkens, A., Li, W., MacLaren, J., Piselli, R., & Yeung, K. (2021). How the EU can achieve legally trustworthy AI: a response to the European Commission's proposal for an Artificial Intelligence Act. *SSRN* 3899991. DOI: 10.2139/ssrn.3899991

Solaiman, B. (2020). Addressing access with artificial intelligence: Overcoming the limitations of deep learning to broaden remote care today. *U. Mem. L. Rev., 51*, 1103.

Stellefson, M., Paige, S. R., Chaney, B. H., & Chaney, J. D. (2020). Evolving role of social media in health promotion: Updated responsibilities for health education specialists. *International Journal of Environmental Research and Public Health, 17*(4), 1153. DOI: 10.3390/ijerph17041153 PMID: 32059561

Tavory, T. (2024). Regulating AI in mental health: Ethics of care perspective. *JMIR Mental Health, 11*(1), e58493. DOI: 10.2196/58493 PMID: 39298759

Ueda, D., Kakinuma, T., Fujita, S., Kamagata, K., Fushimi, Y., Ito, R., Matsui, Y., Nozaki, T., Nakaura, T., Fujima, N., Tatsugami, F., Yanagawa, M., Hirata, K., Yamada, A., Tsuboyama, T., Kawamura, M., Fujioka, T., & Naganawa, S. (2024). Fairness of artificial intelligence in healthcare: Review and recommendations. *Japanese Journal of Radiology, 42*(1), 3–15. DOI: 10.1007/s11604-023-01474-3 PMID: 37540463

Usmani, S. S., Sharath, M., & Mehendale, M. (2022). Future of mental health in the metaverse. *General Psychiatry, 35*(4), e100825. DOI: 10.1136/gpsych-2022-100825 PMID: 36189180

Valentine, L., D'Alfonso, S., & Lederman, R. (2023). Recommender systems for mental health apps: Advantages and ethical challenges. *AI & Society, 38*(4), 1627–1638. DOI: 10.1007/s00146-021-01322-w PMID: 35068708

Wartel, S. G. (2023). *Facilitated Self-Help Guide: A Professional Guide for Technology-Assisted Behavioral Health Care* (Revised Edition). Stephen Wartel.

Wiederhold, B. K., Miller, I., & Wiederhold, M. D. (2018). Augmenting behavioral healthcare: mobilizing services with virtual reality and augmented reality. *Digital Health: Scaling Healthcare to the World*, 123-137.

Wies, B., Landers, C., & Ienca, M. (2021). Digital mental health for young people: A scoping review of ethical promises and challenges. *Frontiers in Digital Health, 3*, 697072. DOI: 10.3389/fdgth.2021.697072 PMID: 34713173

Winfield, A. F., & Jirotka, M. (2017). The case for an ethical black box. In *Towards Autonomous Robotic Systems: 18th Annual Conference, TAROS 2017, Guildford, UK, July 19–21, 2017* [Springer International Publishing.]. *Proceedings, 18*, 262–273.

Chapter 9
Responsible AI in Mental Health:
Ethics, Advocacy, and Innovation

A. Devendran
https://orcid.org/0000-0001-7192-521X
School of Business, Woxsen University, Hyderabad, India

Sai Adarsh Maddu
https://orcid.org/0009-0002-2557-5663
Woxsen University, India

N. KaviPriya
Sri Ramachandra Institute of Higher Education and Research, India

Namita Kotta
Woxsen University, India

V. Bhuvishashri
Woxsen University, India

ABSTRACT

Ethical considerations surrounding technology in mental health applications are complex, multifaceted, and increasingly relevant in both clinical and research settings. Ethical challenges associated with technology's integration into mental health underscore the urgent need for a framework prioritizing safety, fairness, and human dignity. With the rise of technology-driven tools in diagnostics, treatment recommendations, and patient management, the ethical implications of these innovations must be rigorously examined to ensure they serve and protect users—particularly vulnerable individuals seeking mental health care. the chapter begins with an over-

DOI: 10.4018/979-8-3373-0467-0.ch009

view of fundamental ethical challenges in mental health technology. Issues such as algorithmic bias and discrimination are discussed, highlighting how biased datasets can lead to inaccurate diagnoses or inappropriate treatment recommendations for marginalized groups. Addressing these biases is crucial to fostering equitable access to mental health services and avoiding reinforcing societal inequalities through technology-driven systems.

INTRODUCTION

The introduction of AI to mental health care lends itself to a total change in diagnosis, treatment, and support for mental health patients. Mental health care incorporating AI technology promises to make commitment of mental health care, besides care, readily available, personalized, and stigma-free-the latter marking as a sociocultural issue in a number of set-ups. Nonetheless, the quick speed at which AI technologies in such a vulnerable domain are being welcomed has placed substantial ethical concerns to be tackled about what potential benefits of these technological advances can be and how certain considerations can threaten this further in exchange for the rights, safety, and well-being of those being helped by such innovative technologies (IEEE Global Initiative on Ethics of Autonomous and Intelligent Systems, 2019).

In this section, the ethical considerations surrounding AI technology in mental health care will be narrowly examined and the potential benefits, risks, and challenges that come with its implementation will be appraised. Ethical principles including autonomy, beneficence, non-maleficence, and justice and their applicability to the use of AI in mental health care will be reviewed. Discussion will also include privacy issues, consent, bias, and the therapeutic relationship as impacted by AI, as well as the broader societal implications of AI-powered mental health interventions.

1. A GLIMPSE INTO THE ETHICAL CHALLENGES OF AI-DRIVEN MENTAL HEALTH CARE

1.1. Improving Access to Care and Services

One of the most critical applications of AI in mental health care is improved access to health care. A lot of mental health services still languish under geographical, economic, and social barriers, hampering health care they need. The chatbot, e-therapists, and mental health mobile applications of AI can render mental health services to individuals residing in remote or underserved areas, with people too shy to communicate in a real setup, in addition to others.

Take chatbots Woebot and Wysa as an example. They give immediate access to CBT techniques, mood tracking, and emotional support. A tool like this enhances mental health care, allows greater accessibility, and cuts down on the tasks of already overworked therapists.

1.2. Personalization and Early Discovery

AI offers the potential for a real breakthrough in real healthcare by giving ways of highly personalized approaches to illness treatment. With the use of machine learning algorithms and access to huge sets of data, they can study a number of things such as EHR (electronic health record), genetic information, and towards predictive features of individual responses to different treatments. Such precision would help the doctor to steer his interventions in accordance with the patient's requirements, in essence improving outcomes and eliminating the current trial-and-error approach in mental health treatment.

Further, AI tools can also enable early intervention by identifying those who are at risk of initiating the development of mental health conditions before the symptoms actually take their toll. For example, AI algorithms may analyze social media posts, speech patterns, or wearable device data to detect early signs of depression, anxiety, or other mental health issues (Fitzpatrick et al., 2017). Early diagnosis and intervention could assure the end of this downward spiral and ensure a long-term promising prognosis.

1.3. Eradicating Stigma and Enhancing Self-Management

For individuals who bear the stigma of seeking mental health care, AI-affiliated digital platforms can help alleviate the stigma. It so happens that many people may feel more comfortable interacting with an AI-driven platform than with a person in the initial stages of seeking help. Therefore, with the help of AI, they are more likely to approach mental healthcare and engage in self-management of their conditions.

These tools will also empower individuals to actively participate in their mental well-being. Mood tracking applications can help users monitor their emotions and identify what has the power to trigger, with skills to develop methods to term down those activities. This approach of self-management will give the individual greater comfort and assurance of control, as has always been an important aspect of mental health.

Figure 1. AI in Mental Health Care Cycle

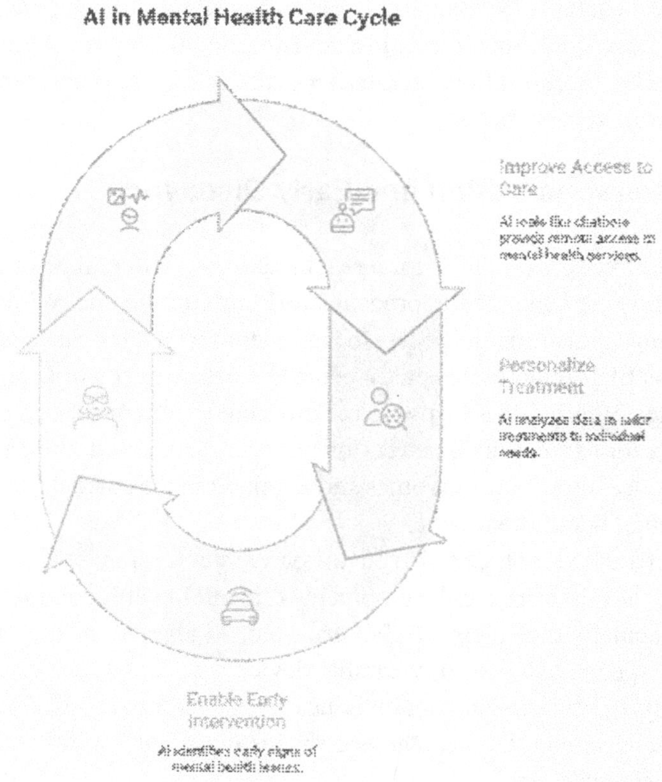

2. ETHICAL PRINCIPLES IN AI-ENHANCED MENTAL HEALTH CARE

As artificial intelligence continues to permeate mental health care practice, underscored by the importance it gives to the ethical principles in bioethics. Among these are autonomy, beneficence, non-maleficence, and justice. Feeling the necessity to unfold with respect to both the ethical principle and any attendant challenges, let us address AI-enhanced mental health care.

2.1. Respect for Autonomy

Respect for autonomy is an essential bioethical principle that underscores individuals' having the right to make their own informed choices. In AI-enhanced mental health care, this means that patients should have the choice of whether or not to use AI tools, understand what the AI does, and have control of their own data.

2.1.1. Informed Consent

When AI is used in health care, we each understand the importance of securing informed consent. Patients must be fully informed about the nature, benefits, risks, and limitations of AI-driven mental health interventions, so they should essentially become aware of what they are getting themselves into. However, obtaining informed consent in the context of AI can be challenged by the complexity of the technology and likely misunderstandings only possible.

For example, patients may not fully grasp how AI algorithms analyze their data or make recommendations. They may also overestimate the capabilities of AI tools, believing that they can replace human therapists entirely. It is the responsibility of health care providers and developers to ensure that patients receive clear, accurate, and comprehensible information about AI tools, enabling them to make informed decisions.

2.1.2. Data Privacy and Control

Respect for autonomy also requires that individuals have control over their personal data. AI-driven mental health tools often depend on the collection and analysis of sensitive data, including medical history, behavioral patterns, and even biometric data from wearable devices (Jobin et al., 2019). Patients must be informed about how their data will be used, stored, and shared, and they should have the option to opt out of data collection if they choose.

Furthermore, it is possible that AI tools can be employed in ways that violate personal autonomy. For example, employers or insurers might use AI-driven mental health assessments to make decisions about hiring; promotions; or coverage. This situation could lead to discrimination or coercion; thus, preventative measures need to be put in place to forestall such misuse and protect privacy rights and self-determination.

2.2. Beneficence and Non-Maleficence

Central to ethical health care practice are the principles of beneficence (doing good) and non-maleficence (averting harm). In the framework of AI-enhanced mental health care, they dictate that AI tools be designed and used to generate positive outcomes and reduce the chances of harms.

2.2.1. Ensuring Accuracy and Consistency

One of the major ethical issues regarding the application of AI in mental health care is coming from the question of accuracy and reliability of AI-operatable assessments and interventions. AI-driven algorithms are as robust or as unreliable as the data that is fed to them. Should this data carry any biases, or is incomplete, or is generally poor, the resultant outputs from these algorithms could very well be misguiding or harmful (Obermeyer et al., 2019).

For example, a tool powered with AI aims to diagnose depression based on speech patterns. However, if it fails to account for linguistic or cultural differences, it may bring about misdiagnosis and a subsequent misfitting treatment. Again, a therapy app developed by AI might offer advice, unsupported by evidence or detrimental to a patient's condition. There is a need to inventory and validate AI technologies so that they are accurate and reliable before putting them into practice in the clinical environment.

2.2.2. Backup Against Over-Reliance on AI

Another ethical concern that surfaces is the possible over-reliance on AI tools at the cost of human judgment and empathy. While AI can provide valuable backsup and augment the capabilities of mental health professionals, it cannot replace the nuanced understanding, emotional connection, and therapeutic alliance that human therapists offer.

An unmeasured trust in AI can lead to impersonalization of mental health care, therefore treating the patient as data points rather than as individuals with their own experiences and needs; this is harmful to the therapeutic relationship, which is the most important lever related to mental health treatment. AI tools need to be encouraged as complements of care rather than substitutes, hence human therapists should still play a central role in care.

2.2.3. Addressing Unforeseen Harms

AI-augmented mental health care has the capability to possibly cause unintended harms. For instance, AI in crisis response may realistically not recognize the severity of the situation to which it's responding and may further worsen the situation by giving advice suitable for the crisis. On the other hand, AI interventions that are not dependent on personal revelations may turn out to be relatively less effective in cases where the respective individuals have a hard time expressing their actual symptoms and emotions.

Mitigation of such risks necessitates the need for built-in safety measures within the AI tool to bring in human therapists whenever the need arises. Ongoing monitoring and evaluations should be undertaken to identify and rectify issues that may worsen with the intention to protect patients within mental health care.

2.3. Justice and Equity

Justice calls for fair and equitable allocation of health resources and benefit from AI-enhanced mental health care. However, AI also stands the risk of reinforcing existing imbalances in access to mental health care, most significantly for the mentally ill, the marginalized, and the underserved.

2.3.1. Addressing Bias and Discrimination

The APIs that train AI models execute upon large data sets. Therefore, if such data sets themselves are subject to bias, the AI may perpetuate or else greatly accentuate such biases. For example, if an AI tool gets trained on data profiles heavily representing a certain particular demographic, then it will be less accurate or effective for working with individuals from other groups and hence promote disparities in mental health care and outcomes.

It is absolutely critical for AI developers to have diverse and inclusive data sets against which they thoroughly test anything that could appear as bias in their algorithms (Obermeyer et al., 2019). Moreover, the mental health professions should remain informed about bias potential inside AI tools and actively work to ensure that equal and ethical care is delivered for all patients.

2.3.2. Ensuring Access for Everyone

Whereas AI poses huge opportunities for improving access to mental health care, it may also establish new barriers. The unequipped with technology or unintelligent on how to use it may be left out from AI-aided mental health intervention. Similarly, they may also simply not afford AI tools, particularly in countries without public health expenditures.

From the vantage point of justice and equity, it ought to manifest that with the happening of AI-enhanced mental health care, inclusiveness to all irrespective of socio-economic status, or geographical location or tech-savvy is vital. This will be a challenge as some level of intervention will need to provide technology access and training to those people who lack vital technologies for these services, with the prevailing policies to ensure thatality cues into the spectrum of affordability and insurance.

3. THE THERAPEUTIC RELATIONSHIP IN AI-ENHANCED MENTAL HEALTH CARE

The therapeutic relationship is considered crucial in effective mental health care. It is based on trust, empathy, and cooperation between therapists and patients. Within such a discussion, the inclusion of AI connotes interesting ethical questions on how to keep the human touch which is the basis for successful therapy.

3.1 AI's Role in the Therapeutic Relationship

AI can be used to help therapeutically in ways that deal with enhancing communication, enhancing self-reflection, and offering resources or aid to patients. So, for example, AI chatbots can help in practicing coping skills between therapy sessions, and virtual reality (VR) can help with exposure therapy or to relax the patient by offering immersive experiences.

However, notes the author, we must not forget that AI lacks the depth of human empathy and understanding which lie at the core of the therapeutic relationship. The AI can only give patients with the needed information and advice; it cannot interpret or respond back to complex feelings and experiences being undergone. Hence, AI can only be used to better the work of the human therapist rather than seek to be the replacement.

3.2. Upholding the Human Connection

One of the ethical dilemmas in using AI in mental health concerns ensuring that it does not undermine the human connection that should exist between therapists and patients. Emphasizing AI devices might lead to a more business-leaning and less amiable manner of caring in which patients might feel they are dealing with a group of inanimate objects as against an empathetic human being.

To maintain the crucial human bond, professionals using AI must consciously participate in the therapy process. This may call for the use of AI to enrich rather than replace physical interactions, ensuring patients feel they are being truly heard, comprehended, and helped while in the hands of their human therapist.

3.3. Ethical Implications of AI-Driven Workforces

The very idea of an AI-driven workforce where patients interact primarily with AI rather than human therapists brings forth new ethical dilemmas. While AI-driven therapy can work wonders for certain individuals and conditions, it may not be helpful or appropriate for all: individuals with complex cases or severe mental health issues may require patient deep understanding and empathy to reach positive outcomes.

Not only that, the use of AI-driven therapy must therefore be closely monitored with regard to any lowering of the quality of care that may damage the therapeutic alliance. Patients, as an option, should be able to turn to human therapy if they feel that the AI-driven therapy has not been able to meet their needs, with mental health professionals made accessible to offer ongoing help when required.

Figure 2. The synergy of AI and human empathy in mental health

4. PRIVACY AND DATA SECURITY IN AI-ENHANCED MENTAL HEALTH CARE

AI usage intrinsically implies that massive data processing tools that are predominantly aimed at mental health carry the practice of data collection and intensive handling and storing of extremely personal data. Holding in fact the anxiety privacy guards and data security are at stake.

4.1. Patient Confidentiality

Patient privacy as the code of the ethics of healthcare is crucial. It furthermore applies specifically to the mental health discipline that inflicts a blur over the stigma of mental illness, causing social and occupational breakdown for patients. AI-driven mental health tools, for this reason, must prioritize robust privacy safeguards that will ensure guarantees for the confidentiality and security of patient private data. Sicherung der Patientendaten

This includes dealing with strong encryptions, access management, and anonymization of data that should protect this data from prosecution. This also extends to grant patients information on the data use with whom they wish to make access of the said information: patients.

4.2. Data Security and Breach Prevention

The greater use of AI in mental health care also presents concerns about data security, with the chance of data breaches. An itinerary for highly confidential mental health data becomes highly damaging in different perspectives-hence differently-from discriminative, take stigmas away from those thus stigmatized, and commotions. The burden lies on the AI pros and health care authorities to appreciate the data-security environment for the prevention of data breaches.

Therefore, these include regular security audits, robust cybersecurity policies, and training for staff on best practices for protecting data. Organizations might also consider policies and procedures for breach response that mandates the affected patients be informed and supported to minimize damages.

4.3. Ethical Use of Data

Aside from privacy and security, another ethical concern is the manner in which data is used in AI-dedicated mental health care. For instance, the risk is always there that data of patients might get employed for some other objectives, not originally intended for when it was collected including marketing or research without the express approval of the patient.

It is really quite important for patients to know the aims of the data changes and to exercise the power of refusal against the union in order to respect dissent. Artificial intelligence and health care organizations should abide by ethical guidelines and legal regulations for the use of patient data, which will ensure that the data is not misused against patients but rather in a way permitted under the rights of the latter.

5. BIAS AND FAIRNESS IN AI-ENHANCED MENTAL HEALTH CARE

The ethical problems concerning bias in AI algorithms are critical, particularly with regard to the mental health system, where biased algorithms may bring about inequalities in the diagnosis, treatment, and outcomes for different populations. It is indeed a main need to recognize sources of bias and advance equity in AI-enhanced mental health care, in order to make secure that every individual obtains an honest and effective medical care.

5.1. Sources that Give Way to Bias in AI Algorithms

Different sources can bring bias into AI algorithms, starting from biased training data, the very underpinning of algorithms, even to subjective decisions of developers in the development of such algorithms. For instance, if training data biased toward a certain demographic group accompanies any AI algorithm, then obviously it would be invalid or ineffective for other groups, too. Moreover, if the design and immanence of the AI tool negatively influence the judgment of developers, it is a real possibility that bias will enter into consideration.

5.2. Consequences of Bias in Mental Health Care

As has been aforementioned, AI-gifted mental health care may build up the dire consequences of bias. Biased algorithms will bring about misdiagnosis, incorrect therapy, or even hindered access to patient care for certain populations. As an example, a tool that misjudges depression in women or people of color, posing so disqualifying, may therefore be suboptimal for the some groups so affected to either be helped by such a tool directly or indirectly.

In connection, removing intolerance into groups and stereotypes would only magnify all the affliction related to mental health. Hence, through the usage of biased data, similar acts may result in maintaining a negative self-identity among certain groups; these AI-aligned tools will only strengthen those stereotypes, hence perpetuating discrimination.

5.3. Promoting Fairness and Equity in AI-Enhanced Mental Health Care

To show fairness and equity along with AI-enhanced mental health care, it is quite crucial to take care of bias for every step involved during the building of an AI project. This must be augmenting the representation of training data that encompasses diversity, testing algorithms for bias, involving multiple stakeholders in the designing and validating of AI tools (World Health Organization, 2020).

Health care providers must also be conscious of the possibility for bias and take care that patients receive equitable treatment (World Health Organization, 2020). This may involve looking at multiple pieces of information in making decisions without relying merely on AI-generated assessments and scrutinizing alerts for possible bias or discrimination laced in the use of AI tools.

6. THE ROLE OF REGULATION AND OVERSIGHT IN AI-ENHANCED MENTAL HEALTH CARE

The ethical dilemma posed by mental health was already notoriously elusive. But AI-enhancing makes the material wayward, miles away, adding filling to arguing about AI regulation. This conundrum renders thinking about AI in mental health a layer dark, a bit more terrifying if wrong handwaves; not to mean holding mental health hostage while looking for that AI care setting faced with some dirt in ethics, in raising principled and sane-flooring before the awkward slipways caregivers lend so pathetically Gramsci and Hegel to do but failed, thus working as stakeholders for yet some divine chance shooting wide of goal in some important issues about AI.

6.1. Ethical Considerations and Regulatory Standards

The role of AI in mental health care should thus blend glamorous ethical guidelines and standard watchwords such as informed consent, data privacy, bias testing, and "good base therapeutic relation." Guidelines should be framed in consultation with related professionals, AI developers, patients (and some sort of ethy backwater, mentally cultivating youngsters among the shores of "me, my, and mine," such as ethicists) to enforce their comprehensiveness, practical viability, and alignment with mental health care values (Mittelstadt, 2019) & (Floridi et al., 2018).

6.2. Legal Framework for AI in Mental Health Care

Yet further to ethical solutions, recourse must be rendered in the shape of a legal framework that oversees the development, testing, and application of AI in mental health care (Gerke et al., 2020). Such a framework should encompass transparency, accountability, and validation and widen the net by activating mechanisms for the monitoring and redress of potential grievances.

For example, if the government stipulated the requirement of rigorous testing and validation-checks to be done before an AI tool is finally to be accepted for realistic usage, the developer would need to come out in the open about the limitations and potential problems with this tool itself. Again, for hindsight, rules that govern the use of such codes for ongoing monitoring and evaluation are necessary to guarantee the usefulness of AI tools supports ethical and safety issues over the days.

6.3. Functions of Professional Bodies and Accreditation Organizations

Professional bodies and accreditation bodies ought to assume the responsibilities in seeing to the pushing forward of the ethical use of AI in mental health care inter alia. Such bodies will have to play the part of developing and disseminating best practices for AI use, educating professionals, and ensuring that accountability is in effect.

For instance, such professional organizations might initiate a certification program that would help regulate AI tools to conform to some ethical and safety standard. These professional bodies could also have a brief to furnish advice as to how AI shall be made to work into clinical practice so that the therapeutic relation is situated to benefit from AI and not be compromised. Likewise, the accreditation authorities could stipulate that organizations charged with mental health care must demonstrate that they are using AI technologies to ensure the safety and well-being of clients as one of the conditions in their accreditation processes (Torous et al., 2018).

7. THE FUTURE OF AI-ENHANCED MENTAL HEALTH CARE: ETHICAL CONSIDERATIONS AND CHALLENGES

As AI technology continues to evolve, the ethical considerations and challenges associated with its use in mental health care will also continue to evolve. In this section, we explore some of the emerging ethical issues that may arise in the future and consider how they might be addressed.

7.1. The Role of AI in Predictive and Preventive Mental Health Care

Predictive and preemptive use of the Artificial Intelligence method in medical fields raises ethical issues too. AI algorithms may cause stigma, discrimination, or unnecessary medicalization for groups identified as "at-risk," to give an example. There is also a scenario in which over-diagnosis comes about due to a classification of the population based on probability, through the pathologization of natural variants in behavior and emotion. It is so much stress to a person branded as at risk when they may never develop mental illness (Musen et al., 2003).

Further, moral issues might rise out of predictive AI applications in mental health care concerning personal autonomy and consensus. Absent direct but informed consent, surveillance of a person by an AI infringes upon his or her privacy and autonomy, setting a chilling example: An employer or educational institution can use AI applications on the students and employees to track them for psychological issues, which could appall the theory and practice of surveillance and coercion Calvo et al., 2020). There must be laid down clear boundaries, ethical guidelines setting that the care will always stay with an individual and that control remains in the hands of the individual.

7.2. The Integration of AI with Emerging Technologies

As the potential of AI grows, increasingly it will be integrated with other nascent technologies such as VR (virtual reality), AR (augmented reality), and BCI (brain-computer interface). These integrations provide avenues through which mental health care can be revolutionized in formulating a range of therapies and even enabling immersive therapeutic approaches, real-time feedback, and novel treatment (Rizzo et al., 2019). VR-related exposure therapy offers promise for phobias and post-traumatic stress disorder (PTSD) and BCIs may even establish a direct interface between the brain and AI systems for severe commutation disabled individuals.

Once again, the integration of AI with these technologies provides for new pool to be fished for ethical challenges. One possible adverse question when raising hands on VR and AR in therapy is toward the potential harm. Does that mean the run of psychological harm or traumatization, or could it be the blurring line between virtual and real-life experiences? Similarly, BCIs raise questions of privacy with its collection of very sensitive neural data; misuse or unintended consequences could always crop up, such as their application for mind control and manipulation.

In order to address these matters, it is important to have ethical considerations and frameworks regarding the integration of AI and emerging technologies for mental health care. Such frameworks need to shield the patient, guarantee informed consent, and promote responsible use of the technology, yet allowing the trees of innovation to grow from the seed of advancement.

7.3. The Role of AI in Global Mental Health

AI can play a key role in bridging the gaps in mental health care worldwide, especially in low-and middle-income countries (LMICs) where access to psychological services is usually constrained. AI-driven tools such as chatbots and mobile applications can become scalable and cost-effective ways of offering mental health support from low-resource contexts (World Health Organization, 2020). For example, AI-based platforms could provide culturally-tailored interventions in various languages, increasing access to mental health care for distinct populations.

However, the use of AI in global mental health care has become mired in ethical matters regarding equity, cultural sensitivity, and sustainability. AI tools developed in high-income countries could prove to be neither suitable nor efficient for application in LMICs due to divergence in culture, language, and healthcare systems. Furthermore, overreliance on AI-driven solutions could drain off the resources that could have been earmarked towards strengthening local mental health infrastructure and workforce capability, rather aggravating the existing inequalities.

To ensure that AI-based mental health care benefits all populations, there is great need to include current stakeholders in the design and implementation of AI-powered tools, concentrating on how their systems will be culturally relevant and contextually relevant. In addition, there should be strategies laid out to help build local capacity and ease integration of AI tools into established systems of mental health care rather than trying to have them as self-operating solutions.

7.4. Ethical Considerations in AI-Driven Research and Innovation

AI is working towards the transformation of mental health research by enabling big data analysis and recognizing new insights into mental health conditions (Vayena & Blasimme, 2018). Algorithms may analyze genetic, neuroimaging, and behavioral data, hence evolving new biomarkers of mental illness and anticipate therapeutic outcome. This could lead to speedier development of personalized and precision mental-healthcare strategies.

AI applied in mental-health research comes with ethical questions about data governance, consent, and exploitation (Floridi, 2018). Who owns data utilized to train AI algorithms, and how should the benefits of research under AI be distributed? There is also a risk of vulnerable groups, such as those with severe mental health statuses, being the targets of disproportionate data collection; exploitation and informed consent, thus, have become a possible concern.

Thus its development requires maintaining the highest ethics standards with regard to AI-driven research in psychiatry: Transparency, openness, and respect for individual rights should inform the way in which data is collected and used. Above all, consideration should be given to the diversity of ethnic prodigies partaking in the AI-driven studies so that the benefits and findings from such research is realized by all.

7.5. The Role of AI in Addressing Mental Health Workforce Shortages

The global shortage of mental health professionals, particularly in underserved areas, is a pressing issue of today's mental health care. By augmenting the capabilities of the existing mental health professionals, AI can provide much-needed support to those individuals who may otherwise remain deprived of care. Applications of AI can, for example, eliminate routine tasks for the counselors-say, diagnostic evaluation, or treatment planning-and hence increase time availability to render highly-personalized care.

At the same time, the integration of AI for tackling the workforce shortage raises pertinent quality-of-care issues and raises the specter of dehumanizing that care. AI efficiencies are bound by the fact that AI is not empathy, intuition, or the therapist-patient relationship central to human mental health care (Topol, 2019). Furthermore, the phobic idea that AI-relying tools may dilute the quality of care they wish to enhance, perhaps even working as a means to cope with physician shortage as opposed to being a provider support.

Reduction of these risks shall lie in the judicious use of AI tools to maintain the sanctity of the human element in mental health care and difficulty incurred during times when the human touch is needed (Calvo et al., 2020).

This social enterprise must strive to tackle the root causes of workforce shortages like lacking adequate training and support for mental health professionals, then look into AI intervention as a supplement and not as an alternative.

REFERENCES

Calvo, R. M., Peters, D., Vold, V. E., & Ryan, R. A. (2020). Supporting human autonomy in AI systems: A framework for ethical enquiry. *Ethics and Information Technology*, *22*(1), 31–42. DOI: 10.1007/s10676-018-9476-2

Fitzpatrick, K., Darcy, A., & Vierhile, M. (2017). Delivering cognitive behavior therapy to young adults with symptoms of depression and anxiety using a fully automated conversational agent (Woebot): A randomized controlled trial. *JMIR Mental Health*, *4*(2), e19. DOI: 10.2196/mental.7785 PMID: 28588005

Floridi, L. (2018). Soft ethics and the governance of the digital. *Philosophy & Technology*, *31*(1), 1–8. DOI: 10.1007/s13347-018-0303-9 PMID: 30322997

Floridi, L., Cowls, J., Beltrametti, M., Chatila, R., Chazerand, P., Dignum, V., Luetge, C., Madelin, R., Pagallo, U., Rossi, F., Schafer, B., Valcke, P., & Vayena, E. (2018). AI4People—An ethical framework for a good AI society: Opportunities, risks, principles, and recommendations. *Minds and Machines*, *28*(4), 689–707. DOI: 10.1007/s11023-018-9482-5 PMID: 30930541

Gerke, S., Minssen, T., & Cohen, G. (2020). Ethical and legal challenges of artificial intelligence-driven healthcare. In *Artificial Intelligence in Healthcare* (pp. 295–336). Academic Press., DOI: 10.1016/B978-0-12-818438-7.00012-5

IEEE Global Initiative on Ethics of Autonomous and Intelligent Systems. (2019). *Ethically aligned design: A vision for prioritizing human well-being with autonomous and intelligent systems* (1st ed.). IEEE. https://standards.ieee.org/content/dam/ieee-standards/standards/web/documents/other/ead1e.pdf

Jobin, A., Ienca, M., & Vayena, E. (2019). The global landscape of AI ethics guidelines. *Nature Machine Intelligence*, *1*(9), 389–399. DOI: 10.1038/s42256-019-0088-2

Mittelstadt, B. (2019). Principles alone cannot guarantee ethical AI. *Nature Machine Intelligence*, *1*(11), 501–507. DOI: 10.1038/s42256-019-0114-4

Musen, M. A. (2003). The evolution of Protégé: An environment for knowledge-based systems development. *International Journal of Human-Computer Studies*, *58*(1), 89–123. DOI: 10.1016/S1071-5819(02)00127-1

Obermeyer, Z., Powers, B., Vogeli, C., & Mullainathan, S. (2019). Dissecting racial bias in an algorithm used to manage the health of populations. *Science*, *366*(6464), 447–453. DOI: 10.1126/science.aax2342 PMID: 31649194

Rizzo, A. S. (2019). Virtual reality exposure therapy for combat-related PTSD: A randomized clinical trial. *IEEE Transactions on Neural Systems and Rehabilitation Engineering*, *27*(12), 2399–2407. DOI: 10.1109/TNSRE.2019.2945555

Topol, A. (2019). High-performance medicine: The convergence of human and artificial intelligence. *Nature Medicine*, *25*(1), 44–56. DOI: 10.1038/s41591-018-0300-7 PMID: 30617339

Torous, J., Nicholas, J., Larsen, M., Firth, J., & Christensen, H. (2018). Clinical review of user engagement with mental health smartphone apps: Evidence, theory and improvements. *Evidence-Based Mental Health*, *21*(3), 116–119. DOI: 10.1136/eb-2018-102891 PMID: 29871870

Vayena, E., & Blasimme, A. (2018). Health research with big data: Time for systemic oversight. *The Journal of Law, Medicine & Ethics*, *46*(1), 119–129. DOI: 10.1177/1073110518766026 PMID: 30034208

World Health Organization (WHO). (2020). *Guidelines on digital interventions for health system strengthening*. WHO. https://www.who.int/publications/i/item/9789241550505

Chapter 10
Adapting AI-Driven Psychotherapy:
Cultural Challenges and Opportunities

Susmita Halder
https://orcid.org/0000-0001-6254-3324
St. Xavier's University Kolkata, India

Akash Mahato
https://orcid.org/0000-0001-9024-4022
Amity University Kolkata, India

ABSTRACT

Psychotherapy has evolved in its approach and imbibing new technologies like Artificial Intelligence (AI). It sounds promising but comes with its own challenges. The present chapter focuses on scope and implications of AI in evidence-based approaches of psychotherapy, primarily in Indian context and discussing the cross-cultural aspect of it. Providing a backdrop of cultural influences in therapy process, it highlights the trend of novel use of AI in psychotherapy practice, its compliance with the client's cultural background, and limitations. It will explore, if the rapid changes in psychosocial milieu, changing patterns of social and personal communications, as well as language complexities could be a barrier in integration of AI in the psychotherapy process. Case studies will be used, to highlight the domains which deem important for integration of AI in the psychotherapy process from a cultural perspective.

DOI: 10.4018/979-8-3373-0467-0.ch010

INTRODUCTION:

Psychotherapy has evolved manifold in its approach, focus as well as scope of implementation across different conditions in past few decades. In principle it remains an informed and intentional application of clinical methods and interpersonal stances derived from established psychological principles for the purpose of assisting people to modify their behaviors, cognitions, emotions, and/ or other personal characteristics in directions that the participants deem desirable. Irrespective of how convincing it may sound; the credibility of psychotherapy has taken years to build up and has toiled to cross cultural barriers for global acceptance. The prominent forms of psychotherapies having credible evidence base, including Cognitive Behavioral Therapy (CBT), Dialectical Behavior Therapy (DBT), Acceptance and Commitment Therapy (ACT), Mindfulness based Cognitive therapy (MBCT), as well as integrated approaches have been customized or modified globally by psychotherapists. These customizations are done primarily to draw the best suitable therapeutic plan for a client without restricting the plan to a single specific therapeutic approach, as well as to meet the cultural demands of the therapy process. These customizations have eventually helped building the evidence base of these psychotherapeutic approaches. However, these modifications may not always have been based on sound psychological theories and may pose a risk of adding erroneous data to the literature. Nevertheless, these adaptations largely help the psychotherapy process. With advancement of technology, psychotherapy process has embraced the advances in technology, including Artificial intelligence (AI). AI probably is the most important invention after internet and appears to be omnipotent having huge implications in health services, including mental health. Literature on potential use of AI in psychotherapy process are emerging with increasing evidence base (Spytska, 2025). Use; rather, attempts to use AI in the psychotherapy process are being reported in higher frequency than ever. From psychodynamic to cognitive-behavioral, followed by acceptance and commitment, mindfulness, and spirituality in psychotherapy approaches, the integration of AI in psychotherapy is likely to be the next wave of psychotherapy.

USE OF AI IN MENTAL HEALTH:

As we speak of adapting AI driven psychotherapy, it is important to understand its background. The early work on use of AI in the health sector dates back to the early 1960's. In 1972, researchers developed Internist-1, a system containing a knowledge base of causal and taxonomic relationships between clinical findings and diagnostic hypotheses, that used a powerful ranking algorithm to reach diagnoses

(Hirani et.al. 2024). However, the earlier models lacked refinement. With the development in the understanding computational power and the availability of more data helped AI scientists build more sophisticated and advanced algorithms which directs towards more precision. In mental health, AI initially restricted to diagnostic application, and assisting in problem diagnosis. To mention in this regard, the DENDRAL project (Smith et.al., 1981) demonstrated the potential of AI used in complex problem-solving, which was later adapted to include diagnostic applications in various fields, including medicine and later in psychiatry.

Today, AI in mental health is no longer just a conceptual possibility but an actively evolving and applied field. Modern AI systems utilize extensive datasets, such as electronic health records and patient-reported outcomes, to aid in diagnosing conditions like depression, anxiety, and schizophrenia. Machine learning (ML) models analyse these data to detect patterns indicative of mental health disorders, often achieving diagnostic accuracy comparable to human clinicians (Iyortsuun, 2023; Madububambachu, Ukpebor, & Ihezue, 2024). Furthermore, AI-driven tools are becoming integral to clinical practice, enabling continuous patient monitoring, predicting treatment responses, and tailoring therapeutic interventions to individual needs. Despite these advancements, challenges remain in fully integrating AI into mental health care.

While AI has shown promise in enhancing diagnostic accuracy and efficiency, challenges remain, particularly regarding the generalizability of AI models across diverse populations and the ethical implications of AI-driven diagnoses. The field continues to grapple with issues related to data privacy, algorithmic bias, and the need for robust validation studies to ensure that AI tools complement rather than replace human judgment. Especially, cultural bias could lead to mismatches in diagnosis and treatment recommendations, affecting the effectiveness and acceptance of AI tools in diverse settings. Moreover, integrating culturally responsive AI in psychotherapy demands ongoing collaboration with local communities and experts to ensure the tools align with cultural beliefs, values, and mental health practices. For instance, culturally adaptive AI could incorporate culturally specific therapeutic practices, such as storytelling or communal healing approaches, and respect culturally bound syndromes, which might be overlooked in a one-size-fits-all AI model. This inclusivity allows AI in psychotherapy to serve as a meaningful support for people across different backgrounds, thereby fostering greater acceptance and trust. Also, designing culturally attuned AI systems in psychotherapy is not only a matter of data inclusion but also requires culturally literate developers who can anticipate and address complex cultural dynamics.

AI-Driven Psychotherapy:

The perception of AI regarding its capabilities and scope is vivid among population, and thus when we speak of AI driven psychotherapy, this may mean different to different layman people (Kotsona, 2024). Psychotherapy primarily had a western origin and took its own time for acceptance across other parts of the world. There will always be apprehension how a procedure which requires human expertise, can be executed by a machine. Thus, the perception of a layperson regarding AI can vary widely (Varghese, Sharma & Patwardhan, 2024).

Some may imagine AI completely replacing a human therapist and doing all the work.

Others may view AI as a supportive tool that helps therapists better understand the therapeutic need of the patient. And some may have no idea at all — a completely blank impression. Probably this could be the reason behind varying degree of trust on use of AI in psychotherapy process. Studies do suggest that study participants, who had highly rated AI based mental health support without knowing its origin, shifted their perception towards human responses upon knowing the origin of the support (Jain, Pareek & Carlbring, 2024).

How AI has been used in psychotherapy process: The reason behind keeping the section heading in present continuous is that integration of AI is an evolving phenomenon. There is vividity in the purpose as well as method employed, though some basic commonality remains in these methods. One of the core principles of AI functioning is identifying patterns in data. This is a fundamental principle behind most AI and machine learning systems. Further AI algorithms are used to assist therapists or professionals in making diagnosis, record keeping, as well as providing basic support independently to the clients. But what is 'data' in psychotherapy process? The answer is almost every information is data, which can be fed, rather can be auto fed to AI based systems. Interestingly AI based systems identify pattern in many other sources which human professionals may ignore or do not consider significant. The table below (Tab 1) gives a generic summary of data types and its source which can be used in AI driven psychotherapy.

Table 1. Data in Psychotherapy process, source and utility

Category of data and source	Use in Psychotherapy
Verbal & Text data: Session transcripts, chat logs, Self-reports, journal entries	Analysing language patterns, mood shifts, thought distortions
Audio & Speech: Voice recordings, tone, pitch, pauses	Detection and validation of emotional state, stress level inference, analysis of engagement in the therapy process
Video & Facial: Facial expressions, gaze tracking, micro expressions which might go unnoticed by human professionals	Detecting non-verbal emotional cues, discomfort, identifying early signs of crisis like self- harm or suicidal behavior
Behavioral & Interaction: App usage logs and pattern, response time, engagement levels	Tracking adherence to therapy, motivation levels, behavioral patterns
Psychometric & Clinical: Screening tool (e.g. PHQ-9, GAD-7), clinical notes, diagnosis history	Progress tracking, monitoring of symptom severity, adaptation to treatment
Physiological: Heart rate, sleep data, stress indicators from wearable devices	Identifying stress triggers, physiological impact of interventions
Environmental & Contextual: Time of interaction, geolocation, weather, life events	Enhancing context-aware responses, predicting mood variations depending on past records

The commonly used AI systems in psychotherapy include Natural language processing (NLP), Machine Learning, Sentiment Analysis, and Generative AI. Like Natural Language processing models find patterns in word usage, syntax, and context to understand and generate human-like text. Sentiment analysis identifies emotional tone (e.g. anxiety, sadness) in text. Generative AI generates free-form empathetic or therapeutic responses which are crucial for mental health chatbots.

Thus, when AI is augmented in the psychotherapy process, it looks into patterns in a patient's facial expressions, emotional tone speech and language, and other relevant data which is matched and analysed by existing data sets to provide input regrading probabilities of treatment outcome, diagnostic possibilities or validating therapist ideas, or simply in automating the process.

A very simple example could be a patient with depressive symptoms, who is hesitant to consult a mental health professional, as he is unclear whom to connect. He looks for help from an AI based chatbot. The chatbot interacts with the patient, and with analysis of the text transcript, it cites risk of suicide and advise the patients managing the immediate crisis and further advises on selecting an appropriate therapist for him as per his geographical location. This process does not undermine the importance of the human therapist but saves time and give desired information to the patient. A similar set of the AI system can log the communication between the therapist and the patient when in real session and even identify patterns in the transcript reflecting subtle suicidal thoughts, or his core beliefs (Figure 1). Again, the AI system only plays a supportive role, but the therapist is definitely saved from efforts of manually writing as well as recapitulating the session (Halder, 2025).

Figure 1. Implications of AI augmentation in psychotherapy process

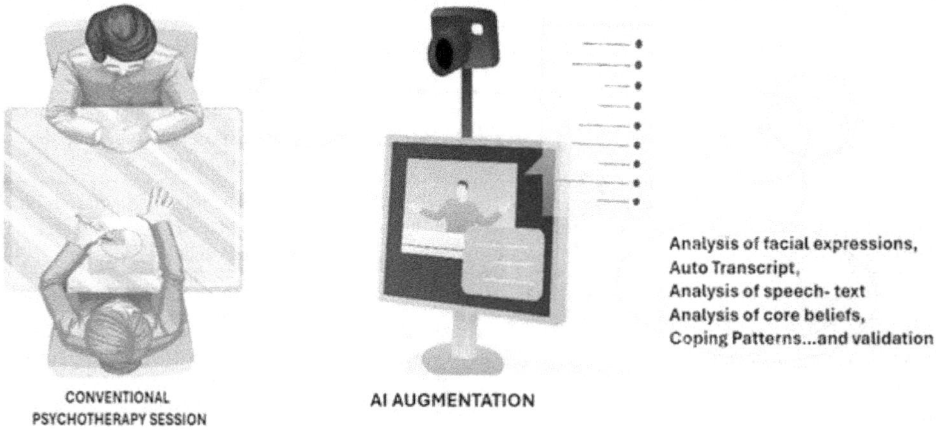

CULTURE AND MENTAL HEALTH

While we speak of incorporating AI in psychotherapeutic process; how easy and smooth would be this augmentation of AI in the psychotherapy process? What could be the potential challenges, especially from cultural perspective? It is important to understand the entwining of culture into our psyche and its influence on our mental health too. Mental health is a critical component of one's overall well-being, influencing how individuals think, feel, and behave in daily life. However, mental health cannot be understood in isolation from cultural contexts, as culture shapes perceptions, expressions, and responses to mental distress. Cultural values, belief systems, traditions, and social norms play a significant role in shaping how individuals conceptualize mental health, seek help, and respond to treatment. Sigmund Freud, often dubbed as the father of modern psychotherapy highlighted the role of conflict between different impulses, drives, and motives and supressed desires in the genesis of 'neurosis'. Freud's contemporary understanding of the suppressed desires may be attributed to then Victorian era's strict sexual morality, which has changed over decades. Even though psychoanalysis appear to have lost contemporary relevance to some extent; it does imply the role and impact of cultural influences in the genesis of psychological symptoms or psychopathology in an individual. Thus, if culture affects the genesis of our symptoms, it is deemed appropriate to be considered when planning its management, including when integrating AI in the process. Understanding of cultural influences in mental health is important for human professionals, thus equally applicable for AI based platforms. The following section delves upon influence of culture on our health behaviours, health seeking behaviour, perception of mental health and related aspects which deem important to be considered when attempting to integrate AI in psychotherapy process.

Cultural variations in perception of mental ill health symptoms and expression of symptoms: Even though international classificatory system like the ICD is uniform in its approach to diagnose mental ill health conditions; a behavior considered as "normal" in one culture may not be a very appropriate or normal behavior in other. Through social learning, culture also shapes our ways of coping we use when under psychological distress. One of the primary cultural influences in mental health is how patients describe their symptoms to their therapists, as there are some well recognized differences in symptom presentation across cultures. There are known variations in symptoms of somatization, the expression of distress through one or more physical symptoms as per ethnicity of the patients. In some Asian and Middle Eastern cultures, emotional restraint is valued, whereas in Latin American and Mediterranean cultures, emotional expressiveness is more common. Understanding these differences helps therapists avoid misinterpreting cultural norms as psychological symptoms. Example, Asian patients are more likely to report their somatic symptoms, while unintentionally not reporting their emotional symptoms. However, they do acknowledge having emotional symptoms when probed further (Lin & Cheung, 1999).

Cultural variations in Language and communication and its impact: Language and communication styles may affect mental health care access. Some cultures discourage open discussions about emotions, making traditional talk therapy less effective. In such cases, alternative therapeutic approaches, such as expressive arts therapy, group interventions, or culturally adapted counselling techniques, may be more suitable. Culturally competent psychotherapy considers language, traditions, and belief systems when designing interventions. For instance, mindfulness-based interventions, which have roots in Eastern philosophy, have been adapted for Western psychotherapy. Similarly, family-based therapy is more effective in cultures that prioritize familial relationships. Language and communication also play role in how AI-based mental health interventions are perceived as well as used by therapists (Zhang & Wang, 2024). AI models must account for linguistic and cultural differences in how people express distress. For example, direct expressions of emotions may be common in Western cultures but less frequent in East Asian societies. Socioeconomic disparities also affect access to AI-driven mental health care. Low-income populations may lack the necessary technological infrastructure, limiting the effectiveness of AI interventions.

Cultural beliefs regarding cause of mental ill health and stigma**:** Cultural beliefs about the causes of mental illness also impact how people seek help. In some societies, mental health issues are attributed to supernatural forces, spiritual imbalances, or moral failures, leading individuals to seek traditional healers rather than professional mental health practitioners. In contrast, biomedical explanations dominate Western mental health discourse, promoting psychiatric interventions and psychotherapy. Cultural stigma surrounding mental illness remains a significant barrier to seeking help. In many cultures, mental health issues are perceived as a sign of weakness or a source of shame, discouraging individuals from openly discussing their struggles. This stigma can lead to social isolation, delayed treatment, and poor health outcomes. Efforts to reduce stigma should incorporate culturally appropriate messaging that resonates with different communities. These propositions on difference in pattern of mental health seeking is supported by relevant data. Disparities in poor utilization and quality of mental health treatments for African Americans, American Indians, Asian Americans, and Latino/as have been documented for many decades, and they continue to persist (Smedley et al. 2003).

Familial systems across cultures and mental health: In collectivist societies, e.g. South Asia, Africa, and Latin America, mental health is often viewed through the lens of family and community well-being rather than individual pathology. This means that mental health concerns are often managed within family structures, with individuals seeking guidance from elders, spiritual leaders, or community members before turning to professional help. For example, in India, families play a crucial role in mental health care, and seeking therapy may only occur if the family collectively deems it necessary. Similarly, in some African cultures, traditional healing practices that involve community rituals and ancestral guidance are commonly used alongside or instead of conventional therapy. In contrast, individualistic cultures, such as those in North America and Western Europe, tend to emphasize personal autonomy and self-expression in mental health treatment. People are encouraged to seek therapy independently, and there is a stronger focus on personal growth and self-improvement. For instance, in the United States, cognitive-behavioral therapy (CBT), which promotes self-awareness and individual coping strategies, is widely practiced and aligns with cultural values of independence. Conversely, this approach may not be as effective in collectivist societies, where interdependence is a fundamental aspect of identity and healing may require involving family or community support systems for better outcome.

Family involvement is a crucial aspect of therapy. Even though family is considered to be cohesive and protective of its members across the globe, communication patterns differ across cultures. Decisions regarding family involvement, group therapy, or structured interventions should also align with cultural norms. In the Indian context, psychotherapy must be approached with a deep understanding of cultural values, family dynamics, and societal expectations. Indian patients often come from collectivist backgrounds where family, community, and tradition play a significant role in shaping emotional expression and mental health perceptions. Many individuals are raised to prioritize social harmony over personal emotions, leading to reluctance in openly discussing mental distress. This necessitates a therapeutic approach that balances cultural sensitivity with the encouragement of self-expression.

Indian families often have strong intergenerational ties, and therapy that incorporates parental engagement can be more effective, especially for younger patients. However, the approach should be flexible—while some families benefit from structured family therapy, others may require a more indirect method where individual therapy subtly integrates family values without direct confrontation. Overall, psychotherapy in India should be adapted to the patient's cultural background, ensuring that therapeutic interventions respect traditions while gently introducing alternative ways of thinking and coping. Integrating cultural sensitivity with evidence-based practices can help create a more effective and accepting mental health framework in the Indian context. There are possibilities of certain roadblocks also on part of the familial structures. While early maladaptive schemas are often related with depressive symptoms (Ghosh & Halder, 2024), it is a possibility as well as observation from clinical practise that patients from India struggle to report significant early childhood experiences that could be significant in shaping their thoughts, ideas and core beliefs.

Culture and Trust factor- Cultural perspectives shape how individuals perceive AI-driven solutions in healthcare, and understanding these variations is crucial for implementing technology-based mental health interventions effectively. Trust in technology is often influenced by historical, social, and economic factors (Ybarra & Turk, 2009; Li, Peng & Rheu, 2024). In societies with a strong tradition of scientific progress and technological innovation, such as the United States, Germany, and Japan, people are generally more open to AI-driven healthcare solutions. In contrast, cultures with a history of scepticism towards technology, government surveillance, or ethical concerns, such as some parts of Africa, Latin America, and Eastern Europe, may exhibit greater hesitation in adopting AI-based mental health interventions. Cultures with a strong emphasis on personal privacy may resist AI-driven mental health monitoring. Some cultures may prefer human therapists over AI interventions due to beliefs about personal connection and empathy. In contrast, cultures with high digital literacy may be more open to virtual therapy solutions.

Religious and philosophical traditions also shape attitudes toward AI. In some Asian cultures influenced by Confucianism and Buddhism, technology is often seen as a tool for progress and harmony. Meanwhile, Western societies influenced by humanistic philosophy may be more cautious about replacing human judgment with machine intelligence, especially in sensitive areas like mental health. AI-powered mental health tools, such as chatbots, predictive analytics, and virtual therapy platforms, offer promising solutions for expanding access to care. These tools can provide early intervention, reduce the burden on healthcare systems, and reach underserved populations. However, cultural factors must be considered when designing and implementing AI solutions (Halder, 2022).

Table 2. Key cultural barriers for AI driven psychotherapy

Key Cultural barriers for AI driven psychotherapy	Potential AI adaptations
Perception of mental ill health symptoms and expression of symptoms	• Large data sets required to identify these cultural influences unique for individual client. • Provision of multilingual support with local dialect. • Generate synthetic conversations with culturally nuanced prompts using advanced LLMs under expert guidance. • Provision of continuous feedback from local users and clinicians.
Language and communication	
Beliefs regarding cause of mental ill health and stigma	

Familial systems

Special conditions requiring cultural competence: While every mental health condition is unique, certain conditions, e.g. grief require cultural competence to engage client for psychotherapy. When addressing grief in individuals from different cultural backgrounds it is crucial to recognize how their communities view mourning, support systems, and emotional expression. Death has distinct philosophical meaning in different cultures, as some communities may have spiritual or ancestral beliefs influencing their grieving process, some culture may emphasize communal support, and some may rely on faith and resilience. Tailoring interventions to respect these perspectives can enhance therapeutic outcomes. On similar lines, grief processing in Indian culture is heavily influenced by religious and spiritual beliefs, whether Hindu, Muslim, Sikh, Christian, or other traditions. Rituals, communal mourning, and ancestral reverence often play a key role in the grieving process. Therapy should acknowledge these aspects, incorporating culturally relevant coping strategies such as guided rituals, storytelling, or spiritual counselling. For women, particularly widows or bereaved mothers, grief may also intersect with societal roles and expectations, requiring an approach that addresses both emotional healing and social empowerment. When working with Indian adolescents, socioeconomic factors, educational pressures, and family obligations must be considered. An underprivileged Indian adolescent may face challenges related to caste, economic hardship, or limited access to mental health resources, making community-based, action-oriented therapy more effective. In contrast, an affluent Indian adolescent may struggle with high parental expectations and academic stress, necessitating an approach that incorporates stress management, career counselling, and emotional resilience training. Thus, a culturally informed approach to mental health care is essential for addressing disparities in access to services, reducing stigma, and ensuring effective interventions.

Culture and Psychotherapy:

Contemporary mental health workers have come to realize that cultural dimensions cannot be ignored in the practice of psychotherapy – a clinical process that is primarily focused on a person's emotional life and utilizes psychological methods to resolve psychological problems and improve the quality of mental life. Psychotherapy, when viewed through a cultural lens, reveals diverse therapeutic approaches that vary in their depth of cultural integration. Some practices are deeply embedded within cultural traditions, utilizing indigenous healing methods rooted in rituals, spiritual beliefs, and communal support. Others are strongly influenced by cultural values but maintain unique frameworks that distinguish them from purely traditional methods. Additionally, there are therapies that, while related to culture, are not fundamentally shaped by it and instead follow widely accepted psychological principles. The effectiveness of psychotherapy often depends on its cultural relevance, as individuals' beliefs, values, and social norms influence their perception of mental health and healing. As societies become more interconnected, intercultural therapy has gained prominence, emphasizing the need for therapists to integrate cultural sensitivity into their practice. A culturally informed approach to psychotherapy not only enhances its effectiveness but also fosters greater acceptance and accessibility across diverse populations.

As stigma around mental health remains prevalent in many Indian communities, it is essential to frame therapy in a way that aligns with culturally accepted concepts such as stress management, holistic well-being, or even spiritual growth. In group therapy settings, Indian patients, particularly those from conservative backgrounds, may struggle with openly discussing personal struggles. Gradual exposure, culturally appropriate self-disclosure, and the inclusion of structured discussion formats may encourage participation while respecting personal boundaries. To improve mental health care globally, it is essential to integrate cultural perspectives into policies, education, and service delivery. This includes training mental health professionals in cultural competence, expanding access to culturally relevant mental health resources, and fostering community-based mental health initiatives. By acknowledging the interplay between culture and mental health, we can develop more inclusive, effective, and accessible mental health care systems.

As we highlight on importance of culture in the psychotherapy process, it might raise a question whether it is best to have a psychotherapist from same cultural and ethnic background as of the patient to have a good treatment outcome and there are interesting findings. Literature suggests ethnic minority clients tend to have more confidence and prefer consultation from a therapist of the same race/ethnicity (Cabral & Smith, 2011; Meyer & Zane, 2014). Interestingly, clients also rated their therapists as more supportive if they perceived their therapists to have similar life experiences, attitudes, values, and personality, regardless of their race or ethnicity (Meyer et al, 2011). This perception was positively linked with working alliance and therapist credibility (i.e., how credible clients perceive their therapists). However, research like these further suggest that a common racial match of the patient and therapist may influence treatment processes in terms of better or long-term retention and follow-up, but may not significantly affect the outcomes of treatment, in terms of symptom or distress reduction. Thus, racial or ethnic match between patient and therapist may help build positive perceptions of the therapists, but other factors (e.g., cultural competence or counselling style of the therapist) may have more direct effects on treatment outcomes (Cabral & Smith, 2011b).

Case presentation:

We present a summarized case of Ms Anna (name changed) highlighting the complexities of using AI in psychotherapy process. Anna is a college girl, who has been struggling with her depressive mood which were fluctuating in nature. In addition, she had turbulent relations with her close friends, on whom she was often dependent for emotional support, but at times she behaved indifferently with them. She occasionally smoked weed and took alcohol and was confronted by her father due to this. She trusted only a few people and was always at loggerhead with her parents. She had consulted a psychologist a year back but did not like her approach and dropped from therapy sessions. She came across an online platform which provided mental health services. It promised complete anonymity and convenience of getting support she desired whenever and wherever she wanted. She interacted with the chatbot on the portal and tried to describe her problems. Initially she was hesitant, to share her troublesome thoughts apprehending if these were being recorded but gradually became comfortable. She liked the comfort of interacting with the platform at her convenience, but her initial relief did not last long. The chatbot appeared to repeat the same questions again and again. In her words, "*it asked questions, in a different way I already answered, and was frustrating. I didn't feel reassured*". Anna tried to talk about her mood, and her problem with his family and habit of taking substance. At times she wanted to vent out, but in turn the bot after expressing sympathetic words, suggested on getting consultation from a psychiatrist and asked if she wanted references. Overall, the bot appeared to be unable to navigate through complex therapeutic requirements of Anna. Again, in her words, "*it was similar to the online bots of customer care of an airlines*". Incidentally Anna could find many advertisement pop ups in her mobile related to mental health services, which were irritating to her. Eventually Anna dropped from the portal. Anna was further referred to another psychologist by her friend. The new therapist took into account her full history including her previous consultations and have detailed psychoeducation regarding her condition and desirable therapeutic processes, some of which might be repetitive. A customized psychotherapeutic formulation was made and discussed with her, which appeared very convincing to her. Anna was able to develop better rapport with her therapist, and she was more regular for her follow ups. She reported gradual improvement in her symptoms as well as functioning and is on regular follow up.

So, do we infer that AI failed in assisting and providing mental health support to Anna? Probably not. We must understand the present level of use and utility of AI based mental health support chatbots and platforms. In this case, we understand the limitations of these platforms in understanding complex human interactions within family and emotional needs of an individual. Example, when Anna is distraught about her father's behaviour towards her; the bot probably is talking about her substance taking behaviour and advising regarding that. In most likely hood these bots or platforms are not yet developed to collect personal and clinical history which helps therapist to formulate therapy plan in real one to one session. Thus, the responses appear monotonous, lacking personal warmth. There might be many more cases like Anna, which somewhere hint that while AI is a definite welcome assistance, there is a need of refinement to understand the vividity in manifestation of mental health symptoms in a single individual. Comorbidity is a rule rather than exception in psychiatry. It is doubtful whether AI can navigate the complexities of comorbidity and analyse appropriate therapeutic plan for patients like Anna.

It may be inferred that existing AI based therapeutic platforms in their current form appear beneficial in giving basic mental health support by engaging clients through empathetic dialogue, providing better insight into their conditions, correcting or modifying thought patterns which could be crucial in emergency situations like suicidal thoughts and also provide desired referral with a mental health professional. Further, for AI to be successfully integrated into mental health care, it must be adapted to different cultural contexts. This can be achieved by developing culturally informed AI models trained on diverse linguistic and cultural datasets to improve accuracy and relevance. Community engagement is crucial, as involving local communities in the design and implementation of AI mental health tools can build trust and increase acceptance. A hybrid approach, combining AI interventions with human therapists, can balance efficiency with cultural sensitivity. Policymakers must also ensure ethical AI use in mental health care, addressing concerns related to bias, privacy, and accessibility. As AI continues to evolve, it has the potential to revolutionize mental health care globally. However, its success depends on culturally sensitive implementation that respects diverse perspectives on mental health, privacy, and human interaction. By integrating cultural considerations into AI-driven mental health solutions, we can create more effective, inclusive, and trusted interventions that benefit people worldwide.

Challenges beyond culture:

Apart from the challenges that cultural variables brings to AI driven psychotherapy, there are other challenges too.*Data sources and challenges:* The incorporation of Artificial Intelligence (AI) into psychiatric practice offers substantial opportunities, particularly in leveraging diverse data sources. However, it also brings considerable challenges, including concerns about data privacy, security, and potential biases.

AI can analyse vast amounts of patient data within Electronic Health Records (EHRs) to identify patterns and predict mental health conditions, improving diagnostic accuracy. However, the complexity of psychiatric data, combined with the variability in how symptoms are documented across different healthcare providers, can introduce inconsistencies. These inconsistencies, if not properly managed, and individual differences are not considered may lead to inaccurate or even wrong predictions or diagnoses. When utilizing patient data in AI-driven mental health care, it is crucial to consider cultural variations that influence data interpretation, privacy expectations, and ethical concerns. The integration of AI with EHRs often faces challenges related to data interoperability, diverse data formats, and large volumes of information. These challenges become even more complex when factoring in cultural differences in how mental health symptoms are expressed, documented, and perceived within different communities.

Another limitation of AI in similar lines comes in terms of availability of large data sets. Even though generative AI attempts to resolve on this, a doubt remains if an AI system developed using data set of western population would be competent enough and equally fit for eastern population, and vice versa. Thus, for generalization and wide scale use of a uniform AI system a very large data set is required, which can balance the cultural differences across patients' profile.

Patient-generated data from sources such as social media, mobile apps, and wearable devices provide valuable insights into mental health. However, cultural norms significantly impact how individuals communicate distress, seek help, and engage with digital platforms. AI models must be trained on diverse datasets to avoid biases that could lead to misinterpretations or misdiagnoses in populations with varying linguistic, behavioral, or social patterns. Standardization efforts should account for cultural differences in symptom expression, healthcare access, and attitudes toward mental illness.

Data privacy and security concerns also vary across cultures, with differing expectations regarding consent, data sharing, and confidentiality. AI systems must be designed to navigate these variations while ensuring compliance with global and local regulations, such as General Data Protection Regulation (GDPR), to protect patient rights. Additionally, ethical AI implementation requires collaboration with culturally diverse stakeholders, including patients, clinicians, and policymakers, to ensure that AI-driven psychiatric tools are both effective and culturally sensitive.

Challenges in AI implementation

The implementation of AI in mental health is fraught with challenges, particularly when considering cultural differences that impact its real-world applicability (Grzenda, 2021). One of the key difficulties lies in the clinical validation of AI models across diverse cultural settings. While AI-driven diagnostic tools have demonstrated promise in controlled studies, their effectiveness in real-world psychiatric practice depends on their ability to generalize across different patient populations (Eskandar, 2024). Mental health symptoms, help-seeking behaviors, and even diagnostic criteria can vary significantly across cultures, making it essential to validate AI models in culturally specific contexts. Without this, there is a risk of misdiagnosis or ineffective interventions due to biases embedded in the training data.

Regulatory and ethical challenges further complicate AI adoption in mental health care (Mennella et.al., 2024). Many existing frameworks do not adequately address cultural nuances in data privacy, consent, and ethical AI use (Nasir et.al., 2024). For instance, some cultures may have distinct perspectives on mental health disclosure, stigma, or reliance on traditional healing practices, which AI models must account for to be both effective and respectful (Gupta et.al.,2023). The opacity of some AI algorithms, especially deep learning models, makes it difficult to ensure transparency and accountability, raising concerns about trust and acceptance in communities where skepticism toward technology is prevalent.

Moreover, AI has the potential to exacerbate healthcare disparities if cultural differences are not carefully considered during model development and implementation (Dankwa-Mullan et.al., 2021). A lack of representation from diverse populations in training datasets can lead to biased outcomes, where AI systems may perform well in some cultural groups but fail in others (Hanna et.al., 2024). To ensure equitable mental health care, AI models must be developed and tested in collaboration with culturally diverse stakeholders, integrating perspectives from patients, clinicians, and policymakers (Morgan & Cogan, 2024). Only through such inclusive approaches can AI truly enhance mental health services across varied cultural landscapes.

Opportunities and future directions

To ensure AI-driven mental health care is both effective and culturally inclusive, future efforts must prioritize the development of models that account for cultural variations in symptom expression, help-seeking behaviors, and treatment preferences. AI should be trained on diverse datasets representing different ethnic, linguistic, and socio-economic backgrounds to minimize biases and improve generalizability. Additionally, collaboration with local mental health professionals and community stakeholders is essential to tailor AI applications to specific cultural contexts. Ethical considerations, such as culturally appropriate consent processes and transparency in AI decision-making, should also be integrated into regulatory frameworks. By embracing these strategies, AI can enhance personalized mental health care while respecting cultural diversity, ensuring its benefits reach all populations equitably.

CONCLUSION

In conclusion, the integration of Artificial Intelligence (AI) in mental health is progressing rapidly, presenting promising opportunities to enhance diagnostic precision, tailor treatments, and improve overall patient outcomes. There is no doubt on AI playing a growing role in diagnostics, treatment, and support systems in mental health including the psychotherapy process. However, trust in technology varies across cultures, influencing its acceptance and effectiveness in different societies. Cultural context does influences how mental health issues are perceived and addressed, which AI needs to respect and adapt to effectively serve varied communities. This involves training AI models to recognize and honour cultural differences in emotional expression, coping mechanisms, and communication styles.

However, the adoption of AI in psychiatric practice faces challenges such as data privacy concerns, algorithmic biases, and the necessity for thorough clinical validation. Nevertheless, case studies involving AI applications in specific psychiatric disorders highlight its ability to revolutionize diagnosis and treatment. Expanding on the cultural aspect of AI in psychotherapy, the ethical implications and potential biases inherent in AI algorithms should also be considered. AI systems trained primarily on data from Western contexts may not capture the nuances of other cultural perspectives, risking misinterpretation of emotional expressions or mental health indicators, and thus need to incorporate cultural perspective. As AI technology advances, its role in mental health is set to grow, offering more personalized, effective, and accessible mental health care, provided that ethical considerations and clinical integration are carefully addressed.

REFERENCES

Cabral, R. R., & Smith, T. B. (2011). Racial/ethnic matching of clients and therapists in mental health services: A meta-analytic review of preferences, perceptions, and outcomes. *Journal of Counseling Psychology*, *58*(4), 537–554. DOI: 10.1037/a0025266 PMID: 21875181

Dankwa-Mullan, I., Scheufele, E. L., Matheny, M. E., Quintana, Y., Chapman, W. W., Jackson, G., & South, B. R. (2021). A proposed framework on integrating health equity and racial justice into the artificial intelligence development lifecycle. *Journal of Health Care for the Poor and Underserved*, *32*(2), 300–317. DOI: 10.1353/hpu.2021.0065

Eskandar, K. (2024). Artificial intelligence in psychiatric diagnosis: Challenges and opportunities in the era of machine learning. *Revista Debates em Psiquiatria*, *14*, 1–16. DOI: 10.25118/2763-9037.2024.v14.1318

Ghosh, D., & Halder, S. (2024). The Role of Social Cognition in Depression and Its Relationship to Early Maladaptive Schemas. *Indian Journal of Psychological Medicine*, *02537176241233632*. DOI: 10.1177/02537176241233632 PMID: 39564327

Grzenda, A. (2021). Artificial intelligence in mental health. *Converg. Ment. Heal. A Transdiscipl. Approach to Innov*, 137.

Gupta, P., Muneshwar, K. N., Juganavar, A., & Shegekar, T. (2023). Beyond the Asylum Walls: Tracing the Tapestry of Mental Health Interventions Across Eras and Cultures. *Cureus*, *15*(11). DOI: 10.7759/cureus.48251 PMID: 38054143

Halder, S. (2022). Application of virtual reality in cognitive rehabilitation: a road ahead. In *Emerging Advancements for Virtual and Augmented Reality in Healthcare* (pp. 217–232). IGI Global. DOI: 10.4018/978-1-7998-8371-5.ch013

Halder, S. (2025). Developing Mental Health Support Chatbots in India: Challenges and Insights. *Annals of Indian Psychiatry*, *9*(1), 99–101. DOI: 10.4103/aip.aip_187_24

Hanna, M., Pantanowitz, L., Jackson, B., Palmer, O., Visweswaran, S., Pantanowitz, J., & Rashidi, H. (2024). Ethical and Bias Considerations in Artificial Intelligence (AI)/Machine Learning. *Modern Pathology*, 100686. PMID: 39694331

Iyortsuun, N. K., Kim, S. H., Jhon, M., Yang, H. J., & Pant, S. (2023, January). A review of machine learning and deep learning approaches on mental health diagnosis. [). MDPI.]. *Health Care*, *11*(3), 285. PMID: 36766860

Jain, G., Pareek, S., & Carlbring, P. (2024). Revealing the source: How awareness alters perceptions of AI and human-generated mental health responses. *Internet Interventions : the Application of Information Technology in Mental and Behavioural Health, 36*, 100745. DOI: 10.1016/j.invent.2024.100745 PMID: 38746873

Kotsona, M. (2024). Virtue in Virtuality: A Cultural Analysis on Artificial Intelligence Companionship and its. *Ethical Dimensions.*

Li, L., Peng, W., & Rheu, M. M. J. (2024). Factors Predicting Intentions of Adoption and Continued Use of Artificial Intelligence Chatbots for Mental Health: Examining the Role of UTAUT Model, Stigma, Privacy Concerns, and Artificial Intelligence Hesitancy. *Telemedicine journal and e-health: the official journal of the American Telemedicine Association, 30*(3), 722–730. DOI: 10.1089/tmj.2023.0313

Lin, K. M., & Cheung, F. (1999). Mental health issues for Asian Americans. *Psychiatric Services (Washington, D.C.), 50*(6), 774–780. DOI: 10.1176/ps.50.6.774 PMID: 10375146

Madububambachu, U., Ukpebor, A., & Ihezue, U. (2024). Machine Learning Techniques to Predict Mental Health Diagnoses: A Systematic Literature Review. *Clinical Practice and Epidemiology in Mental Health, 20*(1), e17450179315688. DOI: 10.2174/0117450179315688240607052117 PMID: 39355197

Mennella, C., Maniscalco, U., De Pietro, G., & Esposito, M. (2024). Ethical and regulatory challenges of AI technologies in healthcare: A narrative review. *Heliyon, 10*(4), e26297. DOI: 10.1016/j.heliyon.2024.e26297 PMID: 38384518

Meyer, O., Zane, N., & Cho, Y. I. (2011). Understanding the psychological processes of the racial match effect in Asian Americans. *Journal of Counseling Psychology, 58*(3), 335–345. DOI: 10.1037/a0023605 PMID: 21574698

Meyer, O. L., & Zane, N. (2013). The influence of race and ethnicity in clients' experiences of mental health treatment. *Journal of Community Psychology, 41*(7), 884–901. DOI: 10.1002/jcop.21580 PMID: 25400301

Morgan, P., & Cogan, N. A. (2024). Using artificial intelligence to address mental health inequalities: Co-creating machine learning algorithms with key stakeholders and citizen engagement. *Journal of Public Mental Health.*

Nasir, S., Khan, R. A., & Bai, S. (2024). Ethical framework for harnessing the power of ai in healthcare and beyond. *IEEE Access : Practical Innovations, Open Solutions, 12*, 31014–31035. DOI: 10.1109/ACCESS.2024.3369912

Smith, D. H., Gray, N. A., Nourse, J. G., & Crandell, C. W. (1981). The DENDRAL project: Recent advances in computer-assisted structure elucidation. *Analytica Chimica Acta*, *133*(4), 471–497. DOI: 10.1016/S0003-2670(01)95414-5

Spytska, L. (2025). The use of artificial intelligence in psychotherapy: Development of intelligent therapeutic systems. *BMC Psychology*, *13*(1), 175. DOI: 10.1186/s40359-025-02491-9 PMID: 40022267

Varghese, M. A., Sharma, P., & Patwardhan, M. (2024). Public Perception on Artificial Intelligence-Driven Mental Health Interventions: Survey Research. *JMIR Formative Research*, *8*, e64380. DOI: 10.2196/64380 PMID: 39607994

Ybarra, C. E., & Turk, T. A. (2009). The evolution of trust in information technology alliances. *The Journal of High Technology Management Research*, *20*(1), 62–74. DOI: 10.1016/j.hitech.2009.02.003

Zhang, Z., & Wang, J. (2024). Can AI replace psychotherapists? Exploring the future of mental health care. *Frontiers in Psychiatry*, *15*, 1444382. DOI: 10.3389/fpsyt.2024.1444382 PMID: 39544371

Chapter 11
Future Trends in AI and Wearable Technology for Mental Health

Aakanksha Mucherla
https://orcid.org/0009-0006-1704-1329
Woxsen University, India

Shreya Mettu
Woxsen University, India

T. Nithya
Woxsen University, India

ABSTRACT

It is estimated that millions of persons are affected by the ongoing global mental epidemic. Wearable technologies, such as AI systems, provide advanced solutions for mental healthcare, which helps in medical diagnosis, monitoring, and therapeutic treatment. This section focuses on the development of mental health AI solutions based on wearable devices by analyzing real-time treatment monitoring and prediction treatment algorithms. The mental health therapy market integrates new technology with emerging trends such as smart fabrics combined with biometric sensors, as well as Augmented and Virtual Reality applications. In addition, to the case studies, the chapter addresses these challenges, including user engagement, AI bias, and legal regulations that must be overcome in order to deploy AI technologies solutions at a larger scale. AI powered wearable technologies can radically change mental health treatment. Chapter scans covers ethical use in case study research with focus on future developments.

DOI: 10.4018/979-8-3373-0467-0.ch011

1. INTRODUCTION

Global attention on mental health has become a major international crisis which heavily impacts individuals throughout different societies. Statistics from Hope Chest (n.d.) show that 970 million people suffer from mental health issues or substance abuse disorders creating a staggering crisis. According to the World Health Organization statistics one quarter of the global population will face mental disorders at some point during their lives and depression together with anxiety disorders represent the most common mental health conditions (SingleCare, n.d.). On a global level depression currently afflicts more than 300 million people and anxiety disorders impact 284 million individuals (IHME, n.d.). According to Project HOPE (n.d.), suicide now ranks as the fourth biggest killer among people between 15 and 29 years old.

A recent study shows that mental illness resulted in an economic expense of $2.5 trillion during 2010 which research forecasts will grow to $6 trillion every year by 2030 (IHME, n.d.). The primary 54% stakeholders of mental health disorders live in lower and middle-income nations which have little access to essential mental care services (Project HOPE, n.d.). The regions face a major care access disparity as psychological treatment reaches only one percent of the population who needs help.

Through the structured assessment of psychological health technology shines as a promising solution for this difficult period. Mental health care receives revolutionary improvements through artificial intelligence and wearable technology advances according to Apexon (n.d.). Modern technology provides real-time tracking capabilities for essential physiological signs which monitor heart rate and sleepqrtacles and mood changes to offer better mental health insights (RoboticsBiz, n.d.). Through AI machine learning processes complex healthcare information allows practitioners to detect conditions earlier while delivering customized treatments and better treatment results (JMIR, 2023). These technologies demonstrate potential to transform mental health care delivery through their ability to heal treatment gaps and improve access across underserved territories and to lower mental health stigma (iMark Infotech, n.d.). Wearable devices connected with artificial intelligence systems present exciting potential to resolve the worldwide mental health emergency by delivering advanced care options which span across all demographics.

2. EVOLUTION OF AI AND WEARABLE TECHNOLOGY IN MENTAL HEALTH

Scientists are engaged in an exciting creative journey toward wearable medical devices. Since spectacles were invented in 1286 wearable technology has transformed how medical practitioners perform monitoring and care tasks. The Holter Monitor became available for continuous cardiac diagnostics in the 1940s along with the modern Akouphone Hearing Aid which delivered electrical sound amplification. The 1956 Clark Electrode represented the genesis of biosensing modernity because it functioned as the first wearable blood oxygen measurement device. Medical systems across the world have introduced person-directed healthcare changes in the twenty-first century by applying continuous glucose monitoring to diabetes management and EEG testing for neurological needs. Through the combination of drug pulse interval monitoring and air intake measurements and cardiac pattern tracking wearable biosensors provide healthcare providers with ongoing surveillance. New applications in mental health diagnosis benefit from machine learning analysis paired with wearable technologies to track emotional patterns for personalized emotional feedback. The integration of adaptive mental health recommendations based on biometric data with Cognitive Behavior Therapy features forms new therapy delivery pathways through wearable technology.

Despite advancements, difficulties still exist. The precise measurement of behavior and biological signals faces limitations because individual attributes interact with data movement instability. User involvement remains fundamental for implementation because adoption success requires both usable interfaces and user comfort. Developers aiming for success must address user data security and privacy requirements alongside securing valid user consent to succeed in this domain. Future mental health care holds promising prospects because of developments in wearable technology systems. Three important leaps in AI technology will become possible through AI insight development which includes individual treatment planning and early diagnosis as well as predictive outcome calculation capabilities. Safe innovations in remote patient monitoring provide expanded healthcare access that delivers precise therapeutic interventions. Crossed capabilities of modern biosensors with user-centered design features and machine learning algorithms bring mental health care toward precise accessible supportive treatment systems.

Industrial development of modern wearables demonstrated their effectiveness in remaking healthcare delivery systems for intricate mental well-being interventions.

3.EMERGING TRENDS IN AI FOR MENTAL HEALTH

Contemporary mental health care treatment experiences radical change through artificial intelligence integration which triggers major operational changes throughout the discipline. Advanced artificial intelligence technologies improve access to mental health care by expanding its availability thereby filling fundamental support and service deficiencies. AI-powered platforms connected to chatbots offer continuous mental health assistance because they operate without time restrictions thus expanding mental health service accessibility through all hours of days to people who normally cannot access traditional care and support. The advancement of global mental health operations through these changes will lead to improved operational efficiency and inclusion for better outcomes in mental health care. AI development possesses the potential to transform mental health systems thus giving more people access to required support at the required time.

3.1 AI-Powered Diagnostics

Through artificial intelligence diagnosis systems mental health diagnosis and understanding is undergoing transformation in current times. Modern artificial intelligence systems grow rapidly to analyze complex databases which support traditional diagnostic work incorporating both physical examinations and standardized surveys. Through artificial Intelligence programming tools experts can uncover unseen patterns missed by humans to produce full mental health assessments.

Tools driven by predictive analytics combine medical records from healthcare systems with patterns in behavioral data collected from sociological media sources to track disease development. The implementation of early detection instruments provides essential support for preventive intervention approaches to produce long-lasting improvements in mental illness severity.

Technical analysis of mental health data using Natural Language Processing offers outstanding potential for medical diagnostic systems. Through ВотышData technology psychiatrists examine mental health indicators with dual assessments of verbal patterns and linguistic components along with interaction styles. Speech samples of limited duration serve as input for Winterlight Labs systems that aim to identify cognitive decline indicators. Tools analyze linguistic changes in speech to identify psychological markers of depression while also tracking anxiety signs and initial indications of early dementia development.

3.2 Personalized Interventions

Predictive AI diagnosis capabilities reach new precision heights annually to provide healthcare professionals sophisticated data about mental healthcare difficulties. Research findings indicate that AI detection algorithms achieve diagnosis accuracy ranging from 63 percent to 92 percent by using different diagnostic methods and varying quality data during training periods. The diagnostic tools demonstrate exceptional abilities in distinguishing various medical conditions including bipolar disorder from major depressive disorder.

Machine learning processing meticulously assembled patient data elements including genetic information and environmental context and neurological factors results in highly tailored treatment protocols. The individual therapeutic approaches made possible by ventilators result in enhanced positive treatment outcomes.

Machine intelligence created communication software which established massive coping platforms for worldwide mental health support. Affordable instant support delivered through accessible digital platforms serves millions of users worldwide to promote equitable inclusive mental healthcare systems. Chatbots track patient mental health progress via therapeutic conversations while supplying established coping techniques.

3.3 Integration with Big Data

Modern mental healthcare is transforming drastically because of artificial intelligence implementation that changes operational methods and boosts service delivery quality. The technological evolution has established new solutions for resolving ongoing mental health care problems involving service access and clinical care availability and delivery efficiency. With the access to more advanced AI tools, these mental health services are now getting much easier for more people, bridging the gaps for cases of support and treatment that are currently not covered. Natural Language Process and Predictive Analytics are used in AI technologies that help in more accurate and timely psychiatric diagnosis. AI systems apply these capabilities for efficient diagnosis of mental health problems so that correct preventative interventions take effect before conditions advance. The way AI systems can come to know massive datasets with the ability to analyze behavioral patterns and in turn comes up with highly personalized treatment plans. Such precision in care levels increases both the pace and the resultability of care, allowing people to get the right kind of interventions that are specific to their requirements. In addition, the use of such AI powered plataforms as chatbots and virtual assistants allow them to have continuous mental health support outside of regular office hours. Such platforms function continuously providing healthcare services to anyone regardless of where they live including those in distant or underprivileged parts of the world. The constant availability of support benefits especially those patients who normally encounter obstacles using conventional mental health resources because of geographic or economic or social factors. Through continuous developmental progress in AI mental health care practitioners will achieve higher operational efficiency and wider accessibility along with enhanced worldwide patient results. AI uses intelligent tools to empower healthcare providers and expanded patient access which drives fundamental changes in worldwide mental health care delivery together with its current practice enhancement.

3.4 Advancements in Natural Language Processing (NLP)

Mobile applications combined with artificial intelligence virtual assistants enable developers to monitor mental health status in real-time. User emotional states evolve over time while AI technologies measure their underlying changes through both speech patterns and linguistic content to provide undetectable mental health care. The system provides users with emergency help at their convenience which reduces stress on the healthcare system.

Through its monitoring system Winterlight Labs tracks changes in the way users speak to identify their depression and anxiety levels. Emotional well-being intervention techniques and live testing capabilities have benefited greatly from modern technological advancements delivering optimized well-being solutions.

4. FUTURE TRENDS IN WEARABLE TECHNOLOGY FOR MENTAL HEALTH

The mental health care landscape experiences a consequential evolution as researchers develop wearable technology. Using artificial intelligence technology these devices have begun to transform the way both monitoring and management and understanding of psychological well-being function. Implantation of biometric alongside psychometric monitoring along with multimodal data collection and modern interfaces including AR and VR within wearable technology provides new levels of mental health insights and interventions.

4.1 Biometric and Psychometric Monitoring

Modern wearable sensors record and study body data to help people check their mental health at any moment. Now these modern devices are able to autogenerate data from many body signals for an early warning of psychological distress. For instance, stress detection technologies such as electrodermal activity (EDA) sensor based on skin conductivity monitoring skin conductivity, which is converted by skin as emotional arousal to levels. When combined with heart rate variability (HRV), respiratory patterns and also the brainwave, these metrics present a deeper level about a person's emotional state. Such features worn by one tend to be tracked sleep patterns and emotional fluctuations in order to detect early signs of mental health conditions. Both interrupted sleep and constantly high stress markers are common indicators that may indicate anxiety. These are the first and often the only measures to be showcased when a diagnosis is started. These platforms integrate data from different sources to give an overall view of the mental health of a person. Beyond that, they deliver real time technological feedback allowing the users to self regulate their own emotional health. Continuous monitoring with individualized insights helps those who can spot the symptoms early on to prevent worsening of symptoms.

4.2 Multimodal Wearables

Multiple sensing technologies coming together enable greater capabilities for wearable mental health devices. Modern wearable technology scans EEG sensor data through machine learning algorithms and combines this data with heart rate readings and movement monitoring to develop complete mental health profiles. Different sensor technologies integrated together yield improved understanding of how well people maintain emotional and mental stability. Wearables tracking EEG signals along with motor patterns can identify depression by measuring stable movement behavior. Multiple levels of sensor analysis create these devices able to offer personalized recommendations for relaxation practices and wellness techniques. The method demonstrates accurate matching functions which leads to improved efficiency in mental health treatment.

4.3 Miniaturization and Smart Fabrics

Wearable technology development allows smaller discreet devices that made smart fabrics possible for mental health monitoring. The advanced clothing now includes small sensing components to track mental health status without anyone noticing. The wearable fabrics function like sensors because they gather information about heart rate breathing trends and body temperature from inside the fabric. Smart fabrics help patients feel less self-conscious due to their mental health tracking capabilities. Customers seeking monitoring solutions appreciate these garments because they blend seamlessly with their everyday apparel and do not attract attention. Wearable technology stays hidden under clothing so users keep their privacy protected as they collect thorough mental health updates in real time. Users find smart fabrics easy to wear and unobtrusive so they work well for everyday activities. People can stay monitored with ease because they can wear smart fabrics for everyday activities which reports their mental health status early before health issues become dangerous. Smart fabrics create new opportunities to advance mental healthcare through user-friendly monitoring technology that people will more willingly accept. Modern technology helps create support systems that let more people seek mental health help while meeting their personal needs and lowering treatment-related worries.

4.4 Augmented Reality (AR) and Virtual Reality (VR) Integration

VR and AR technology brings new mental healthcare advances by developing virtual therapeutic spaces that aid PTSD patients and survivors of extreme anxiety disorders. Users benefit from these virtual reality capabilities as they enter organized digital realms for psychological treatment during their sessions.

Virtual reality technology creates virtualve environments of public spaces to help anxious patients progressively encounter triggers across controlled simulation scenarios. PTSD patients gain distinct opportunities through accurate Virtual Reality replication to process emotions within controlled virtual settings. VR-based immersive therapy leads to superior treatment engagement which helps patients recover more quickly.

4.5 Brain-Computer Interfaces and Neural Integration

Wearable mental health technology features Brain-Computer Interfaces (BCIs) as an emerging powerful system for such technology. The creation of these systems provides an immediate connection between neurological signals in the brain and external digital equipment so real-time neurological tracking and device responses become possible. BCIs developed for neurorehabilitation started being studied for their potential role in mental health through identification and response of anxiety symptoms together with depression symptoms and cognitive tiredness. Wearable EEG headbands function as non-invasive BCIs because they both monitor brainwaves and make detections about neural signals that connect to emotional states. The signals sent to artificial intelligence and machine learning models become useful for identifying psychological distress at its earliest stages. The systems analyze data in real time to generate behavioral therapeutics which include respiratory exercises combined with mindfulness methods and caregiver notifications based on the processing of neural signals. Neurofeedback training gains support from BCI systems because users can monitor their brain activity in real-time. This approach provides successful results when treating ADHD together with PTSD in addition to generalized anxiety disorders. Wearable systems together with BCI technology demonstrate promising potential for continuous personalized mental healthcare even though they are currently developing their accessibility features alongside affordability and ethical standards. BCIs will transform mental health tracking and therapeutic practice by becoming more accessible and easier to use which will result in an advanced neural-informed healthcare era.

5. CHALLENGES AND ETHICAL CONSIDERATIONS

Wearable mental health technologies have experienced rapid advancements which offer extraordinary potential for tracking psychological health conditions alongside better mental health service provision. The explosive development of innovative technologies creates significant moral and implementation issues requiring immediate discovery and resolution to sustain ethical innovation. These technologies need complete implementation by addressing security concerns regarding sensitive data simultaneously with reducing artificial intelligence (AI) bias through effective resolution of related challenges.

5.1 Data Privacy and Security

Although wearable mental health technologies include several benefits of real time psychological monitoring as well as personalised care, they also entail considerable privacy and security issues. Inside these devices, feeling delicate and continuously streaming where mental health data, for example, emotional states, stress levels, sleep patterns, etc. is collected. Physical information about an individual's psychological well-being can be seen very differently from healthy records, which are usually collected during visits with a doctor and stored in centralized systems, than from wearables that allow constant and very detailed perspective of how one is doing psychologically. As a result, there is a unified psychological profile that can be exposed at risk if the right protection is not in place. Misuse of such data collection is possible. First, the information collected is quite sensitive including mood fluctuations, stress measurements etc., which make the data an attractive target for access by the unauthorized parties. Exploitation or harm through this data breach, and potential mental health data breach (MHD) or mismanagement could include identity theft, targeted advertising, or discrimination on the basis of mental health condition. With the risk given, it's important to be able to trust that your personal data is not going to be handled with the lack of security and respect that your personal data deserves. If this information not protected, it could harm individuals and severely damage the trust necessary to get widespread uptake of these technologies.

In fact, a study reveals that the worry over personal health information explosure to unathorized parties accounts for 41% of wearble health devise usres. The percentage is this significant as it shows how deep the mistrust between the user and how personal data is entrusted when dealing with the sensitive matters of mental health. People are becoming aware that their mental health data is more prone to exploitation than other forms of health data due to its continual nature and the possibility of having intimate disclosures related to emotional and psychological status. If these technologies are not guarded properly there is a serious risk of the same unauthorized access and misuse and hamper acceptance of these technologies. Yet, in order for wearable mental health technologies to be successful, developers and manufacturers needs to tackle these concerns by having robust security measures in place. One thing that is important is making sure that data is from the time it is collected all the way to the transmission to storage; so, that it is encrypted end to end. The encryption carries out in this is such a way that even data is intercepted or read by the unauthorized entity, it is not readable. Additionally, data needs to be stored securely, such as encrypted cloud server with strong access control mechanism to avoid data breach. In addition, data sharing agreements must be established with users that clearly spell out the terms with the understanding that someone can have full transparency and control at all times over use, sharing, and storage of their data. The data should be processed by someone who has specific permission to see it, for what purpose, and according to some conditions.

Since there are existing laws like the General Data Protection Regulation (GDPR) for us to follow when it comes to protecting user privacy, we started from there. Since GDPR has established the global standards for the protection of data, people have come with the right to be informed, right to rectify and delete their personal data. In addition, it imposes hard and fast rules on data security and consents. Unfortunately, if laws do not continually evolve to meet the speed of development of wearable technologies and AI based mental health solutions, GDPR serves as the bare minimum of what's necessary in protecting people's data. The wearables health technologies pose new challenges that existing regulations maybe not fully addressed including the instant data collection and the incorporation of the AI in the processing of the sensitive mental health information. Requisite attention must be given to updating legal framework to specifically instruct on developing, deploying, and regulating wearable mental health technologies. Newer laws will make it more clear how the user's consent to have their data collected, the data to be anonymize or to be de identified, and which safeguards must be in place. In addition, these laws should insist that users are well aware of how their data will be used and allow users to choose not to share their data if they are uncomfortable with the data sharing practices.

5.2 Bias in AI Algorithms

In order for wearable mental health devices to generate individualized mental well being insights together with the proper interventions, they require the use of AI algorithms that can analyze complex data. AI systems are applied in wearable mental health devices, but the technology based solution develops bias from the unbalanced training sample data collections. For example, trained algorithms drawn primarily from a specific demographic database tend to falter in reading mental health signals from minority populations. The bias may continue to exist health care disparities already in play, particularly in the area of discriminatory mental health practices, especially directed at people with disabilities or chronic medical problems. Developers and researchers need to use various training datasets, and conduct frequent audits on them, to provide equal treatment. Comprehensive teamwork between technology specialists and psychologists, ethicists, and so forth is required in order to achieve risk reduction.

5.3 User Compliance and Engagement

Regular usage and correct usage of a mental health wearable is determined by building trust and creating user engagement, therefore, the effectiveness of these depends on it. To make a significant statement in mental health care, wearable technologies require that people are confident in their ability to use the device consistently and correctly. According to research findings, 87 percent of people are willing to share their wearable health data with medical professionals only if they trust the system. This illustrates how user trust is crucial for those technologies being widely used. Transparency is one of the major factors in putting this trust in place. All users should be properly informed about how their personal data will be collected, processed, and used. The main barrier of digital health tool acceptance, especially regarding sensitive information of the mental health, is privacy. As such, mental health wearables must be clear and accessible about how they use their data if they want to become widely accepted. Users will feel confident to form a trust in these devices if they know that their data is being handled securely and with responsibility. The design of the device itself further matters: transparency is not enough if the design is frightening or impedes the user from achieving his goals or from feeling comfortable doing so. The real challenge with mental health wearables is that they must be easy to use (i.e. they need to not be intrusive in users' daily lives). Complicated devices or those devices that require a long period of effort to be driven are less likely to garner long term engagement. In order for wearables to become seamlessly integrated into everyday routines, they have to be able to work as things seamlessly as possible with everyday routines. Intuitive technology makes users more likely to use technology into their routines without being burdened or distracted. Devices that are less obtrusive, and easy to use, increase the chance that the users will use them regularly.

A solution is to provide flexible features that satisfy the tastes of each person. By enabling easy setting of personalised notifications, reminders and alerts, the wearable meets a user's specific needs and preferences. They promote the device's utility and help its adaptability to the user's lifestyle. It's also common for a mental health wearable to nudge a user to practice mindfulness when their stress levels are elevated or to take a breathing exercise when they're being stressed, for example. What makes it helpful is the ability to customize the device to suit one's own needs. Wearables also have to have clear and simple exit mechanisms. It is important that users can disable the device whenever they want without having unnecessary barriers. As user-centric design, there should be the option of an individual to unsubscribe from notifications or even to completely disable the device if they want. Control is essential on this level of trust; users will be more responsive to new technologies if they retain some control over how they are used. Manufacturers embed these features into wearable mental health devices to make for an easy to use experience and help build confidence in the device, and encourage long term usage. Users are more likely to succeed with using the devices over time if users have control over their mental health data and understand how the technology works. It is a way of ongoing interaction that helps persons to know more of one's mental wellness and hence able to cope with mental health in a better way.

5.4 Ethical Dilemmas

More extended ethical questions than privacy preservation arise with mental health wearables. A large ethical issue emerges when psychological data can be confiscated as an object. Psychological profiles have the potential for market entities such as employers and insurance companies as well as other entities to make personally discriminatory employment decisions or discriminate in privates insurance coverage costs. We need to create new legal safeguards to protect personal autonomy in the age of electronic data, while at the same time trying to keep abuse of that data at bay. A fundamental ethical requirement of earned patient knowledge about data utilization. With absolute clarity of what data is intended for, which storage mechanisms are being used, and how it is distributed, users will need to be involved in the development of any new service if possible. However, mental health data consent moves forward when growing taught protocols are set up that explain complex technical terms in easy to understand language which the patient can understand and consent to making the proper decisions about their data. Now it is a subtle management problem because of the marriage of advancing technology and ethical patient care guidelines. However, wearable technology has big potential to contribute positively to proactive mental healthcare if designers first put user health first and not just building technological capability. Developers need to prioritize three key aspects: psychological safety improvements alongside precise data control functions and diverse design inclusivity standards.

This trend will take ethical considerations into an increasingly difficult to manage domain of the development of wearable mental health technology. Mental health technology success can be measured to the accolades of genuine psychological support share with proven respect for human dignity and unimpeachable privacy standards. Managing the issues of data privacy in wearable mental health technology and minimizing bias through user participation through effective management of data by combining positive ethical commitments are important components to a successful approach to wearable mental health technology. The future of mental health care will be accessible through wearable technologies because these efforts protect patient privacy and will promote successful mental health treatment systems.

6. CASE STUDIES AND EXAMPLES

Artificial intelligence (AI) combined with wearable techs is the coming together of new generation designed solutions that help assist not just your mental health but make it proactive which has completely changed the outlook of mental health care. By this integration, real time monitoring, early intervention, and tailor made therapeutic strategies yielding to transformative client and real world clinical outcomes are delivered. Among many other examples, one is using smartwatches that integrate the power of AI to detect early symptoms anxiety and depression as indicated by metrics such as heart rate variability, sleep patterns and daily activities. Can help in indicating anomalies in a user's normal physiological patterns and recommending interventions like mindfulness exercises or seeking professional help – if necessary. They include cases such as wearable EEG headbands combined with AI used to track brainwaves during meditation and sleep. Feedback is given to users to help improve emotional regulation and sleep quality that is important to good mental health. Also, therapists and healthcare providers are using smart rings and AI-based mental health apps to constantly track up of their patients and change their treatment plans accordingly. Innovations do not only make care more progressive inclusive but also more personalized according to individual needs.

6.1 Innovative AI Solutions

Online platforms with artificial intelligence capabilities show exceptional ability to help people overcome mental health problems in both a precise and compassionate manner. At the forefront stands Talkspace as an online therapy platform which combines Artificial Intelligence to make mental health care more accessible. Real-time conversation data analysis through AI helps Detect individuals severely distressed based on patterns of speech with an 83% precision mark. Through this feature therapists can provide rapid assistance to patients who need crisis support (ITRex Group, n.d.). Spring Health demonstrates that Artificial Intelligence provides methods to tailor mental health interventions. The technological power of machine learning enables Spring Health to process millions of data points for creating employee-specific treatment plans. Organizations implementing this method experience higher recovery rates measuring twice as high as common techniques alongside accelerated treatment schedules. The shift from standardization with Spring Health demonstrates how artificial intelligence can achieve personalized medical care according to VinBrain (n.d.). Industrial applications of AI natural language processing now include vital mental health support tools facilitating conversations through chatbots. Woebot together with Wysa and Youper use artificial intelligence algorithms to provide evidence-supported therapeutic help using cognitive behavioral therapy and mindfulness foundation. The research demonstrates that these conversational agents lower depression symptoms by 64% and they stay continuously available to help people with mental health difficulties (ResearchGate, 2023).

6.2 Wearable Technology Impact

Real-time monitoring of mental health has become possible through wearable devices because these devices track physiological and emotional states in real-time. The app-based Meru Health program represents an important example of mental health care because it combines therapy with mindfulness and biofeedback to handle ailments such as depression and anxiety and burnout. Through utilising wearable sensors Meru Health obtains biometric data such as heart rate variability which enables it to generate actionable information regarding user stress and emotional conditions. The combination of evidence-based therapy with technological monitoring establishes a holistic mental health system which goes further than traditional therapeutic approaches (AIM Research, n.d.).

Research demonstrates how wearable technology acts as an essential tool for suicide prevention programs. The combination of AI-driven models working with wearable devices enables suicide attempt predictions with a 92% accuracy rate that extends across one week. Risk patterns that evade clinical detection become identifiable through the analysis of electronic health records and clinical notes and social media interactions by these systems. Wearable technology shows its essential role for future-focused mental healthcare through its predictive capabilities (PMC, 2023).

6.3 Combined AI and Wearable Initiatives

With the help of the artificial intelligence that integrated with wearable technology, we discovered the new dimension of mental health care, which is data driven, individualized interventions that work not only boosting attention, but also helping people to connect in and out. Digital therapeutic evidence-based interventions are available today as software platforms because the initial innovation led to their development. Varieties of therapeutic tools have been created by research teams for the handling of multiple mental health issues such as PTSD and anxiety together with depression and substance abuse disorders. For instance, the wearable sensor can include heart rate variability or sleep patterns to alter the cognitive behavioral therapy (CBT) modules or add some mindfulness exercises in the digital therapeutic app that offers real time change. These systems use automatic data sharing to relate therapeutic treatments to patient biological measurements thus improving patient-handling while boosting compliance rates. Users' mental health receives insight through digital phenotyping systems operated by some digital platforms that examine wireless device activities including phone behavior and typing speed and social interaction patterns. Wearable technology integration allows healthcare professionals to build connected care systems that develop fluid biological-to-therapy communication networks accessible to users. The growing popularity of digital therapeutic technologies enables flexible service delivery extending medical care support to communities who need it most especially in areas without access to regular medical facilities. The future trends in personal medicine is the basis in which this form of an all encompassing, adaptive mental health support is integrated with AI driven wearables. The grand opening is happening when regulatory frameworks are established and clinical validation is growing: Digital therapeutics have positioned themselves to be at the front of tech enabled mental healthcare among the rest of the AI, wearables and personalized service around more integrated and proactive psychological well being delivery.

7. FUTURE OUTLOOK AND OPPORTUNITIES

Digital technology tools help doctors deliver top-notch care more easily and produce fresh mental health product solutions for everyone. Through IoT and AI technology wearable devices give us the chance to take better proactive decisions in mental health care. These advanced systems allow doctors to monitor patients more effectively and forecast their needs to keep mental health treatments on the path of prevention instead of reaction. Taking action ahead of crisis helps people with better results while making mental healthcare delivery work better. The advances in healthcare technology are transforming standard care methods to allow users quality customized support in different locations. The ongoing development of new technology systems creates a mental health care system that helps today's society in better ways.

7.1 AI-Powered Personalized Therapy

Advanced mental healthcare depends on AI-based therapy sessions as they create a major leap toward personal and data-guided patient care for mental health. The progress of Artificial Intelligence improves its ability to spot emotional changes and identifying verbal hints that people miss during their work. Modern AI systems track mental health conditions by assessing how people express themselves in online conversations and evaluate their spoken voice, spoken habits, and communication speed. Timely detection of mental health issues creates the right opportunity to act earlier in their development. The best feature of AI in medical treatment settings is its automatic processing of important data at the moment it happens. This feature of system enhancement permits ongoing adjustments to therapy methods for better patient outcomes. Every therapeutic encounter makes progress in determining how someone's mental health changes. This insight allows healthcare providers to improve treatment progressively. When AI works with powerful mental health support features such as VR exposure therapy and AI-powered journaling tools it delivers better results to patients. VR technology produces safety-adjusted scenarios for therapy exercises and AI journaling platforms assist emotional self-assessment by giving intelligent advice and result feedback. AI has the ability to re-condition how mental health is delivered through customization of mental health care to each patients unique need. These new methods boost the standards and speed of medical services while extending help to a greater number of people faster. Through its assistance AI systems will create better ways to serve psychological health needs to everyone without limitations.

7.2 Integration with IoT and Home-Based Systems

In mental health, the continuous monitoring of physiological and psychological metrics is revolutionizing the way the patients adopt mental health care as the users of the system will be connected to those they care about in one's own home, forming an interconnected smart home systems. The systems that result are based on wearable devices, applications on mobile phones, environmental sensors, all interconnected to run a comprehensive platform to watch in real time data like heart rate variability, sleep pattern, and emotional response. A combination of these technologies makes it possible for them to understand better, without them even being in the users' living space, a deeper insight into a user's mental state. Speech patterns in a smart homes with voice recognition can be detected by the voice indicators of stress. If signs of emotional distress are seen, the system can immediately provide guidance to help such as asking the student to perform deep breathing exercises. Proactive responses to users' stress help control the problems and stop them from growing into severe mental health issues. In case of emergency, smart home systems are lifesavers due to their role as alert mechanisms. Up to this point they can alert medical personnel and family members in real time so that they can respond quickly. This is a very important layer of security for people in crisis mental health, especially for those who are alone or who are in some other vulnerable condition. The smart homes encourage a normalized and stigma free approach to emotional wellness by integration mental health care into their everyday. Mental health support that becomes part of our daily routines can change the perception of public to make mental care not extraordinary but an ordinary and important part of personal health.

7.3 Research and Innovation in Mental Health Technologies

Research efforts today concentrate on developing AI-powered wearable technology systems which better help users manage their mental health issues. Key areas of innovation include:

Empathetic AI Systems - Future technology research seeks to build AI systems which excel at human-type communication yet keep objective data accuracy.

Seamless Wearable Technologies - The next generation of wearables based on textile-embedded sensors and microscopic biosensors will deliver continuous non-invasive examination of psychological states. Protected devices allow data collection while releasing stress-reducing interventions to users experiencing high stress levels (PolarSeal, n.d.).

Predictive Analytics for Risk Mitigation - AI predictive models have evolved their sophistication by evaluating combinations of medical diagnosis records together with behavioral information and environmental elements. By analyzing early mental health indicators these systems detect risks before symptoms develop thus allowing preventive healthcare and decreasing community costs of untreated mental disorders (Google AI, n.d.).

Global Accessibility - Technology-enabled mental health solutions deliver accessible yet scalable treatment approaches beyond the reach of limited resource areas. Targeting underserved geographical areas Project HOPE (n.d.) shows how AI chatbots along with mobile applications can spread psychological treatment access across society.

8. CONCLUSION

Artificial intelligence (AI) combined with wearable devices create revolutionary advancements which drive both technological advances and direct humanitarian effects in mental healthcare. The technological advancements have moved basic monitoring instruments into advanced systems which deliver both real-time and custom-made psychological assistance for patients (Apexon, n.d.). Diagnostic systems fueled by artificial intelligence detect psychological cues with unmatched accuracy to support early interventions that prove unattainable before (Augnito, n.d.). The technology of wearables grew from simple monitoring devices into health platforms that track mental health indicators without disruption while showing personalized data segments (ACL Digital, n.d.). Technology advancements help early mental health identification and produce custom intervention methods that focus on complete psychological care.

The development of advanced diagnostics poses critical challenges for implementation. Effective precise regulatory policies need implementation in order to defend psychological data sensitivity and support individual rights through human dignity preservation (MDPI, n.d.-b). Multiple research fields must work together to ensure creativity stays within ethical boundary lines.

The obstacles do exist but the potential advantages extend toward major levels. Similar innovations between AI and wearable technology work to decentralize mental health support while making therapy solutions accessible to underprivileged demographics thus lowering the barriers patients face to access psychiatric assistance (Project HOPE, n.d.). These technological systems exist to potentiate therapeutic relationships instead of displacing interpersonal connection. New psychological systems must combine advanced computational features with emotional sensibility to serve as resources that improve mental healthcare delivery (Nexocode, n.d.).

Research teams should work side-by-side with clinical practitioners and government officials to create solutions which meet specific patient needs and promote inclusive design standards. Silicon Valley has distinguished itself by allocating financial resources toward developing strong ethical frameworks and privacy safeguards which let these technologies grow responsibly (Healthcare Technology Report, n.d.). Mental health care will develop into an accessible healthcare system which unites compassion and intelligent practices. Interdisciplinary cooperation together with responsible innovation will enable us to build worldwide mental health care systems which deliver personalized proactive support to all individuals thus making them able to develop psychological resilience and well-being (AIM Research, n.d.)

9. REFERENCES

Apexon. (2023). *Future of wearable health technology innovations*. https://www.apexon.com/blog/evolution-of-wearable-devices/

Augnito. (2023). *AI in mental health: Trends and statistics*. https://augnito.ai/resources/ai-in-mental-health-trends-and-statistcs/

Builder.ai. (2023). *Case studies: AI-driven mental health support applications*. https://www.builder.ai/case-studies/moodit

ESRGroups. (2023). *Wearable technology and health monitoring*. Retrieved from https://journal.esrgroups.org/jes/article/view/5589

Health, X. R. (2023). *AI therapy and mental health applications*. Retrieved from [https://www.xr.health/blog/ai-therapy/](https://www.xr.health

HopeChest. (2023). *Global mental health statistics*. Retrieved from https://www.hopechest.org/global-mental-health-statistics/

Institute for Health Metrics and Evaluation (IHME). (2023). *Mental health data and analysis*. Retrieved from https://www.healthdata.org/research-analysis/health-risks-issues/mental-health

Jain, N. (2023). *Global mental health crisis and the role of technology*. LinkedIn. Retrieved from https://www.linkedin.com/pulse/global-mental-health-crisis-role-technology-neeraj-jain-okbsc/

Journal of Medical Internet Research. (2023). *Global mental health and AI applications*. https://www.jmir.org/2023/1/e42672/

Ministry of General Affairs. (2023). *Mental health worldwide facts and figures*. https://www.government.nl/topics/mhpss/mhpss-worldwide-facts-and-figures

Nmeters. (2023). *The future of wearables: Technological innovations*. https://9meters.com/technology/the-future-of-wearables-exploring-next-generation-technological-innovations

PolarSeal. (2023). *Evolution of wearables*. https://www.polarseal.net/blog/evolution-of-wearables/

Project, H. O. P. E. (2023). *The global mental health crisis: 10 numbers to note*. https://www.projecthope.org/news-stories/story/the-global-mental-health-crisis-10-numbers-to-note/

PubMed Central (PMC). (2023). *Wearable technology in health and mental health.* https://pmc.ncbi.nlm.nih.gov/articles/PMC6631918/

ResearchGate. (2023). *The role of artificial intelligence in mental health: A review.* https://www.researchgate.net/publication/376134724_The_Role_of_Artificial_Intelligence_in_Mental_Health_A_Review

Robotics Biz. (2023). *A complete history of wearable technology: A timeline.* https://roboticsbiz.com/a-complete-history-of-wearable-technology-a-timeline-updated/

Sharpener. (2023). *The future of wearables: Exploring next-generation innovations.* https://gosharpener.com/blogs/552205/The-Future-of-Wearable-Technology

Simform. (2023). *AI-driven mental health applications: A case study.* https://www.simform.com/case-studies/ai-driven-mental-health-support-application/

SingleCare. (2023). *Mental health statistics.* https://www.singlecare.com/blog/news/mental-health-statistics/

T-Rex Group. (2023). *AI and mental health: Trends and examples.* https://itrexgroup.com/blog/ai-mental-health-examples-trends/

VinBrain. (2023). *AI in mental health: A new revolution in diagnosis and treatment.* https://vinbrain.net/ai-in-mental-health-a-new-revolution-in-diagnosis-and-treatment

Wikipedia. (2023). *Wearable technology.* https://en.wikipedia.org/wiki/Wearable_technology

Compilation of References

Abbas, A., Schultebraucks, K., & Galatzer-Levy, I. R. (2021). Digital Measurement of mental health: Challenges, promises, and future directions. *Psychiatric Annals*, *51*(1), 14–20. DOI: 10.3928/00485713-20201207-01

Abd-alrazaq, A., AlSaad, R., Aziz, S., Ahmed, A., Denecke, K., Househ, M., Farooq, F., & Sheikh, J. (2023a). Wearable Artificial Intelligence for Anxiety and Depression: Scoping Review. *Journal of Medical Internet Research*, *25*, e42672. Advance online publication. DOI: 10.2196/42672 PMID: 36656625

Abd-Alrazaq, A., AlSaad, R., Shuweihdi, F., Ahmed, A., Aziz, S., & Sheikh, J. (2023b). Systematic review and meta-analysis of performance of wearable artificial intelligence in detecting and predicting depression. *NPJ Digital Medicine*, *6*(1), 84. Advance online publication. DOI: 10.1038/s41746-023-00828-5 PMID: 37147384

Addepally, S. A., & Purkayastha, S. (2017). Mobile-application based cognitive behavior therapy (CBT) for identifying and managing depression and anxiety. In *Digital Human Modeling. Applications in Health, Safety, Ergonomics, and Risk Management: Health and Safety: 8th International Conference, DHM 2017, Held as Part of HCI International 2017, Vancouver, BC, Canada, July 9-14, 2017* [Springer International Publishing.]. *Proceedings*, *8*(Part II), 3–12.

Adnan, M., Xiao, B., Ali, M. U., Bibi, S., Yu, H., Xiao, P., Zhao, P., Wang, H., & An, X. (2024). Human inventions and its environmental challenges, especially artificial intelligence: New challenges require new thinking. *Environmental Challenges*, *16*, 100976. DOI: 10.1016/j.envc.2024.100976

Agarwal, S., Jalan, M., Wilcox, H. C., Sharma, R., Hill, R., Pantalone, E., ... & Robinson, K. A. (2022). Evaluation of mental health mobile applications. PMID: 35675475

Aghdam, L. & Science Arena Publications. (2017). Effectiveness of cognitive-behavioral therapy on the reduction of marital disaffection of women filing for divorce. *International Journal of Philosophy and Social-Psychological Sciences, 3–6*, 1–7. https://sciarena.com/storage/models/article/jpN8UTmaxPHN2fkw0xysoAlVAXwJnoJHOJKsGOnRIy4oGoBVLmtQhwawN5Pu/effectiveness-of-cognitive-behavioral-therapy-on-the-reduction-of-marital-disaffection-of-women-fi.pdf

Aguirre, C.. (2021). Addressing bias in AI-driven mental health interventions. *Journal of Medical Internet Research, 23*(9), e27894.

AI chatbots are here to help with your mental health, despite limited evidence they work. (March 24th, 2024). *The Hindu*. Retrieved from https://www.thehindu.com/sci-tech/technology/ai-chatbots-are-here-to-help-with-your-mental-health-despite-limited-evidence-they-work/article67984641.ece

Ajayi, R. (2025). AI-powered innovations for managing complex mental health conditions and addiction treatments. *International Research Journal of Modernization in Engineering Technology and Science, 7*(1), 87.

Aktan, M. E., Turhan, Z., & Dolu, İ. (2022). Attitudes and perspectives towards the preferences for artificial intelligence in psychotherapy. *Computers in Human Behavior, 133*, 107273. DOI: 10.1016/j.chb.2022.107273

Alfano, L., Malcotti, I., & Ciliberti, R. (2023). Psychotherapy, artificial intelligence and adolescents: Ethical aspects. *Journal of Preventive Medicine and Hygiene, 64*(4), E438–E442. DOI: 10.15167/2421-4248/jpmh2023.64.4.3135 PMID: 38379752

Alharthi, A. S., Ameen, M. S., Alshahrani, M. A., & Alghamdi, A. S. (2024). Artificial intelligence in mental health care: Current trends and challenges. *ScienceDirect*. https://www.sciencedirect.com/science/article/pii/S2949916X24000525

Almufareh, M. F., Tehsin, S., Humayun, M., & Kausar, S. (2023). Intellectual disability and technology: An artificial intelligence perspective and framework. *Journal of Disability Research, 2*(4), 58–70. DOI: 10.57197/JDR-2023-0055

Alonso, J., Liu, Z., Evans-Lacko, S., Sadikova, E., Sampson, N., Chatterji, S., Abdulmalik, J., Aguilar-Gaxiola, S., Al-Hamzawi, A., Andrade, L. H., Bruffaerts, R., Cardoso, G., Cia, A., Florescu, S., de Girolamo, G., Gureje, O., Haro, J. M., He, Y., de Jonge, P., & Thornicroft, G.WHO World Mental Health Survey Collaborators. (2018). Treatment gap for anxiety disorders is global: Results of the World Mental Health Surveys in 21 countries. *Depression and Anxiety, 35*(3), 195–208. DOI: 10.1002/da.22711 PMID: 29356216

American Psychiatric Association. (2013). *Diagnostic and statistical manual of mental disorders* (5th ed.). APA Publishing.

American Psychological Association (APA). (2023). Artificial intelligence in mental health care. https://www.apa.org/practice/artificial-intelligence-mental-health-care

American Psychological Association. (2021). *APA guidelines for ethical conduct in the use of digital technologies in psychological practice*. https://www.apa.org/ethics/digital-guidelines

American Speech-Language-Hearing Association. (n.d.). Confidentiality. https://www.asha.org/practice/ethics/confidentiality/

Anderson, P. L., Zimand, E., Hodges, L. F., & Rothbaum, B. O. (2020). Virtual reality exposure therapy for social anxiety. *Journal of Anxiety Disorders*, *42*, 1–7.

Andersson, G., Carlbring, P., Titov, N., & Lindefors, N. (2019). Internet Interventions for Adults with Anxiety and Mood Disorders: A Narrative Umbrella Review of Recent Meta-Analyses. *Canadian Journal of Psychiatry*, *64*(7), 465–470. DOI: 10.1177/0706743719839381 PMID: 31096757

Apexon. (2023). *Future of wearable health technology innovations*. https://www.apexon.com/blog/evolution-of-wearable-devices/

Arendt, I. T. P., Riisager, L. H. G., Larsen, J. E., Christiansen, T. B., & Moeller, S. B. (2021). Distinguishing between rumination and intrusive memories in PTSD using a wearable self-tracking instrument: A proof-of-concept case study. *Cognitive Behaviour Therapist*, *14*, e15. Advance online publication. DOI: 10.1017/S1754470X2100012X

Arora, A. (2020). Conceptualising artificial intelligence as a digital healthcare innovation: an introductory review. *Medical Devices: Evidence and Research*, 223-230.

Arvind, B. A., Gururaj, G., Loganathan, S., Amudhan, S., Varghese, M., Benegal, V., Rao, G. N., Kokane, A. M., B S, C., P K, D., Ram, D., Pathak, K., R K, L. S., Singh, L. K., Sharma, P., Saha, P. K., C, R., Mehta, R. Y., & T M, S. (2019). Prevalence and socioeconomic impact of depressive disorders in India: Multisite population-based cross-sectional study. *BMJ Open*, *9*(6), e027250. DOI: 10.1136/bmjopen-2018-027250 PMID: 31253618

Asad, M. M., Naz, A., Churi, P., & Tahanzadeh, M. M. (2021). Virtual reality as a pedagogical tool to enhance experiential learning: A systematic literature review. *Education Research International*, *2021*(1), 7061623. DOI: 10.1155/2021/7061623

Assuncao, G., Patrao, B., Castelo-Branco, M., & Menezes, P. (2022). An overview of emotion in artificial intelligence. *IEEE Transactions on Artificial Intelligence, 3*(6), 867–886. DOI: 10.1109/TAI.2022.3159614

Augnito. (2023). *AI in mental health: Trends and statistics*. https://augnito.ai/resources/ai-in-mental-health-trends-and-statistcs/

Australian National University. (n.d.). Confidentiality and privacy. https://services.anu.edu.au/research-support/research-ethics-integrity-compliance/key-ethical-concerns/confidentiality-and-privacy

Auto, M. L. Framework. (n.d.). Google AutoML: Machine learning made easy. https://cloud.google.com/automl

Baeza-Yates, R. (2024, Mar). Introduction to Responsible AI. In *Proceedings of the 17th ACM International Conference on Web Search and Data Mining* (pp. 1114-1117). DOI: 10.1145/3616855.3636455

Bajwa, J., Munir, U., Nori, A., & Williams, B. (2021). Artificial intelligence in healthcare: Transforming the practice of medicine. *Future Healthcare Journal, 8*(2), e188–e194. DOI: 10.7861/fhj.2021-0095 PMID: 34286183

Baker, R., Owens, M., Thomas, S., Whittlesea, A., Abbey, G., Gower, P., Tosunlar, L., Corrigan, E., & Thomas, P. W. (2011). Does CBT facilitate emotional processing? *Behavioural and Cognitive Psychotherapy, 40*(1), 19–37. DOI: 10.1017/S1352465810000895 PMID: 21392415

Bakker, D., Kazantzis, N., Rickwood, D., & Rickard, N. (2016). Mental health smartphone apps: Review and evidence-based recommendations for future developments. *JMIR Mental Health, 3*(1), e4984. DOI: 10.2196/mental.4984 PMID: 26932350

Balconi, M., Fronda, G., & Crivelli, D. (2019). Effects of technology-mediated mindfulness practice on stress: Psychophysiological and self-report measures. *Stress (Amsterdam, Netherlands), 22*(2), 200–209. DOI: 10.1080/10253890.2018.1531845 PMID: 30472901

Bandura, A. (1977). *Social learning theory*. Prentice-Hall.

Barlow, D. H. (2002). *Anxiety and its disorders: The nature and treatment of anxiety and panic*. Guilford Press.

Beck, A. T. (1970). Cognitive therapy: Nature and relation to behavior therapy. *Behavior Therapy, 1*(2), 184–200. DOI: 10.1016/S0005-7894(70)80030-2 PMID: 27993332

Beck, A. T. (1976). *Cognitive therapy and the emotional disorders*. International Universities Press.

Beck, A. T., & Clark, D. A. (1997). An information processing model of anxiety: Automatic and strategic processes. *Behaviour Research and Therapy*, *35*(1), 49–58. DOI: 10.1016/S0005-7967(96)00069-1 PMID: 9009043

Beck, N. A. (2017). *Mobile apps for cognitive restructuring: A review and comparative analysis*. Pepperdine University.

Beg, M. J. (2024). Responsible AI Integration in Mental Health Research: Issues, Guidelines, and Best Practices. *Indian Journal of Psychological Medicine*, 02537176241302898. PMID: 39650770

Beg, M. J., & Verma, M. K. (2024). Exploring the Potential and Challenges of Digital and AI-Driven Psychotherapy for ADHD, OCD, schizophrenia, and Substance Use Disorders: A Comprehensive Narrative review. *Indian Journal of Psychological Medicine*, 02537176241300569. Advance online publication. DOI: 10.1177/02537176241300569 PMID: 39677516

Beg, M. J., Verma, M., & Verma, M. K. (2024). Artificial intelligence for psychotherapy: A review of the current state and future directions. *Indian Journal of Psychological Medicine, 02537176241260819*. Advance online publication. DOI: 10.1177/02537176241260819

Birk, R. H., & Samuel, G. (2022). Digital Phenotyping for Mental Health: Reviewing the challenges of using data to monitor and predict mental health problems. *Current Psychiatry Reports*, *24*(10), 523–528. DOI: 10.1007/s11920-022-01358-9 PMID: 36001220

Borghare, P. T., Methwani, D. A., & Pathade, A. G. (2024). A comprehensive review on harnessing wearable technology for enhanced depression treatment. *Cureus*. Advance online publication. DOI: 10.7759/cureus.66173 PMID: 39233951

Borthakur, R. (2021). U.S. Patent Application No. 16/630,451.

Botella, C., Baños, R. M., Villa, H., Perpiñá, C., & García-Palacios, A. (2017). Virtual reality in the treatment of phobias. *Cyberpsychology & Behavior*, *7*(3), 283–290. DOI: 10.1089/1094931041291295

Botella, C., Fernandez-Alvarez, J., Guillén, V., García-Palacios, A., & Baños, R. M. (2017). Virtual reality exposure-based therapy in anxiety disorders: A systematic review. *Behaviour Research and Therapy*, *87*, 172–183. DOI: 10.1016/j.brat.2016.10.003

Boucher, E. M., Harake, N. R., Ward, H. E., Stoeckl, S. E., Vargas, J., & Stoeckl, S. M. (2021). Artificial intelligence and cognitive behavioral therapy: An integrative approach to address challenges in mental health care delivery. *Journal of Medical Internet Research, 23*(5), e26980. DOI: 10.2196/26980

Brandmeyer, T., Delorme, A., & Wahbeh, H. (2013). The neuroscience of meditation: Classification, phenomenology, correlates, and mechanisms. *Progress in Brain Research, 244*, 1–29. DOI: 10.1016/bs.pbr.2018.10.020 PMID: 30732832

Brantley, J. (2005). Mindfulness-based stress reduction. In *Acceptance and mindfulness-based approaches to anxiety: Conceptualization and treatment* (pp. 131–145). Springer US., DOI: 10.1007/0-387-25989-9_5

Braun, V., & Clarke, V. (2006). Using thematic analysis in psychology. *Qualitative Research in Psychology, 3*(2), 77–101. DOI: 10.1191/1478088706qp063oa

Brewer, R., Guberman, J., & Piper, A. M. (2020). Designing for disability: Evaluating the state of accessibility design in mental health apps. *Proceedings of the ACM on Human-Computer Interaction, 4*(CSCW2), 1–25. https://doi.org/DOI: 10.1145/3415184

Brewin, C. R., Andrews, B., & Valentine, J. D. (2000). Meta-analysis of risk factors for posttraumatic stress disorder in trauma-exposed adults. *Journal of Consulting and Clinical Psychology, 68*(5), 748–766. DOI: 10.1037/0022-006X.68.5.748 PMID: 11068961

Broderick, J. E., Keefe, F. J., Schneider, S., Junghaenel, D. U., Bruckenthal, P., Schwartz, J. E., Kaell, A. T., Caldwell, D. S., McKee, D., & Gould, E. (2016). Cognitive behavioral therapy for chronic pain is effective, but for whom? *Pain, 157*(9), 2115–2123. DOI: 10.1097/j.pain.0000000000000626 PMID: 27227692

Brown, A. (2023). Regulatory frameworks for AI in healthcare. *MDPI*. https://www.mdpi.com/2673-2688/5/1/19

Brown, T. (2023). Ethical considerations in AI development. *Nature Humanities and Social Sciences Communications*, 10(1), 15. https://www.nature.com/articles/s41599-023-02079-x

Brown, M. (2023). AI governance: Themes, knowledge gaps, and future agendas. *International Journal of Information Technology : an Official Journal of Bharati Vidyapeeth's Institute of Computer Applications and Management, 29*(5), 51–65. https://www.emerald.com/insight/content/doi/10.1108/intr-01-2022-0042/full/html

Bruggeman, K. J., & Wurster, S. W. (2018). The hiatus system: virtual healing spaces: low dose mindfulness based stress reduction virtual reality application. In *ACM siggraph 2018 appy hour* (pp. 1-2). https://doi.org/DOI: 10.1145/3213779.3213785

Bryant, E. F. (2015). *The yoga sutras of Patanjali: A new edition, translation, and commentary*. North Point Press.

Builder.ai. (2023). *Case studies: AI-driven mental health support applications*. https://www.builder.ai/case-studies/moodit

Byrne, S. M., Fursland, A., Allen, K. L., & Watson, H. (2011). The effectiveness of enhanced cognitive behavioural therapy for eating disorders: An open trial. *Behaviour Research and Therapy*, *49*(4), 219–226. DOI: 10.1016/j.brat.2011.01.006 PMID: 21345418

Cabral, R. R., & Smith, T. B. (2011). Racial/ethnic matching of clients and therapists in mental health services: A meta-analytic review of preferences, perceptions, and outcomes. *Journal of Counseling Psychology*, *58*(4), 537–554. DOI: 10.1037/a0025266 PMID: 21875181

Calvo, R. M., Peters, D., Vold, V. E., & Ryan, R. A. (2020). Supporting human autonomy in AI systems: A framework for ethical enquiry. *Ethics and Information Technology*, *22*(1), 31–42. DOI: 10.1007/s10676-018-9476-2

Cannon, W. B. (1932). The Wisdom of the Body WW Norton & Company. Inc., New York.

Cannon, W. B. (1932). *The wisdom of the body*. W.W. Norton & Company, Inc.

Carl, E., Stein, A. T., Levihn-Coon, A., Pogue, J. R., Rothbaum, B., Emmelkamp, P., & Powers, M. B. (2019). Virtual reality exposure therapy for anxiety and related disorders: A meta-analysis. *Journal of Anxiety Disorders*, *61*, 27–36. DOI: 10.1016/j.janxdis.2018.08.003 PMID: 30287083

Carter, R.III, Carter, K. S., Holliday, J., Holliday, A., & Harrison, C. K. (2018). A Community-based Stress Management Program: Using Wearable Devices to Assess Whole Body Physiological Responses in Non-laboratory Settings. *Journal of Visualized Experiments*, *55816*(131), •••. DOI: 10.3791/55816 PMID: 29443019

Cerasa, A., Gaggioli, A., Marino, F., Riva, G., & Pioggia, G. (2022). The promise of the metaverse in mental health: The new era of MEDverse. *Heliyon*, *8*(11), e11762. DOI: 10.1016/j.heliyon.2022.e11762 PMID: 36458297

Chand, S., Ravi, C., Chakkamparambil, B., Prasad, A., & Vora, A. (2018). CBT for depression: What the evidence says. *Current Psychiatry*, *17*(9), 14–23.

Chen, Z., Lu, Y., & Wang, W. Y. (2023). Empowering psychotherapy with large language models: Cognitive distortion detection through diagnosis of thought prompting. *arXiv preprint arXiv:2310.07146*. https://doi.org//arXiv.2310.07146DOI: 10.18653/v1/2023.findings-emnlp.284

Cheung, F., Lucas, G. M., & Gratch, J. (2019). Personalizing mental health interventions through adaptive conversational agents. *Frontiers in Psychology, 10*, 2647. DOI: 10.3389/fpsyg.2019.02647

Chikersal, P., Belgrave, D., Doherty, G., Enrique, A., Palacios, J. E., Richards, D., & Thieme, A. (2020, April). Understanding client support strategies to improve clinical outcomes in an online mental health intervention. In *Proceedings of the 2020 CHI conference on human factors in computing systems* https://doi.org/DOI: 10.1145/3313831.3376341

Chiovato, A., Demarzo, M., & Notargiacomo, P. (2021). Evaluation of mindfulness state for the students using a wearable measurement system. *Journal of Medical and Biological Engineering, 41*(5), 1–14. DOI: 10.1007/s40846-021-00658-4 PMID: 34608381

Choy, Y., Fyer, A. J., & Lipsitz, J. D. (2007). Treatment of specific phobia in adults. *Clinical Psychology Review, 27*(3), 266–286. DOI: 10.1016/j.cpr.2006.10.002 PMID: 17112646

Chrousos, G. P. (2009). Stress and disorders of the stress system. *Nature Reviews. Endocrinology, 5*(7), 374–381. DOI: 10.1038/nrendo.2009.106 PMID: 19488073

Chung, A. H., Gevirtz, R. N., Gharbo, R. S., Thiam, M. A., & Ginsberg, J. P. (2021). Pilot study on reducing symptoms of anxiety with a heart rate variability biofeedback wearable and remote stress management coach. *Applied Psychophysiology and Biofeedback, 46*(4), 347–358. DOI: 10.1007/s10484-021-09519-x PMID: 34308526

Clark, D. M., & Wells, A. (1995). A cognitive model of social phobia. In Heimberg, R. G., Liebowitz, M. R., Hope, D. A., & Schneier, F. R. (Eds.), *Social phobia: Diagnosis, assessment, and treatment* (pp. 69–93). Guilford Press.

Cogito. (2025, February 14). *AI Training Data Company | Cogito Tech*. Cogitotech. https://www.cogitotech.com/

Collins, C., Dennehy, D., Conboy, K., & Mikalef, P. (2021). Artificial intelligence in information systems research: A systematic literature review and research agenda. *International Journal of Information Management, 60*, 102383. DOI: 10.1016/j.ijinfomgt.2021.102383

Columbia Center for New Media Teaching and Learning. (n.d.). Ethical issues in research. https://ccnmtl.columbia.edu/projects/cire/pac/foundation/

Copeland, W. E., Keeler, G., Angold, A., & Costello, E. J. (2007). Traumatic events and posttraumatic stress in childhood. *Archives of General Psychiatry*, *64*(5), 577–584. DOI: 10.1001/archpsyc.64.5.577 PMID: 17485609

Corrigan, P. W., Druss, B. G., & Perlick, D. A. (2014). The impact of mental illness stigma on seeking and participating in mental health care. *Psychological Science in the Public Interest*, *15*(2), 37–70. DOI: 10.1177/1529100614531398 PMID: 26171956

Craske, M. G., & Stein, M. B. (2016). Anxiety disorders. *Lancet*, *388*(10063), 3048–3059. DOI: 10.1016/S0140-6736(16)30381-6 PMID: 27349358

Creed, T. A., Salama, L., Slevin, R., Tanana, M., Imel, Z., Narayanan, S., & Atkins, D. C. (2022). Enhancing the quality of cognitive behavioral therapy in community mental health through artificial intelligence generated fidelity feedback (Project AFFECT): A study protocol. *BMC Health Services Research*, *22*(1), 1177. DOI: 10.1186/s12913-022-08519-9 PMID: 36127689

Czaja, S. J., Boot, W. R., Charness, N., & Rogers, W. A. (2019). *Designing for older adults: Principles and creative human factors approaches* (3rd ed.). CRC Press., DOI: 10.1201/b22189

D'Alfonso, S. (2020). AI in mental health. *Current Opinion in Psychology*, *36*, 112–117. DOI: 10.1016/j.copsyc.2020.04.005 PMID: 32604065

Danks, D. (2023). Ethical AI design: Principles and practices. *ACM Transactions on Human-AI Interaction*, 4(3), 1–22. https://dl.acm.org/doi/fullHtml/10.1145/3544548.3581252

Dankwa-Mullan, I., Scheufele, E. L., Matheny, M. E., Quintana, Y., Chapman, W. W., Jackson, G., & South, B. R. (2021). A proposed framework on integrating health equity and racial justice into the artificial intelligence development lifecycle. *Journal of Health Care for the Poor and Underserved*, *32*(2), 300–317. DOI: 10.1353/hpu.2021.0065

Davidson, R. J., Kabat-Zinn, J., Schumacher, J., Rosenkranz, M., Muller, D., Santorelli, S. F., Urbanowski, F., Harrington, A., Bonus, K., & Sheridan, J. F. (2003). Alterations in brain and immune function produced by mindfulness meditation. *Psychosomatic Medicine*, *65*(4), 564–570. DOI: 10.1097/01.PSY.0000077505.67574.E3 PMID: 12883106 de Almeida, P. G. R., dos Santos, C. D., & Farias, J. S. (2021). Artificial intelligence regulation: A framework for governance. *Ethics and Information Technology*, *23*(3), 505–525. DOI: 10.1007/s10676-021-09593-z

De Freitas, J., Uğuralp, A. K., Oğuz-Uğuralp, Z., & Puntoni, S. (2024). Chatbots and mental health: Insights into the safety of generative AI. *Journal of Consumer Psychology*, *34*(3), 481–491. DOI: 10.1002/jcpy.1393

Dehbozorgi, R., Zangeneh, S., Khooshab, E., Nia, D. H., Hanif, H. R., Samian, P., Yousefi, M., Hashemi, F. H., Vakili, M., Jamalimoghadam, N., & Lohrasebi, F. (2025). The application of artificial intelligence in the field of mental health: A systematic review. *BMC Psychiatry*, *25*(1), 132. Advance online publication. DOI: 10.1186/s12888-025-06483-2 PMID: 39953464

Denecke, K., May, R., & Deng, Y. (2021). SERMO: An AI-based mobile application to improve emotion regulation through mindfulness and cognitive behavioral therapy. *Frontiers in Artificial Intelligence*, *4*, 654789. DOI: 10.3389/frai.2021.654789

DeVault, D., Artstein, R., Benn, G., Dey, T., Fast, E., Gainer, A., & Morency, L. P. (2014, May). SimSensei Kiosk: A virtual human interviewer for healthcare decision support. In *Proceedings of the 2014 international conference on Autonomous agents and multi-agent systems* (pp. 1061-1068).

Dezyit. (2024). *Scaling mental health support globally: AI success stories from developing countries*. https://www.dezyit.com/post/scaling-mental-health-support-globally-ai-success-stories-from-developing-countries

Dhiman, V. K. (2024). *The emergence of AI in mental health: A transformative journey*. World Journal of Advanced Research and Reviews., DOI: 10.30574/wjarr.2024.22.1.1298

Difede, J., Cukor, J., Jayasinghe, N., Patt, I., Spielman, L., Giosan, C., & Hoffman, H. G. (2014). Virtual reality exposure therapy for the treatment of posttraumatic stress disorder: A review of the literature. *Annals of the New York Academy of Sciences*, *1208*(1), 68–78. DOI: 10.1111/j.1749-6632.2010.05696.x

Dimova, A. (2024). *Mental Health Advocacy on Social Media: A Multimodal Discourse Analysis* (Doctoral dissertation, Queen Mary University of London).

Dobson, K. S. (2016). New frontiers in Cognitive-Behavioral Therapy for Depression. *International Journal of Cognitive Therapy*, *9*(2), 107–123. DOI: 10.1521/ijct.2016.9.2.107

Dobson, K. S., & Dozois, D. J. (Eds.). (2021). *Handbook of cognitive-behavioral therapies*. Guilford Publications.

Donnermann, M., Schaper, P., & Lugrin, B. (2020, August). Integrating a social robot in higher education–a field study. In *2020 29th IEEE international conference on robot and human interactive communication (RO-MAN)* (pp. 573-579). IEEE. DOI: 10.1109/RO-MAN47096.2020.9223602

Doorn, K. A., Kamsteeg, C., Bate, J., & Aafjes, M. (2020). A scoping review of machine learning in psychotherapy research. *Psychotherapy Research, 31*(1), 92–116. DOI: 10.1080/10503307.2020.1808729 PMID: 32862761

Durán, J. M., & Jongsma, K. R. (2021). Who is afraid of black box algorithms? On the epistemological and ethical basis of trust in medical AI. *Journal of Medical Ethics, 47*(5), 329–335. DOI: 10.1136/medethics-2020-106820 PMID: 33737318

Earkick, Your Free Personal AI Therapist. (n.d.). *Earkick*. Retrieved from https://earkick.com/

Eley, T. C., Gregory, A. M., & Clark, D. M. (2015). Genetic and environmental influences on phobias. *Behaviour Research and Therapy, 64*, 44–50. DOI: 10.1016/j.brat.2014.10.006

El-Sappagh, S. H., Nazih, W., Alharbi, M., & AbuHmed, T. (2025). Responsible Artificial Intelligence for Mental Health Disorders: Current Applications and Future Challenges. *Journal of Disability Research, 4*(1), 20240101. DOI: 10.57197/JDR-2024-0101

Emmelkamp, P. M. G., Krijn, M., Hulsbosch, A. M., De Vries, S., Schuemie, M. J., & Van Der Mast, C. A. (2002). Virtual reality treatment versus exposure in vivo: A comparative evaluation in acrophobia. *Behaviour Research and Therapy, 40*(5), 509–516. DOI: 10.1016/S0005-7967(01)00023-7 PMID: 12038644

Engel, J., & Prentice, D. (2013). The ethics of interprofessional collaboration. *Nursing Ethics, 20*(4), 426–435. DOI: 10.1177/0969733012468466 PMID: 23329780

Eskandar, K. (2024). Artificial intelligence in psychiatric diagnosis: Challenges and opportunities in the era of machine learning. *Revista Debates em Psiquiatria, 14*, 1–16. DOI: 10.25118/2763-9037.2024.v14.1318

Espejo, G., Reiner, W., & Wenzinger, M. (2023). Exploring the role of artificial intelligence in mental healthcare: Progress, pitfalls, and promises. *Cureus, 15*. Advance online publication. DOI: 10.7759/cureus.44748 PMID: 37809254

ESRGroups. (2023). *Wearable technology and health monitoring*. Retrieved from https://journal.esrgroups.org/jes/article/view/5589

Etkin, A., & Wager, T. D. (2007). Functional neuroimaging of anxiety: A meta-analysis of emotional processing in PTSD, social anxiety disorder, and specific phobia. *The American Journal of Psychiatry, 164*(10), 1476–1488. DOI: 10.1176/appi.ajp.2007.07030504 PMID: 17898336

Eyre, H. A., Ellsworth, W., Fu, E., Manji, H. K., & Berk, M. (2020). Responsible innovation in technology for mental health care. *The Lancet. Psychiatry, 7*(9), 728–730. DOI: 10.1016/S2215-0366(20)30192-9 PMID: 32828153

Eysenck, H. J. (1952). The effects of psychotherapy: An evaluation. *Journal of Consulting Psychology, 16*(5), 319–324. DOI: 10.1037/h0063633 PMID: 13000035

Fitzpatrick, K. K.. (2017). Delivering cognitive behavior therapy to young adults with symptoms of depression and anxiety using a fully automated conversational agent (Woebot). *Journal of Medical Internet Research, 19*(6), e7785. PMID: 28588005

Fitzpatrick, K. K., Darcy, A., & Vierhile, M. (2017). Delivering cognitive behavior therapy to young adults with symptoms of depression and anxiety using a fully automated conversational agent (Woebot): A randomized controlled trial. *JMIR Mental Health, 4*(2), e7785. DOI: 10.2196/mental.7785 PMID: 28588005

Flavián, C., Ibáñez-Sánchez, S., & Orús, C. (2019). The impact of virtual, augmented and mixed reality technologies on the customer experience. *Journal of Business Research, 100*, 547–560. DOI: 10.1016/j.jbusres.2018.10.050

Floridi, L. (2018). Soft ethics and the governance of the digital. *Philosophy & Technology, 31*(1), 1–8. DOI: 10.1007/s13347-018-0303-9 PMID: 30322997

Floridi, L., Cowls, J., Beltrametti, M., Chatila, R., Chazerand, P., Dignum, V., Luetge, C., Madelin, R., Pagallo, U., Rossi, F., Schafer, B., Valcke, P., & Vayena, E. (2018). AI4People—An ethical framework for a good AI society: Opportunities, risks, principles, and recommendations. *Minds and Machines, 28*(4), 689–707. DOI: 10.1007/s11023-018-9482-5 PMID: 30930541

Fredrikson, M., Annas, P., Fischer, H., & Wik, G. (1996). Gender and age differences in the prevalence of specific fears and phobias. *Behaviour Research and Therapy, 34*(1), 33–39. DOI: 10.1016/0005-7967(95)00048-3 PMID: 8561762

Freeman, C. (1990). Cognitive Behavioural Therapy for Psychiatric Problems: A Practical Guide. Edited by Keith Hawton, Paul M. Salkovskis, Joan Kirk and David M. Clark. Oxford: Oxford University Press. 1989. 456 pp. £40 (hb), £15 (pb). *British Journal of Psychiatry, 157*(3), 468–468. DOI: 10.1192/S000712500006356X

Freeman, D., Haselton, P., Freeman, J., Spanlang, B., Kishore, S., Albery, E., & Slater, M. (2017). Automated psychological therapy using immersive virtual reality for treatment of fear of heights: A single-blind, parallel-group, randomised controlled trial. *The Lancet. Psychiatry*, *5*(8), 625–632. DOI: 10.1016/S2215-0366(18)30226-8 PMID: 30007519

Freeman, D., Reeve, S., Robinson, A., Ehlers, A., Clark, D., Spanlang, B., & Slater, M. (2017). Advances in virtual reality and implications for clinical psychology. *Psychological Medicine*, *47*(16), 2395–2400. DOI: 10.1017/S003329171700040X PMID: 28325167

Freeman, D., Reeve, S., Robinson, A., Ehlers, A., Clark, D., Spanlang, B., & Slater, M. (2017). Virtual reality in the treatment of persecutory delusions: Randomised controlled experimental study testing how to reduce delusional conviction. *The British Journal of Psychiatry*, *210*(6), 429–436. DOI: 10.1192/bjp.bp.116.193564 PMID: 28302702

Freud, S. (1920). *Beyond the pleasure principle*. International Psycho-Analytical Press.

Fulmer, R., Joerin, A., Gentile, B., Lakerink, L., & Rauws, M. (2018). Using psychological artificial intelligence (Tess) to relieve symptoms of depression and anxiety: Randomized controlled trial. *JMIR Mental Health*, *5*(4), e9782. DOI: 10.2196/mental.9782 PMID: 30545815

Fusar-Poli, P., Manchia, M., Koutsouleris, N., Leslie, D., Woopen, C., Calkins, M. E., & Andreassen, O. A. (2022). Ethical considerations for precision psychiatry: A roadmap for research and clinical practice. *European Neuropsychopharmacology*, *63*, 17–34. DOI: 10.1016/j.euroneuro.2022.08.001 PMID: 36041245

Galič, M. (2018). AI and the ethics of automating consent. https://www.researchgate.net/publication/325979872_AI_and_the_Ethics_of_Automating_Consent

Gamble, A. (2020). Artificial intelligence and mobile apps for mental healthcare: A social informatics perspective. *Aslib Journal of Information Management*, *72*(4), 509–523. DOI: 10.1108/AJIM-11-2019-0316

Garcia-Palacios, A., Botella, C., Hoffman, H. G., & Fabregat, S. (2002). Comparing acceptance and refusal rates of virtual reality exposure versus in vivo exposure by patients with specific phobias. *Cyberpsychology & Behavior*, *5*(3), 219–224. DOI: 10.1089/109493102760147117 PMID: 12123244

Garcia-Palacios, A., Hoffman, H. G., Kwong See, S., Tsai, A., & Botella, C. (2007). Redefining therapeutic success with virtual reality exposure therapy. *Cyberpsychology & Behavior*, *4*(3), 341–348. DOI: 10.1089/109493101300210231 PMID: 11710258

Garland, E. L., Gullapalli, B. T., Prince, K. C., Hanley, A. W., Sanyer, M., Tuomenoksa, M., & Rahman, T. (2023). Zoom-based Mindfulness-Oriented Recovery Enhancement plus just-in-time mindfulness practice triggered by wearable sensors for opioid craving and chronic pain. *Mindfulness, 14*(6), 1329–1345. DOI: 10.1007/s12671-023-02137-0 PMID: 37362184

Garriga, R., Mas, J., Abraha, S., Nolan, J., Harrison, O., Tadros, G., & Matic, A. (2022). Machine learning model to predict mental health crises from electronic health records. *Nature Medicine, 28*(6), 1240–1248. DOI: 10.1038/s41591-022-01811-5 PMID: 35577964

Gera, P. (2024). Integrating Mindfulness: Enhancing Art and Design Student Well-being through AI Voice Assistants and Wearable Technology (Doctoral dissertation, OCAD University).

Gerke, S., Minssen, T., & Cohen, G. (2020). Ethical and legal challenges of artificial intelligence-driven healthcare. In *Artificial Intelligence in Healthcare* (pp. 295–336). Academic Press., DOI: 10.1016/B978-0-12-818438-7.00012-5

Gerke, S., Yeung, S., & Cohen, I. G. (2020). Ethical and Legal Aspects of Ambient Intelligence in Hospitals. *Journal of the American Medical Association, 323*(7), 601–602. DOI: 10.1001/jama.2019.21699 PMID: 31977033

Ghosh, D., & Halder, S. (2024). The Role of Social Cognition in Depression and Its Relationship to Early Maladaptive Schemas. *Indian Journal of Psychological Medicine, 02537176241233632*. DOI: 10.1177/02537176241233632 PMID: 39564327

Ghosh, K., Nanda, S., Hurt, R. T., Schroeder, D. R., West, C. P., Fischer, K. M., Bauer, B. A., Fokken, S. C., Ganesh, R., Hanson, J. L., Lindeen, S. A., Pruthi, S., & Croghan, I. T. (2023). Mindfulness using a wearable brain sensing device for health care professionals during a pandemic: A pilot program. *Journal of Primary Care & Community Health, 14*, 21501319231162308. DOI: 10.1177/21501319231162308 PMID: 36960553

Ghosh, K., Singh, A., & Ghosh, S. (2023). Neurofeedback: A comprehensive review on system design, methodology and clinical applications. *Biomedical Signal Processing and Control, 78*, 104024. DOI: 10.1016/j.bspc.2022.104024

Gloster, A. T., Wittchen, H. U., Einsle, F., & Lang, T. (2011). Agoraphobia. *Psychopathology, 44*(6), 335–345. DOI: 10.1159/000325225 PMID: 21534651

Goharinejad, S., Alizadehsani, R., Khosravi, A., & Nahavandi, S. (2022). A comprehensive review of applications of EEG and fNIRS in brain-computer interface and neurofeedback. *Neuroscience and Biobehavioral Reviews*, *138*, 104712. DOI: 10.1016/j.neubiorev.2022.104712

Goleman, D. (2003). Finding happiness: Cajole your brain to lean to the left. *New York Times, 4*(3).

Gooding, P., & Kariotis, T. (2021). Ethics and law in research on algorithmic and data-driven technology in mental health care: Scoping review. *JMIR Mental Health*, *8*(6), e24668. DOI: 10.2196/24668 PMID: 34110297

Goodwin, T. R., Maldonado, R., & Harabagiu, S. M. (2017). Automatic recognition of symptom severity from psychiatric evaluation records. *Journal of Biomedical Informatics*, *75*, S71–S84. DOI: 10.1016/j.jbi.2017.05.020 PMID: 28576748

Goyal, M., Singh, S., Sibinga, E. M., Gould, N. F., Rowland-Seymour, A., Sharma, R., Berger, Z., Sleicher, D., Maron, D. D., Shihab, H. M., Ranasinghe, P. D., Linn, S., Saha, S., Bass, E. B., & Haythornthwaite, J. A. (2014). Meditation programs for psychological stress and well-being: A systematic review and meta-analysis. *JAMA Internal Medicine*, *174*(3), 357–368. DOI: 10.1001/jamainternmed.2013.13018 PMID: 24395196

Grodniewicz, J. P., & Hohol, M. (2023). Waiting for a digital therapist: Three challenges on the path to psychotherapy delivered by artificial intelligence. *Frontiers in Psychiatry*, *14*, 1190084. DOI: 10.3389/fpsyt.2023.1190084 PMID: 37324824

Grzenda, A. (2021). Artificial intelligence in mental health. *Converg. Ment. Heal. A Transdiscipl. Approach to Innov*, 137.

Gual-Montolio, P., Jaén, I., Martínez-Borba, V., Castilla, D., & Suso-Ribera, C. (2022). Using artificial intelligence to enhance ongoing psychological interventions for emotional problems in real- or close to real-time: A systematic review. *International Journal of Environmental Research and Public Health*, *19*(13), 7737. Advance online publication. DOI: 10.3390/ijerph19137737 PMID: 35805395

Guidance, W. H. O. (2021). *Ethics and governance of artificial intelligence for health*. World Health Organization.

Guo, C., & Saxton, G. D. (2014). Tweeting social change: How social media are changing nonprofit advocacy. *Nonprofit and Voluntary Sector Quarterly*, *43*(1), 57–79. DOI: 10.1177/0899764012471585

Gupta, A. (2023). *From ethical AI frameworks to tools: A review of approaches for implementation. Springer Nature*. https://link.springer.com/article/10.1007/s43681-023-00258-9

Gupta, M., Malik, T., & Sinha, C. (2022). Delivery of a mental health intervention for chronic pain through an artificial intelligence–enabled app (Wysa): Protocol for a prospective pilot study. *JMIR Research Protocols, 11*(3), e36910. DOI: 10.2196/36910 PMID: 35314423

Gupta, P., Muneshwar, K. N., Juganavar, A., & Shegekar, T. (2023). Beyond the Asylum Walls: Tracing the Tapestry of Mental Health Interventions Across Eras and Cultures. *Cureus, 15*(11). DOI: 10.7759/cureus.48251 PMID: 38054143

Hadjistavropoulos, T. (2023). Domains of contemporary and future ethical dilemmas. *Canadian Psychology, 64*(3), 178–188. DOI: 10.1037/cap0000356

Hahn, A., Stein, P., Windischberger, C., Weissenbacher, A., Spindelegger, C., Moser, E., & Kasper, S. (2010). Reduced resting-state functional connectivity between amygdala and orbitofrontal cortex in social anxiety disorder. *NeuroImage, 56*(3), 881–889. DOI: 10.1016/j.neuroimage.2011.02.064 PMID: 21356318

Halder, S. (2022). Application of virtual reality in cognitive rehabilitation: a road ahead. In *Emerging Advancements for Virtual and Augmented Reality in Healthcare* (pp. 217–232). IGI Global. DOI: 10.4018/978-1-7998-8371-5.ch013

Halder, S. (2025). Developing Mental Health Support Chatbots in India: Challenges and Insights. *Annals of Indian Psychiatry, 9*(1), 99–101. DOI: 10.4103/aip.aip_187_24

Hamilton, A. (2023, May 5). *The Rise of Online Therapy: History and Challenges - masc*. Masc. https://mascmedical.com/healthcare-staffing-agencies-online-therapy-history-and-challenges

Hammen, C. (2005). Stress and depression. *Annual Review of Clinical Psychology, 1*(1), 293–319. DOI: 10.1146/annurev.clinpsy.1.102803.143938 PMID: 17716090

Hanh, T. N. (2006). *True love: A practice for awakening the heart*. Shambhala Publications.

Hanna, M., Pantanowitz, L., Jackson, B., Palmer, O., Visweswaran, S., Pantanowitz, J., & Rashidi, H. (2024). Ethical and Bias Considerations in Artificial Intelligence (AI)/Machine Learning. *Modern Pathology*, 100686. PMID: 39694331

Haque, M. D. R., Rahman, M. M., & Islam, M. R. (2023). AI chatbots in mental health: A comprehensive review. *Journal of Medical Systems, 47*(1), 5. DOI: 10.1007/s10916-022-01810-3 PMID: 38127210

Hariri, A. R., & Holmes, A. (2006). Genetics of emotional regulation: The role of the serotonin transporter in neural function. *Trends in Cognitive Sciences, 10*(4), 182–191. DOI: 10.1016/j.tics.2006.02.011 PMID: 16530463

Hawker, D. S. J., & Boulton, M. J. (2000). Twenty years' research on peer victimization and psychosocial maladjustment: A meta-analytic review of cross-sectional studies. *Journal of Child Psychology and Psychiatry, and Allied Disciplines, 41*(4), 441–455. DOI: 10.1111/1469-7610.00629 PMID: 10836674

Health, X. R. (2023). *AI therapy and mental health applications.* Retrieved from [https://www.xr.health/blog/ai-therapy/](https://www.xr.health

Herath, H. M. K. K. M. B., Jayasekara, A. G. B. P., & Madhusanka, B. G. D. A. (2024). *Vision Attentive Technology: AI Approaches for Functional Mobility Assessments in Elderly Healthcare* (1st ed.). CRC Press., DOI: 10.1201/9781032622088

Heron, K. E., & Smyth, J. M. (2010). Ecological momentary interventions: Incorporating mobile technology into psychosocial and health behaviour treatments. *British Journal of Health Psychology, 15*(1), 1–39. DOI: 10.1348/135910709X466063 PMID: 19646331

Hettema, J. M., Neale, M. C., & Kendler, K. S. (2001). A review and meta-analysis of the genetic epidemiology of anxiety disorders. *The American Journal of Psychiatry, 158*(10), 1568–1578. DOI: 10.1176/appi.ajp.158.10.1568 PMID: 11578982

Hickey, S., Kavanagh, E., & McGrath, D. (2021). Wearable technology in the mindfulness space: A systematic review. *PLoS One, 16*(11), e0259623. PMID: 34793519

Hirsch, T., Merced, K., Narayanan, S., Imel, Z. E., & Atkins, D. C. (2017, June). Designing contestability: Interaction design, machine learning, and mental health. In *Proceedings of the 2017 Conference on Designing Interactive Systems* (pp. 95-99).https://doi.org/DOI: 10.1145/3064663.3064703

Hofmann, S. G., Asnaani, A., Vonk, I. J., Sawyer, A. T., & Fang, A. (2010). The efficacy of cognitive behavioral therapy: A review of meta-analyses. *Cognitive Therapy and Research, 36*(5), 427–440. DOI: 10.1007/s10608-012-9476-1 PMID: 23459093

Hohman, F., Head, A., Caruana, R., DeLine, R., & Drucker, S. M. (2019, May). Gamut: A design probe to understand how data scientists understand machine learning models. In *Proceedings of the 2019 CHI conference on human factors in computing systems* (pp. 1-13). https://doi.org/DOI: 10.1145/3290605.3300809

Holistic, A. I. (2024). *Ethical concerns in AI healthcare applications.* https://www.holisticai.com/blog/healthcare-laws-us

Hölzel, B. K., Lazar, S. W., Gard, T., Schuman-Olivier, Z., Vago, D. R., & Ott, U. (2011). How does mindfulness meditation work? Proposing mechanisms of action from a conceptual and neural perspective. *Perspectives on Psychological Science*, *6*(6), 537–559. DOI: 10.1177/1745691611419671 PMID: 26168376

HopeChest. (2023). *Global mental health statistics*. Retrieved from https://www.hopechest.org/global-mental-health-statistics/

Howard, J. (2019). Artificial intelligence: Implications for the future of work. *American Journal of Industrial Medicine*, *62*(11), 917–926. DOI: 10.1002/ajim.23037 PMID: 31436850

Huang, T., Xu, H., Wang, H., Huang, H., Xu, Y., Li, B., Hong, S., Feng, G., Kui, S., Liu, G., Jiang, D., Li, Z.-C., Li, Y., Ma, C., Su, C., Wang, W., Li, R., Lai, P., & Qiao, J. (2023). Artificial intelligence for medicine: Progress, challenges, and perspectives. *The Innovation Medicine*, *1*(2), 100030. DOI: 10.59717/j.xinn-med.2023.100030

Hunkin, H., King, D. L., & Zajac, I. T. (2021). Perceived acceptability and feasibility of digital interventions for reducing problematic smartphone use: A qualitative study. *BMC Psychiatry*, *21*(1), 1–12. DOI: 10.1186/s12888-021-03420-2 PMID: 33388029

IEEE Global Initiative on Ethics of Autonomous and Intelligent Systems. (2019). *Ethically aligned design: A vision for prioritizing human well-being with autonomous and intelligent systems* (1st ed.). IEEE. https://standards.ieee.org/content/dam/ieee-standards/standards/web/documents/other/ead1e.pdf

Imel, Z. E., Caperton, D. D., Tanana, M., & Atkins, D. C. (2017). Technology-enhanced human interaction in psychotherapy. *Journal of Counseling Psychology*, *64*(4), 385–393. DOI: 10.1037/cou0000213 PMID: 28318277

Indian Council of Medical Research (ICMR). (2023). Ethical guidelines for artificial intelligence applications in biomedical research. https://www.icmr.gov.in/icmrobject/uploads/Guidelines/1724842648_ethical_guidelines_application_artificial_intelligence_biomed_rsrch_2023.pdf

Inkster, B., Sarda, S., & Subramanian, V. (2018). An empathy-driven, conversational artificial intelligence agent (Wysa) for digital mental well-being: Real-world data evaluation mixed-methods study. *JMIR mHealth and uHealth*, *6*(11), e12106. DOI: 10.2196/12106 PMID: 30470676

Inside Precision Medicine. (2023). Medical artificial intelligence: A new frontier in precision medicine. https://www.insideprecisionmedicine.com/topics/informatics/medical-artificial-intelligence-a-new-frontier-in-precision-medicine/

Institute for Health Metrics and Evaluation (IHME). (2023). *Mental health data and analysis*. Retrieved from https://www.healthdata.org/research-analysis/health-risks-issues/mental-health

Integrating Technology into Mental Healthcare. (n.d.). https://www.psychotherapy.net/article/integrating-technology

ISACA. (2016). An ethical approach to data privacy protection. https://www.isaca.org/resources/isaca-journal/issues/2016/volume-6/an-ethical-approach-to-data-privacy-protection

Iyortsuun, N. K., Kim, S. H., Jhon, M., Yang, H. J., & Pant, S. (2023, January). A review of machine learning and deep learning approaches on mental health diagnosis. [). MDPI.]. *Health Care*, *11*(3), 285. PMID: 36766860

Jacobson, N. C., & Feng, B. (2022). Digital phenotyping of generalized anxiety disorder: Using artificial intelligence to accurately predict symptom severity using wearable sensors in daily life. *Translational Psychiatry*, *12*(1), 336. DOI: 10.1038/s41398-022-02038-1 PMID: 35977932

Jain, N. (2023). *Global mental health crisis and the role of technology*. LinkedIn. Retrieved from https://www.linkedin.com/pulse/global-mental-health-crisis-role-technology-neeraj-jain-okbsc/

Jain, D. (2023). Regulation of Digital Healthcare in India: Ethical and Legal Challenges. *Healthcare (Basel)*, *11*(6), 911. DOI: 10.3390/healthcare11060911 PMID: 36981568

Jain, G., Pareek, S., & Carlbring, P. (2024). Revealing the source: How awareness alters perceptions of AI and human-generated mental health responses. *Internet Interventions : the Application of Information Technology in Mental and Behavioural Health*, *36*, 100745. DOI: 10.1016/j.invent.2024.100745 PMID: 38746873

Jang, S., Kim, J. J., Kim, S. J., Hong, J., Kim, S., & Kim, E. (2021). Mobile app-based chatbot to deliver cognitive behavioral therapy and psychoeducation for adults with attention deficit: A development and feasibility/usability study. *International Journal of Medical Informatics*, *150*, 104440. DOI: 10.1016/j.ijmedinf.2021.104440 PMID: 33799055

Janiesch, C., Zschech, P., & Heinrich, K. (2021). Machine learning and deep learning. *Electronic Markets*, *31*(3), 685–695. DOI: 10.1007/s12525-021-00475-2

Jansen, C.-P., Gordt-Oesterwind, K., & Schwenk, M. (2022). Wearable Motion Sensors in Older Adults: On the Cutting Edge of Health and Mobility Research. *Sensors (Basel)*, *22*(3), 973. DOI: 10.3390/s22030973 PMID: 35161719

Jauhar, S., McKenna, P. J., Radua, J., Fung, E., Salvador, R., & Laws, K. R. (2014). Cognitive–behavioural therapy for the symptoms of schizophrenia: Systematic review and meta-analysis with examination of potential bias. *The British Journal of Psychiatry*, *204*(1), 20–29. DOI: 10.1192/bjp.bp.112.116285 PMID: 24385461

Jeevanandam, N. (2023). Five interesting Generative AI-powered smart wearables in 2023. IndiaAI. Retrieved from https://indiaai.gov.in/article/five-interesting-generative-ai-powered-smart-wearables-in-2023

Jiang, M., Zhao, Q., Li, J., Wang, F., He, T., Cheng, X., Yang, B. X., Ho, G. W. K., & Fu, G. (2024). A Generic review of Integrating Artificial intelligence in cognitive Behavioral therapy. *arXiv (Cornell University)*. https://doi.org//arxiv.2407.19422DOI: 10.48550

Jobin, A., Ienca, M., & Vayena, E. (2019). The global landscape of AI ethics guidelines. *Nature Machine Intelligence*, *1*(9), 389–399. DOI: 10.1038/s42256-019-0088-2

Johnson, L. (2023). Balancing act: Exploring the interplay between human judgment and artificial intelligence in problem-solving. *ResearchGate*. https://www.researchgate.net/publication/379328371_Balancing_Act_Exploring_the_Interplay_Between_Human_Judgment_and_Artificial_Intelligence_in_Problem-solving_Creativity_and_Decision-making

Johnson, M. (2023). *Ethical concerns in AI mental health screening. Therapy Helpers Blog*. https://therapyhelpers.com/blog/ai-assisted-mental-health-screening-ethical/

Johnson, S. (2023). *Integrating ethics in AI development: A qualitative study on the adoption of ethical principles. BMC Medical Ethics*. https://bmcmedethics.biomedcentral.com/articles/10.1186/s12910-023-01000-0

Johnson, K. B., Wei, W. Q., Weeraratne, D., Frisse, M. E., Misulis, K., Rhee, K., Zhao, J., & Snowdon, J. L. (2021). Precision medicine, AI, and the future of personalized health care. *Clinical and Translational Science*, *14*(1), 86–93. DOI: 10.1111/cts.12884 PMID: 32961010

Johnson, P. (2023). A systematic literature review of human-centered, ethical, and responsible AI. *arXiv Preprint*. https://arxiv.org/abs/2302.05284

Jones, L. (2023). Artificial intelligence in healthcare: Challenges and opportunities. *SAGE Journals*. https://journals.sagepub.com/doi/full/10.1177/20552076231218141

Jones, M. (2023). Algorithmic bias in AI-driven customer management. SSRN *Electronic Journal*. https://papers.ssrn.com/sol3/papers.cfm?abstract_id=4466681

Jorie, A. I. (n.d.). Harnessing the power of AI in mental health: Unveiling a new frontier in treatment and support. https://www.jorie.ai/post/harnessing-the-power-of-ai-in-mental-health-unveiling-a-new-frontier-in-treatment-and-support

Journal of Medical Internet Research. (2023). *Global mental health and AI applications*. https://www.jmir.org/2023/1/e42672/

Jurblum, M., & Selzer, R. (2024). Potential promises and perils of artificial intelligence in psychotherapy - The AI Psychotherapist (APT). *Australasian Psychiatry*. Advance online publication. DOI: 10.1177/10398562241286312 PMID: 39344214

Kabat-Zinn, J. (1990). *Full catastrophe living: Using the wisdom of your body and mind to face stress, pain, and illness*. Delta.

Kabat-Zinn, J. (1994). *Wherever you go, there you are: Mindfulness meditation in everyday life*. Hyperion.

Kabat-Zinn, J. (2003). Mindfulness-based interventions in context: Past, present, and future. *Clinical Psychology : a Publication of the Division of Clinical Psychology of the American Psychological Association, 10*(2), 144–156. DOI: 10.1093/clipsy.bpg016

Kabat-Zinn, J., Lipworth, L., Burncy, R., & Sellers, W. (1986). Four-year follow-up of a meditation-based program for the self-regulation of chronic pain: Treatment outcomes and compliance. *The Clinical Journal of Pain, 2*(3), 159–774. DOI: 10.1097/00002508-198602030-00004

Kasula, B. Y. (2023). Ethical Considerations in the Adoption of Artificial Intelligence for Mental Health Diagnosis. *International Journal of Creative Research In Computer Technology and Design, 5*(5), 1–7.

Kendell, R. E. (1980). Diagnostic and statistical manual of mental disorders. *The American Journal of Psychiatry, 137*(12), 1630–1631. DOI: 10.1176/ajp.137.12.1630

Kendler, K. S., Neale, M. C., Kessler, R. C., Heath, A. C., & Eaves, L. J. (1992). The genetic epidemiology of phobias in women: The interrelationship of agoraphobia, social phobia, situational phobia, and simple phobia. *Archives of General Psychiatry, 49*(4), 273–281. DOI: 10.1001/archpsyc.1992.01820040025003 PMID: 1558461

Keshav Desiraju India Mental Health Observatory. (March 7th, 2024). The interim Union Budget FY 2024-25: where does mental health stand? Retrieved from https://cmhlp.org/imho/blog/the-interim-union-budget-fy-2024-25-where-does-mental-health-stand/

Khamassi, M., Velentzas, G., Tsitsimis, T., & Tzafestas, C. (2018). Robot fast adaptation to changes in human engagement during simulated dynamic social interaction with active exploration in parameterized reinforcement learning. *IEEE Transactions on Cognitive and Developmental Systems, 10*(4), 881–893. DOI: 10.1109/TCDS.2018.2843122

Kim, J. J., & Fanselow, M. S. (1992). Modality-specific retrograde amnesia of fear. *Science, 256*(5057), 675–677. DOI: 10.1126/science.1585183 PMID: 1585183

Kleinknecht, R. A., Dinnel, D. L., Kleinknecht, E. E., Hiruma, N., & Harada, N. (1997). Cultural factors in social anxiety: A comparison of social phobia symptoms and Taijin Kyofusho. *Journal of Anxiety Disorders, 11*(2), 157–177. DOI: 10.1016/S0887-6185(97)00004-2 PMID: 9168340

Kodal, A., Fjermestad, K., Bjelland, I., Gjestad, R., Öst, L. G., Bjaastad, J. F., Haugland, B. S. M., Havik, O. E., Heiervang, E., & Wergeland, G. J. (2018). Long-term effectiveness of cognitive behavioral therapy for youth with anxiety disorders. *Journal of Anxiety Disorders, 53*, 58–67. DOI: 10.1016/j.janxdis.2017.11.003 PMID: 29195188

Kok, J. N., Boers, E. J., Kosters, W. A., Van der Putten, P., & Poel, M. (2009). Artificial intelligence: definition, trends, techniques, and cases. *Artificial intelligence, 1*(270-299), 51.

König, P., Zwiky, E., Küttner, A., Uhlig, M., & Redlich, R. (2024). Brain functional effects of cognitive behavioral therapy for depression: A systematic review of task-based fMRI studies. *Journal of Affective Disorders*. Advance online publication. DOI: 10.1016/j.jad.2024.09.084 PMID: 39299583

Koolhaas, J. M., Bartolomucci, A., Buwalda, B., de Boer, S. F., Flügge, G., Korte, S. M., Meerlo, P., Murison, R., Olivier, B., Palanza, P., Richter-Levin, G., Sgoifo, A., Steimer, T., Stiedl, O., van Dijk, G., Wöhr, M., & Fuchs, E. (2011). Stress revisited: A critical evaluation of the stress concept. *Neuroscience and Biobehavioral Reviews, 35*(7), 1743–1761. DOI: 10.1016/j.neubiorev.2011.02.003 PMID: 21316391

Kotsona, M. (2024). Virtue in Virtuality: A Cultural Analysis on Artificial Intelligence Companionship and its. *Ethical Dimensions*.

Koutsouleris, N., Hauser, T. U., Skvortsova, V., & De Choudhury, M. (2022). From promise to practice: Towards the realisation of AI-informed mental health care. *The Lancet. Digital Health, 4*(11), e829–e840. DOI: 10.1016/S2589-7500(22)00153-4 PMID: 36229346

Kumar, D. (2023). AI ethics in mental health care: Continuous monitoring and improving outcomes. *Journal of Healthcare, MDPI, 13*(7), 381. https://www.mdpi.com/2076-0760/13/7/381

Kuziemsky, C., Maeder, A. J., John, O., Gogia, S. B., Basu, A., Meher, S., & Ito, M. (2019). Role of artificial intelligence within the telehealth domain. *Yearbook of medical informatics, 28*(01), 035-040.

LaFargue, M. (2010). *The Tao of the Tao Te Ching: A translation and commentary.* State University of New York Press.

Lazar, S. W., Kerr, C. E., Wasserman, R. H., Gray, J. R., Greve, D. N., Treadway, M. T., ... & Fischl, B. (2005). Meditation experience is associated with increased cortical thickness. *neuroreport, 16*(17), 1893-1897. https://doi.org/DOI: 10.1097/01.wnr.0000186598.66243.19

Lazarus, A. A. (1977). Has behavior therapy outlived its usefulness? *The American Psychologist, 32*(7), 550–554. DOI: 10.1037/0003-066X.32.7.550 PMID: 889160

Lazarus, R. S., & Folkman, S. (1984). *Stress, Appraisal, and Coping.* Springer Publishing Company.

LeBeau, R. T., Glenn, D., Liao, B., Wittchen, H. U., Beesdo, K., & Craske, M. G. (2010). Specific phobia: Diagnostic issues and epidemiology. *The Psychiatric Clinics of North America, 33*(2), 235–248. DOI: 10.1016/j.psc.2010.04.006

Lee, J. (2024). *Trust in AI: Progress, challenges, and future directions in healthcare. Nature Humanities and Social Sciences Communications*, 11(1), 88. https://www.nature.com/articles/s41599-024-04044-8

Lee, E. E., Torous, J., De Choudhury, M., Depp, C. A., Graham, S. A., Kim, H. C., Paulus, M. P., Krystal, J. H., & Jeste, D. V. (2021). Artificial intelligence for mental health care: Clinical applications, barriers, facilitators, and artificial wisdom. *Biological Psychiatry: Cognitive Neuroscience and Neuroimaging, 6*(9), 856–864. DOI: 10.1016/j.bpsc.2021.02.001 PMID: 33571718

Lee, Y.. (2023). Transformer-based Multi-modal Learning for Bipolar Prediction. *IEEE Transactions on Affective Computing.*

Levine, S. (2005). Developmental determinants of sensitivity and resistance to stress. *Psychoneuroendocrinology, 30*(10), 939–946. DOI: 10.1016/j.psyneuen.2005.03.013 PMID: 15958281

Levine, S., & Ursin, H. (1991). What is stress? In Brown, M. R., Koob, G. F., & Rivier, C. (Eds.), *Stress: Neurobiology and neuroendocrinology* (pp. 3–21). Marcel Dekker, Inc.

Levine, S., & Ursin, H. (1991). What is stress? In Brown, M. R., Koob, G. F., & Rivier, C. (Eds.), *Stress: Neurobiology and Neuroendocrinology* (pp. 3–21). Marcel Dekker, Inc.

LexisNexis. (2023). Balancing AI and human judgment: Ethical considerations in the legal profession. https://www.lexisnexis.com/blogs/za/b/legal/posts/balancing-ai-and-human-judgment-ethical-considerations-in-the-legal-profession

Li, L., Peng, W., & Rheu, M. M. J. (2024). Factors Predicting Intentions of Adoption and Continued Use of Artificial Intelligence Chatbots for Mental Health: Examining the Role of UTAUT Model, Stigma, Privacy Concerns, and Artificial Intelligence Hesitancy. *Telemedicine journal and e-health: the official journal of the American Telemedicine Association, 30*(3), 722–730. DOI: 10.1089/tmj.2023.0313

Liang, L., Zheng, Y., Ge, Q., & Zhang, F. (2022). Exploration and Strategy Analysis of Mental Health Education for Students in Sports Majors in the Era of Artificial Intelligence. *Frontiers in Psychology, 12*, 762725. DOI: 10.3389/fpsyg.2021.762725 PMID: 35308079

Li, H., Zhang, R., Lee, Y. C., Kraut, R. E., & Mohr, D. C. (2023). Systematic review and meta-analysis of AI-based conversational agents for promoting mental health and well-being. *NPJ Digital Medicine, 6*(1), 236. DOI: 10.1038/s41746-023-00979-5 PMID: 38114588

LIME (Local Interpretable Model-Agnostic Explanations). (2016). Ribeiro, M. T., Singh, S., & Guestrin, C. "Why should I trust you?": Explaining the predictions of any classifier. *Proceedings of the 22nd ACM SIGKDD International Conference on Knowledge Discovery and Data Mining*. DOI: 10.1145/2939672.2939778

Lim, S. M., Shiau, C. W. C., Cheng, L. J., & Lau, Y. (2022). Chatbot-delivered psychotherapy for adults with depressive and anxiety symptoms: A systematic review and meta-regression. *Behavior Therapy*, *53*(2), 334–347. DOI: 10.1016/j.beth.2021.09.007 PMID: 35227408

Lin, K. M., & Cheung, F. (1999). Mental health issues for Asian Americans. *Psychiatric Services (Washington, D.C.)*, *50*(6), 774–780. DOI: 10.1176/ps.50.6.774 PMID: 10375146

Litwack, E. B. (2024). *Chatbots, robots, and the ethics of automating psychotherapy*. Athens Journal of Philosophy., DOI: 10.30958/ajphil.3-3-1

Liu, F., Ju, Q., Zheng, Q., & Peng, Y. (2024). Artificial intelligence in mental health: Innovations brought by artificial intelligence techniques in stress detection and interventions of building resilience. *Current Opinion in Behavioral Sciences*, *60*, 101452. DOI: 10.1016/j.cobeha.2024.101452

Li, X. (2023). Evaluation and Analysis of Elderly Mental Health Based on Artificial Intelligence. *Occupational Therapy International*, *7077568*, 1–11. Advance online publication. DOI: 10.1155/2023/7077568 PMID: 36817324

Löchner, J., Carlbring, P., Schuller, B., Torous, J., & Sander, L. B. (2025). Digital interventions in mental health: An overview and future perspectives. *Internet Interventions : the Application of Information Technology in Mental and Behavioural Health*, *100824*, 100824. Advance online publication. DOI: 10.1016/j.invent.2025.100824 PMID: 40330743

Loies, S. E. (2025). Combating Mental Health Stigma With Artificial Intelligence: Improving Societal Understanding. In *Supporting Psychological and Emotional Wellbeing Among Entrepreneurs* (pp. 75-92). IGI Global.

Long, N., Lei, Y., Peng, L., Xu, P., & Mao, P. (2022). A scoping review on monitoring mental health using smart wearable devices. *Mathematical Biosciences and Engineering*, *19*(8), 7899–7919. DOI: 10.3934/mbe.2022369 PMID: 35801449

Lovejoy, C. A. (2019). Technology and mental health: The role of artificial intelligence. *European Psychiatry : the Journal of the Association of European Psychiatrists*, *55*, 1–3. DOI: 10.1016/j.eurpsy.2018.08.004 PMID: 30384105

Lustgarten, S. D., & Elhai, J. D. (2018). Technology use in mental health practice and research: Legal and ethical risks. *Clinical Psychology : a Publication of the Division of Clinical Psychology of the American Psychological Association*, *25*(2), 20. DOI: 10.1111/cpsp.12234

Lutz, C., & Tamò-Larrieux, A. (2021). The privacy implications of conversational agents in mental health: A systematic review. *Journal of Medical Internet Research*, *23*(6), e27854. DOI: 10.2196/27854

Luxton, D. D. (2020). Ethical implications of conversational agents in mental health care. *Journal of Technology in Behavioral Science*, *5*(4), 346–354. DOI: 10.1007/s41347-020-00144-8

Madububambachu, U., Ukpebor, A., & Ihezue, U. (2024). Machine Learning Techniques to Predict Mental Health Diagnoses: A Systematic Literature Review. *Clinical Practice and Epidemiology in Mental Health*, *20*(1), e17450179315688. DOI: 10.2174/0117450179315688240607052117 PMID: 39355197

Maha, C. C., Kolawole, T. O., & Abdul, S. (2024). Revolutionizing community health literacy: The power of digital health tools in rural areas of the US and Africa. *GSC Advanced Research and Reviews, 19*(02), 286–296. DOI: 10.30574/gscarr.2024.19.2.0189

Maheu, M. M., Pulier, M. L., Wilhelm, F. H., McMenamin, J. P., & Brown-Connolly, N. E. (2004). *The mental health professional and the new technologies: A handbook for practice today*. Routledge. DOI: 10.4324/9781410611444

Maleki Varnosfaderani, S., & Forouzanfar, M. (2024). The Role of AI in Hospitals and Clinics: Transforming Healthcare in the 21st Century. *Bioengineering (Basel, Switzerland), 11*(4), 337. DOI: 10.3390/bioengineering11040337 PMID: 38671759

Malgaroli, M., Hull, T. D., & Schultebraucks, K. (2021). Digital Health and Artificial intelligence for PTSD: Improving treatment delivery through personalization. *Psychiatric Annals, 51*(1), 21–26. DOI: 10.3928/00485713-20201203-01

Manjunatha, N., Jayasankar, P., Suhas, S., Rao, G. N., Gopalkrishna, G., Varghese, M., & Benegal, V. (2022). Prevalence and its correlates of anxiety disorders from India's National Mental Health Survey 2016. *Indian Journal of Psychiatry, 64*(2), 138–142. DOI: 10.4103/indianjpsychiatry.indianjpsychiatry_964_21 PMID: 35494323

Maples-Keller, J. L., Yasinski, C., Manjin, M., & Rothbaum, B. O. (2017). Virtual reality-enhanced extinction of phobias and post-traumatic stress. *Neurotherapeutics; the Journal of the American Society for Experimental NeuroTherapeutics, 14*(3), 554–563. DOI: 10.1007/s13311-017-0534-y PMID: 28512692

Marks, I. M. (1975). *Living with fear*. McGraw-Hill.

Martinez-Martin, N. (2022). Minding the AI: Ethical challenges and practice for AI mental health care tools. In *Artificial intelligence in brain and mental health: Philosophical, ethical & policy issues* (pp. 111–125). Springer International Publishing.

Martin, J. K., Pescosolido, B. A., & Tuch, S. A. (2000). Of fear and loathing: The role of'disturbing behavior,'labels, and causal attributions in shaping public attitudes toward people with mental illness. *Journal of Health and Social Behavior, 41*(2), 208–223. DOI: 10.2307/2676306

Masoumian Hosseini, M., Masoumian Hosseini, S. T., Qayumi, K., Hosseinzadeh, S., & Sajadi Tabar, S. S. (2023). Smartwatches in healthcare medicine: Assistance and monitoring; a scoping review. *BMC Medical Informatics and Decision Making, 23*(1), 248. DOI: 10.1186/s12911-023-02350-w PMID: 37924029

Matsumoto, K., Sutoh, C., Asano, K., Seki, Y., Urao, Y., Yokoo, M., Takanashi, R., Yoshida, T., Tanaka, M., Noguchi, R., Nagata, S., Oshiro, K., Numata, N., Hirose, M., Yoshimura, K., Nagai, K., Sato, Y., Kishimoto, T., Nakagawa, A., & Shimizu, E. (2018). Internet-based cognitive behavioral therapy with real-time therapist support via videoconference for patients with obsessive-compulsive disorder, panic disorder, and social anxiety disorder: Pilot single-arm trial. *Journal of Medical Internet Research, 20*(12), e12091. DOI: 10.2196/12091 PMID: 30559094

Matusiewicz, A. K., Hopwood, C. J., Banducci, A. N., & Lejuez, C. W. (2010). The effectiveness of cognitive behavioral therapy for personality disorders. *Psychiatria Clinica, 33*(3), 657–685. DOI: 10.1016/j.psc.2010.04.007 PMID: 20599139

McGorry, P. (2020). Mental health and COVID-19: Are we really all in this together? *The Medical Journal of Australia, 213*(10), 454–455. DOI: 10.5694/mja2.50834 PMID: 33107074

McLeod, S. (2015). Psychodynamic approach. *Simply Psychology*. Retrieved from https://www.simplypsychology.org/psychodynamic.html

Meghrajani, V. R., Marathe, M., Sharma, R., Potdukhe, A., Wanjari, M. B., & Taksande, A. B. (2023). A Comprehensive Analysis of Mental Health Problems in India and the Role of Mental Asylums. *Cureus, 15*(7), e42559. DOI: 10.7759/cureus.42559 PMID: 37637646

Mehta, A., Niles, A. N., Vargas, J. H., Marafon, T., Couto, D. D., & Gross, J. J. (2021). Acceptability and Effectiveness of Artificial Intelligence Therapy for Anxiety and Depression (YOUPER): Longitudinal observational study. *Journal of Medical Internet Research, 23*(6), e26771. DOI: 10.2196/26771 PMID: 34155984

Mehta, U. M., Venkatasubramanian, G., & Chandra, P. S. (2021). The "mind" of the mindful therapist: Mindfulness meditation, neuroscience, and therapeutic presence. *Asian Journal of Psychiatry, 56*, 102547. DOI: 10.1016/j.ajp.2020.102547

Mennella, C., Maniscalco, U., De Pietro, G., & Esposito, M. (2024). Ethical and regulatory challenges of AI technologies in healthcare: A narrative review. *Heliyon, 10*(4), e26297. DOI: 10.1016/j.heliyon.2024.e26297 PMID: 38384518

Mental health needs have multiplied. Support hasn't. (n.d.). Until now. *Woebot Health*. https://woebothealth.com/faq/

Meszaros, J., Minari, J., & Huys, I. (2022). The future regulation of artificial intelligence systems in healthcare services and medical research in the European Union. *Frontiers in Genetics, 13*, 927721. DOI: 10.3389/fgene.2022.927721 PMID: 36267404

Meyer, O. L., & Zane, N. (2013). The influence of race and ethnicity in clients' experiences of mental health treatment. *Journal of Community Psychology*, *41*(7), 884–901. DOI: 10.1002/jcop.21580 PMID: 25400301

Meyer, O., Zane, N., & Cho, Y. I. (2011). Understanding the psychological processes of the racial match effect in Asian Americans. *Journal of Counseling Psychology*, *58*(3), 335–345. DOI: 10.1037/a0023605 PMID: 21574698

Milic, J., Zrnic, I., Grego, E., Jovic, D., Stankovic, V., Djurdjevic, S., & Sapic, R. (2025). The Role of Artificial Intelligence in Managing Bipolar Disorder: A New Frontier in Patient Care. *Journal of Clinical Medicine*, *14*(7), 2515. DOI: 10.3390/jcm14072515 PMID: 40217964

Millstine, D. M., Chen, C. Y., & Bauer, B. A. (2019). Complementary and integrative medicine in the management of headache. *BMJ (Clinical Research Ed.)*, *364*, k5235. DOI: 10.1136/bmj.k5235 PMID: 28512119

Miner, A. S., Shah, N., Bullock, K. D., Arnow, B. A., Bailenson, J., & Hancock, J. (2019). Key considerations for Incorporating Conversational AI in Psychotherapy. *Frontiers in Psychiatry*, *10*, 746. Advance online publication. DOI: 10.3389/fpsyt.2019.00746 PMID: 31681047

Ministry of General Affairs. (2023). *Mental health worldwide facts and figures*. https://www.government.nl/topics/mhpss/mhpss-worldwide-facts-and-figures

Mirmomeni, M., Fazio, T., von Cavallar, S., & Harrer, S. (2021). From wearables to THINKables: artificial intelligence-enabled sensors for health monitoring. In *Wearable Sensors* (pp. 339–356). Academic Press. DOI: 10.1016/B978-0-12-819246-7.00012-7

Mitchell, M. D., Gehrman, P., Perlis, M., & Umscheid, C. A. (2012). Comparative effectiveness of cognitive behavioral therapy for insomnia: A systematic review. *BMC Family Practice*, *13*(1), 1–11. DOI: 10.1186/1471-2296-13-40 PMID: 22631616

Mittelstadt, B. (2020). AI ethics and data governance. SSRN *Electronic Journal*. https://papers.ssrn.com/sol3/papers.cfm?abstract_id=3529576

Mittelstadt, B. (2019). Principles alone cannot guarantee ethical AI. *Nature Machine Intelligence*, *1*(11), 501–507. DOI: 10.1038/s42256-019-0114-4

Mohr, D. C., Tomasino, K. N., Lattie, E. G., Palac, H. L., Kwasny, M. J., Weingardt, K., Karr, C. J., Kaiser, S. M., Rossom, R. C., Bardsley, L. R., Caccamo, L., Stiles-Shields, C., & Schueller, S. M. (2017). IntelliCare: An eclectic, skills-based app suite for the treatment of depression and anxiety. *Journal of Medical Internet Research*, *19*(1), e10. DOI: 10.2196/jmir.6645 PMID: 28057609

Morgan, P., & Cogan, N. A. (2024). Using artificial intelligence to address mental health inequalities: Co-creating machine learning algorithms with key stakeholders and citizen engagement. *Journal of Public Mental Health.*

Mowrer, O. H. (1947). On the dual nature of learning: A reinterpretation of "conditioning" and "problem-solving.". *Harvard Educational Review, 17*(2), 102–148.

Müller, V. C. (2016). Fundamental issues of artificial intelligence. In *Synthese Library/Synthese library*. https://doi.org/DOI: 10.1007/978-3-319-26485-1

Musen, M. A. (2003). The evolution of Protégé: An environment for knowledge-based systems development. *International Journal of Human-Computer Studies, 58*(1), 89–123. DOI: 10.1016/S1071-5819(02)00127-1

Muthuraman, P. (2024, August). Anxiety Monitoring and Digital Meditation Aids. In *2024 International Conference on Electrical Electronics and Computing Technologies (ICEECT)* (Vol. 1, pp. 1-6). IEEE.

Naeem, M., & Coronato, A. (2022). An AI-Empowered Home-Infrastructure to Minimize Medication Errors. *Journal of Sensor and Actuator Networks, 11*(1), 13. DOI: 10.3390/jsan11010013

Naik, S., & Reddy, B. (2025). A framework for the governance of AI in healthcare for India: An exploratory model. *Indian Journal of Medical Ethics.* Advance online publication. DOI: 10.20529/IJME.2025.037

Nasir, S., Khan, R. A., & Bai, S. (2024). Ethical framework for harnessing the power of ai in healthcare and beyond. *IEEE Access : Practical Innovations, Open Solutions, 12*, 31014–31035. DOI: 10.1109/ACCESS.2024.3369912

National Academies of Sciences. (2016). Division of Behavioral, Social Sciences, Board on Behavioral, Sensory Sciences, & Committee on the Science of Changing Behavioral Health Social Norms. *Ending discrimination against people with mental and substance use disorders: The evidence for stigma change.* National Academies Press.

Nazareth, S., Nussbaum, R. L., Siglen, E., & Wicklund, C. A. (2020). Chatbots & artificial intelligence to scale genetic information delivery. *Journal of Genetic Counseling, 30*(1), 7–10. DOI: 10.1002/jgc4.1359 PMID: 33191601

Neff, K. D., Tóth-Király, I., & Colosimo, K. (2020). Self-compassion is best measured as a global construct and is overlapping with but distinct from neuroticism: A response to Pfattheicher, Geiger, Hartung, Weiss, and Schindler (2017). *European Journal of Personality, 34*(5), 810–823. DOI: 10.1002/per.2248

Nemeroff, C. B. (2003). The role of GABA in anxiety disorders. *The Journal of Clinical Psychiatry, 64*(Suppl 3), 3–7. PMID: 12662127

Niles, A. N., Axelsson, E., Andersson, E., Hedman-Lagerlöf, E., Carlbring, P., Andersson, G., ... & Ljotsson, B. (2021). Internet-based cognitive behavior therapy for depression, social anxiety disorder, and panic disorder: Effectiveness and predictors of response in a teaching clinic. *Behaviour Research and Therapy, 136*, 103767. https://doi.org/ rights and contentDOI: 10.1016/j.brat.2020.103767Get

Nmeters. (2023). *The future of wearables: Technological innovations*. https://9meters.com/technology/the-future-of-wearables-exploring-next-generation-technological-innovations

North, M. M., North, S. M., & Coble, J. R. (1997). Virtual reality therapy: An innovative paradigm. *Cyberpsychology & Behavior, 1*(1), 39–47. DOI: 10.1089/cpb.1998.1.39

O'Donnell, L. A., Brydon, D. M., & González-Prendes, A. (2022). Technological Advances in Cognitive-Behavioral Therapy and Clinical Practice: Challenges in an Evolving Field. *American Journal of Psychotherapy, 75*(4), 186–190. DOI: 10.1176/appi.psychotherapy.20220003 PMID: 36052453

O'Neil, C. (2016). *Weapons of Math Destruction: How Big Data Increases Inequality and Threatens Democracy*. Crown Publishing Group.

Obermeyer, Z., Powers, B., Vogeli, C., & Mullainathan, S. (2019). Dissecting racial bias in an algorithm used to manage the health of populations. *Science, 366*(6464), 447–453. DOI: 10.1126/science.aax2342 PMID: 31649194

Öhman, A., & Mineka, S. (2001). Fears, phobias, and preparedness: Toward an evolved module of fear and fear learning. *Psychological Review, 108*(3), 483–522. DOI: 10.1037/0033-295X.108.3.483 PMID: 11488376

Olatunji, B. O., Davis, M. L., Powers, M. B., & Smits, J. A. (2013). Cognitive-behavioral therapy for obsessive-compulsive disorder: A meta-analysis of treatment outcome and moderators. *Journal of Psychiatric Research, 47*(1), 33–41. DOI: 10.1016/j.jpsychires.2012.08.020 PMID: 22999486

Olendzki, A. (2009). Mindfulness and meditation. In *Clinical handbook of mindfulness* (pp. 37–44). Springer New York., DOI: 10.1007/978-0-387-09593-6_3

Palmer, C. E., Marshall, E., Millgate, E., Warren, G., Ewbank, M. P., Cooper, E., Lawes, S., Bouazzaoui, M., Smith, A., Hutchins-Joss, C., Young, J., Margoum, M., Healey, S., Marshall, L., Mehew, S., Cummins, R., Tablan, V., Catarino, A., Welchman, A. E., & Blackwell, A. D. (2024). Combining AI and human support in mental health: a digital intervention with comparable effectiveness to human-delivered care. medRxiv *(Cold Spring Harbor Laboratory)*. DOI: 10.1101/2024.07.17.24310551

Pandey, A. K., Gelin, R., & Robot, A. M. P. S. H. (2018). Pepper: The first machine of its kind. *IEEE Robotics & Automation Magazine*, *25*(3), 40–48. DOI: 10.1109/MRA.2018.2833157

Parekh, A. E., Shaikh, O. A., Simran, N., Manan, S., & Hasibuzzaman, M. A. (2023). Artificial intelligence (AI) in personalized medicine: AI-generated personalized therapy regimens based on genetic and medical history: short communication. *Annals of Medicine and Surgery (London)*, *85*(11), 5831–5833. DOI: 10.1097/MS9.0000000000001320 PMID: 37915639

Patel, K. (2024). AI and medical ethics. *PubMed Central*. https://pmc.ncbi.nlm.nih.gov/articles/PMC11064747/

Patel, V., Saxena, S., De Silva, M., & Samele, C. (2013). Transforming lives, enhancing communities: innovations in mental health. *WISH Mental Health Report*.

Pathak, V. N., & Judge, J. & Anjalee. (2023). *Indigenous Phobia VR Device* (Patent No. 1234567). India: Indian Patent Office.

Pavlopoulos, A., Rachiotis, T., & Maglogiannis, I. (2024). An overview of tools and technologies for anxiety and depression management using AI. *Applied Sciences (Basel, Switzerland)*, *14*(19), 9068. DOI: 10.3390/app14199068

Pawade, D., Karia, S., & De Sousa, A. (2022). Neurofeedback: A comprehensive review on system design, methodology and clinical applications. *Asian Journal of Psychiatry*, *67*, 102937. DOI: 10.1016/j.ajp.2021.102937

Pesapane, F., Codari, M., & Sardanelli, F. (2018). Artificial intelligence in medical imaging: Threat or opportunity? Radiologists again at the forefront of innovation in medicine. *European Radiology Experimental*, *2*(1), 1–10. DOI: 10.1186/s41747-018-0061-6 PMID: 30353365

Peters, D., Vold, K., Robinson, D., & Calvo, R. A. (2020). Responsible AI—Two frameworks for ethical design practice. *IEEE Transactions on Technology and Society*, *1*(1), 34–47. DOI: 10.1109/TTS.2020.2974991

Pickens, J. N. PhD. (2021, October 13). Click “less worrying” or “improved sleep” to fine-tune your virtual therapy? *Psychology Today*. https://www.psychologytoday.com/us/blog/psychology-through-technology/202110/the-evolution-digital-therapy-technology

Piwek, L., Ellis, D. A., Andrews, S., & Joinson, A. (2016). The rise of consumer health wearables: Promises and barriers. *PLoS Medicine*, *13*(2), e1001953. DOI: 10.1371/journal.pmed.1001953 PMID: 26836780

Polar Seal. (2023). *Evolution of wearables*. https://www.polarseal.net/blog/evolution-of-wearables/

Powell, W. A., Corbett, N., & Powell, V. (2019). The rise of the virtual human. In *Natural language processing*. (pp. 1459–1488). https://doi.org/DOI: 10.4018/978-1-7998-0951-7.ch070

Prochaska, J. J., Vogel, E. A., Chieng, A., Kendra, M., Baiocchi, M., Pajarito, S., & Robinson, A. (2021). A therapeutic relational agent for reducing problematic substance use (Woebot): Development and usability study. *Journal of Medical Internet Research*, *23*(3), e24850. DOI: 10.2196/24850 PMID: 33755028

Project, H. O. P. E. (2023). *The global mental health crisis: 10 numbers to note*. https://www.projecthope.org/news-stories/story/the-global-mental-health-crisis-10-numbers-to-note/

Proudfoot, J., Goldberg, D., Mann, A., Everitt, B., Marks, I., & Gray, J. A. (2003). Computerized, interactive, multimedia cognitive-behavioural program for anxiety and depression in general practice. *Psychological Medicine*, *33*(2), 217–227. DOI: 10.1017/S0033291702007225 PMID: 12622301

PubMed Central (PMC). (2023). *Wearable technology in health and mental health*. https://pmc.ncbi.nlm.nih.gov/articles/PMC6631918/

Puiatti, A., Mudda, S., Giordano, S., & Mayora, O. (2011, August). *Smartphone-centred wearable sensors network for monitoring patients with bipolar disorder. In 2011 Annual International Conference of the IEEE Engineering in Medicine and Biology Society*. IEEE., DOI: 10.1109/IEMBS.2011.6090613

Raj, G., Sharma, A. K., & Arora, Y. (2024). Analyzing the Effect of Digital Technology on Mental Health. In *Strategies for E-Commerce Data Security: Cloud, Blockchain, AI, and Machine Learning* (pp. 54-82). IGI Global. DOI: 10.4018/979-8-3693-6557-1.ch003

Ramana, Y. V. S., & Peri, P. (2024). AI-Based Mindfulness: The Future of Healthcare and Customer Service. In *Comparative Analysis of Digital Consciousness and Human Consciousness: Bridging the Divide in AI Discourse* (pp. 54-64). IGI Global. DOI: 10.4018/979-8-3693-2015-0.ch003

Rampton, V., Mittelman, M., & Goldhahn, J. (2020). Implications of artificial intelligence for medical education. *The Lancet. Digital Health*, *2*(3), e111–e112. DOI: 10.1016/S2589-7500(20)30023-6 PMID: 33328081

Rebenitsch, L., & Owen, C. (2016). Review on cybersickness in applications and visual displays. *Virtual Reality (Waltham Cross)*, *20*(2), 101–125. DOI: 10.1007/s10055-016-0285-9

ResearchGate. (2023). *The role of artificial intelligence in mental health: A review.* https://www.researchgate.net/publication/376134724_The_Role_of_Artificial_Intelligence_in_Mental_Health_A_Review

Richards, D. (2024). Artificial intelligence and psychotherapy: A counterpoint. *Counselling & Psychotherapy Research.* Advance online publication. DOI: 10.1002/capr.12758

Riches, S., Schlosser, M., & Gaebler, M. (2023). Immersive virtual reality in the treatment of mental disorders: A systematic review of its efficacy. *Frontiers in Psychology*, *14*, 1122334. DOI: 10.3389/fpsyg.2023.1122334

Riisager, L. G., Christiansen, T. B., Moeller, S. B., Huniche, L., & Larsen, J. E. (2024, May). Snapping Out of It: How a Wearable for Self-Tracking Assisted Psychotherapy Bridges the Gap Between Thoughts and the World. In *Extended Abstracts of the CHI Conference on Human Factors in Computing Systems* (pp. 1-7). https://doi.org/DOI: 10.1145/3613905.3637112

Riva, G., Baños, R. M., Botella, C., Mantovani, F., & Gaggioli, A. (2019). Transforming experience: The potential of augmented reality and virtual reality for enhancing personal and clinical change. *Frontiers in Psychiatry*, *10*, 164. DOI: 10.3389/fpsyt.2019.00164 PMID: 27746747

Riva, G., Baños, R. M., Botella, C., Wiederhold, B. K., & Gaggioli, A. (2019). Virtual reality in psychotherapy: From imagination to immersion. *Journal of Clinical Psychology*, *75*(3), 456–470. DOI: 10.1002/jclp.22799

Rizzo, A. S. (2019). Virtual reality exposure therapy for combat-related PTSD: A randomized clinical trial. *IEEE Transactions on Neural Systems and Rehabilitation Engineering*, *27*(12), 2399–2407. DOI: 10.1109/TNSRE.2019.2945555

Rizzo, A., & Koenig, S. T. (2017). Is clinical virtual reality ready for primetime? *Neuropsychology*, *31*(8), 877–899. DOI: 10.1037/neu0000405 PMID: 29376669

Robotics Biz. (2023). *A complete history of wearable technology: A timeline*. https://roboticsbiz.com/a-complete-history-of-wearable-technology-a-timeline-updated/

Rodgers, M. M., Pai, V. M., & Conroy, R. S. (2014). Recent advances in wearable sensors for health monitoring. *IEEE Sensors Journal*, *15*(6), 3119–3126. DOI: 10.1109/JSEN.2014.2357257

Rodriguez, V. H., Medrano, C. T., & Plaza, I. (2018). Wearable sensors for measuring movement in short sessions of mindfulness sitting meditation: A pilot study. *Journal of Healthcare Engineering*, *2018*(1), 7275049. DOI: 10.1155/2018/7275049 PMID: 29854363

Rollwage, M., Habicht, J., Juechems, K., Carrington, B., Viswanathan, S., Stylianou, M., Hauser, T. U., & Harper, R. (2023). Using conversational AI to facilitate mental health assessments and improve clinical efficiency within psychotherapy services: Real-world observational study. *JMIR AI*, *2*, e44358. Advance online publication. DOI: 10.2196/44358 PMID: 38875569

Romero, L. M., Dickens, M. J., & Cyr, N. E. (2009). The reactive scope model: A new model integrating homeostasis, allostasis, and stress. *Hormones and Behavior*, *55*(3), 375–389. DOI: 10.1016/j.yhbeh.2008.12.009 PMID: 19470371

Rothbaum, B. O., Hodges, L., Smith, S., Lee, J. H., & Price, L. (2000). A controlled study of virtual reality exposure therapy for the fear of flying. *Journal of Consulting and Clinical Psychology*, *68*(6), 1020–1026. DOI: 10.1037/0022-006X.68.6.1020 PMID: 11142535

Rubin, M., Arnon, H., Huppert, J., & Perry, A. (2024). Considering the role of human empathy in AI-driven therapy. *JMIR Mental Health*, *11*, e56529. Advance online publication. DOI: 10.2196/56529 PMID: 38861302

Ruskin, P. E., Silver-Aylaian, M., Kling, M. A., Reed, S. A., Bradham, D. D., Hebel, J. R., Barrett, D., Knowles, F.III, & Hauser, P. (2004). Treatment outcomes in Depression: Comparison of remote treatment through telepsychiatry to In-Person Treatment. *The American Journal of Psychiatry*, *161*(8), 1471–1476. DOI: 10.1176/appi.ajp.161.8.1471 PMID: 15285975

Saarni, S. I., Hofmann, B., Lampe, K., Lühmann, D., Mäkelä, M., Velasco-Garrido, M., & Autti-Rämö, I. (2008). Ethical analysis to improve decision-making on health technologies. *Bulletin of the World Health Organization*, *86*(8), 617–623. DOI: 10.2471/BLT.08.051078 PMID: 18797620

Sadeh-Sharvit, S., & Hollon, S. D. (2020). Leveraging the power of nondisruptive technologies to optimize mental health treatment: Case study. *JMIR Mental Health, 7*(11), e20646. DOI: 10.2196/20646 PMID: 33242025

Saeb, S., Zhang, M., Kwasny, M. J., Karr, C. J., Kordy, H., & Mohr, D. C. (2015). Mobile phone sensor correlates of depressive symptom severity in daily-life behaviour: An exploratory study. *Journal of Medical Internet Research, 17*(7), e175. DOI: 10.2196/jmir.4273 PMID: 26180009

Saeidnia, H. R., Hashemi Fotami, S. G., Lund, B., & Ghiasi, N. (2024). Ethical considerations in artificial intelligence interventions for mental health and well-being: Ensuring responsible implementation and impact. *Social Sciences (Basel, Switzerland), 13*(7), 381. DOI: 10.3390/socsci13070381

Salkovskis, P. M. (1991). The importance of behavior in the maintenance of anxiety and panic: A cognitive account. *Behavioural Psychotherapy, 19*(1), 6–19. DOI: 10.1017/S0141347300011472

Samoili, S., Lopez, C. M., Delipetrev, B., Martinez-Plumed, F., Gomez, G. E., & De, P. G. (2021). *AI Watch. Defining Artificial Intelligence 2.0*. JRC Publications Repository., DOI: 10.2760/019901

Santomauro, D. F., Herrera, A. M. M., Shadid, J., Zheng, P., Ashbaugh, C., Pigott, D. M., & Ferrari, A. J. (2021). Global prevalence and burden of depressive and anxiety disorders in 204 countries and territories in 2020 due to the COVID-19 pandemic. *Lancet, 398*(10312), 1700–1712. DOI: 10.1016/S0140-6736(21)02143-7 PMID: 34634250

Schneider, S., & Kummert, F. (2017, August). Exploring embodiment and dueling bandit learning for preference adaptation in human-robot interaction. In *2017 26th IEEE International Symposium on Robot and Human Interactive Communication (RO-MAN)* (pp. 1325-1331). https://doi.org/DOI: 10.1109/ROMAN.2017.8172476

Schueller, S. M.. (2018). Designing AI for mental health: The role of user experience. *Psychiatric Services (Washington, D.C.), 69*(12), 1201–1203.

Schueller, S. M., Neary, M., O'Loughlin, K., & Adkins, E. C. (2018). Discovery of and interest in health apps among those with mental health needs: Survey and focus group study. *Journal of Medical Internet Research, 20*(6), e10141. DOI: 10.2196/10141 PMID: 29891468

Schuemie, M. J., van der Straaten, P., Krijn, M., & van der Mast, C. A. (2001). Research on presence in virtual reality: A survey. *Cyberpsychology & Behavior, 4*(2), 183–201. DOI: 10.1089/109493101300117884 PMID: 11710246

Scribbr. (n.d.). Research ethics. https://www.scribbr.com/methodology/research-ethics/

Seabrook, E., Kelly, R., Foley, F., Theiler, S., Thomas, N., Wadley, G., & Nedeljkovic, M. (2020). Understanding how virtual reality can support mindfulness practice: Mixed methods study. *Journal of Medical Internet Research, 22*(3), e16106. DOI: 10.2196/16106 PMID: 32186519

Seligman, M. E. (1971). Phobias and preparedness. *Behavior Therapy, 2*(3), 307–320. DOI: 10.1016/S0005-7894(71)80064-3 PMID: 27816071

Selye, H. (1936). A syndrome produced by diverse nocuous agents. *Nature, 138*(3479), 32–32. DOI: 10.1038/138032a0

Selye, H. (1950). *Stress: The physiology and pathology of exposure to stress*. Acta Medica Publication.

Selye, H. (1976). *Stress in health and disease*. Butterworths Scientific.

Semmler, S., & Rose, Z. (2017). Artificial intelligence: Application today and implications tomorrow. *Duke L. & Tech. Rev., 16*, 85.

Sequeira, L., Perrotta, S., LaGrassa, J., Merikangas, K., Kreindler, D., Kundur, D., Courtney, D., Szatmari, P., Battaglia, M., & Strauss, J. (2020). Mobile and wearable technology for monitoring depressive symptoms in children and adolescents: A scoping review. *Journal of Affective Disorders, 265*, 314–324. DOI: 10.1016/j.jad.2019.11.156 PMID: 32090755

Shaban-Nejad, A., Michalowski, M., & Buckeridge, D. L. (2018). Health intelligence: How artificial intelligence transforms population and personalized health. *NPJ Digital Medicine, 1*(1), 53. DOI: 10.1038/s41746-018-0058-9 PMID: 31304332

Shajari, S., Kuruvinashetti, K., Komeili, A., & Sundararaj, U. (2023). The emergence of AI-based wearable sensors for digital health technology: A review. *Sensors (Basel), 23*(23), 9498. DOI: 10.3390/s23239498 PMID: 38067871

Shamseer, L., Moher, D., Clarke, M., Ghersi, D., Liberati, A., Petticrew, M., Shekelle, P., & Stewart, L. A. (2015). Preferred reporting items for systematic review and meta-analysis protocols (PRISMA-P) 2015: Elaboration and explanation. *BMJ (Clinical Research Ed.), 349*(jan02 1), g7647. Advance online publication. DOI: 10.1136/bmj.g7647 PMID: 25555855

Sharma, V., Goyal, M., & Malik, D. (2017). An intelligent behaviour shown by chatbot system. *International Journal of New Technology and Research, 3*(4), 263312.

Sharpener. (2023). *The future of wearables: Exploring next-generation innovations*. https://gosharpener.com/blogs/552205/The-Future-of-Wearable-Technology

Shatte, A. B., Hutchinson, D. M., & Teague, S. J. (2019). Machine learning in mental health: A scoping review of methods and applications. *Psychological Medicine*, *49*(9), 1426–1448. DOI: 10.1017/S0033291719000151 PMID: 30744717

Siegel, R. D. (2010). *The mindfulness solution: Everyday practices for everyday problems*. Guilford Press.

Simform. (2023). *AI-driven mental health applications: A case study*. https://www.simform.com/case-studies/ai-driven-mental-health-support-application/

Simon, J., Bass, T., Boelman, V., & Mulgan, G. (2017). Digital democracy: the tools transforming political engagement.

SingleCare. (2023). *Mental health statistics*. https://www.singlecare.com/blog/news/mental-health-statistics/

Sinha, R. (2024). The role and impact of new technologies on healthcare systems. *Discover Health Systems*, *3*(1), 96. Advance online publication. DOI: 10.1007/s44250-024-00163-w

Smith, J. (2021). Ethical frameworks for AI deployment. *PubMed Central*. https://pmc.ncbi.nlm.nih.gov/articles/PMC8404542/

Smith, J. (2023). Addressing ethical concerns in AI-driven healthcare. *PubMed Central*. https://pmc.ncbi.nlm.nih.gov/articles/PMC10048681/

Smith, J. (2024). Ethical implications of AI adoption. *PubMed Central*. https://pmc.ncbi.nlm.nih.gov/articles/PMC11181958/

Smith, K. (2023). Ethical dilemmas in AI-driven decision-making. *Nature Humanities and Social Sciences Communications*, 10(1), 22. https://www.nature.com/articles/s41599-023-01787-8

Smith, L. (2024). *Chatbot tragedy highlights AI risks in mental health care. Le Monde*. https://www.lemonde.fr/pixels/article/2024/08/06/comment-l-ia-bouscule-le-milieu-de-la-sante-mentale-plutot-que-de-payer-une-nouvelle-seance-chez-le-psy-j-allais-sur-chatgpt_6270640_4408996.html

Smith, D. H., Gray, N. A., Nourse, J. G., & Crandell, C. W. (1981). The DENDRAL project: Recent advances in computer-assisted structure elucidation. *Analytica Chimica Acta*, *133*(4), 471–497. DOI: 10.1016/S0003-2670(01)95414-5

Smith, E. N., Santoro, E., Moraveji, N., Susi, M., & Crum, A. J. (2020). Integrating wearables in stress management interventions: Promising evidence from a randomized trial. *International Journal of Stress Management*, *27*(2), 172–182. https://psycnet.apa.org/doi/10.1037/str0000137. DOI: 10.1037/str0000137

Smith, J., Brown, T., & Anderson, K. (2021). Multi-modal feature engineering for anxiety prediction using wearable and smartphone data. *Computers in Human Behavior*, *124*, 106925. DOI: 10.1016/j.chb.2021.106925

Smith, K. (2023). Exploring patient perspectives on AI in healthcare. *BMC Health Services Research*, *23*(1), 100. https://bmchealthservres.biomedcentral.com/articles/10.1186/s12913-023-10098-2 PMID: 36717808

Smuha, N. A., Ahmed-Rengers, E., Harkens, A., Li, W., MacLaren, J., Piselli, R., & Yeung, K. (2021). How the EU can achieve legally trustworthy AI: a response to the European Commission's proposal for an Artificial Intelligence Act. *SSRN* 3899991. DOI: 10.2139/ssrn.3899991

Sokolovskaya, M. (2024). *Enhancing supervisory practice in Positive and Transcultural Psychotherapy through Artificial Intelligence*. The Global Psychotherapist., DOI: 10.52982/lkj253

Sokołowska, E., Sokołowska, B., Chrapusta, S. J., & Sulejczak, D. (2025). Virtual environments as a novel and promising approach in (neuro)diagnosis and (neuro) therapy: A perspective on the example of autism spectrum disorder. *Frontiers in Neuroscience*, *18*, 1461142. DOI: 10.3389/fnins.2024.1461142 PMID: 39886337

Solaiman, B. (2020). Addressing access with artificial intelligence: Overcoming the limitations of deep learning to broaden remote care today. *U. Mem. L. Rev.*, *51*, 1103.

Sorvaniemi, M., Ojanen, E., & Santamäki, O. (2005). Telepsychiatry in emergency consultations: A follow-up study of sixty patients. *Telemedicine Journal and e-Health*, *11*(4), 439–441. DOI: 10.1089/tmj.2005.11.439 PMID: 16149889

Spytska, L. (2025). The use of artificial intelligence in psychotherapy: Development of intelligent therapeutic systems. *BMC Psychology*, *13*(1), 175. DOI: 10.1186/s40359-025-02491-9 PMID: 40022267

Statista. (2023). Number of social media users worldwide. Retrieved from https://www.statista.com

Stellefson, M., Paige, S. R., Chaney, B. H., & Chaney, J. D. (2020). Evolving role of social media in health promotion: Updated responsibilities for health education specialists. *International Journal of Environmental Research and Public Health*, *17*(4), 1153. DOI: 10.3390/ijerph17041153 PMID: 32059561

Suleiman-Martos, N., Albendín-García, L., Gómez-Urquiza, J. L., Vargas-Román, K., Ramirez-Baena, L., Ortega-Campos, E., & Cañadas-De la Fuente, G. A. (2020). Prevalence and predictors of burnout in midwives: A systematic review and meta-analysis. *International Journal of Environmental Research and Public Health*, *17*(2), 641. DOI: 10.3390/ijerph17020641 PMID: 31963831

Sullivan, G. M., Coplan, J. D., Kent, J. M., & Gorman, J. M. (1999). The noradrenergic system in pathological anxiety: A focus on panic with relevance to generalized anxiety and phobias. *Biological Psychiatry*, *46*(9), 1205–1218. DOI: 10.1016/S0006-3223(99)00246-2 PMID: 10560026

Sunflower Lab. (2023). AI vs human decision-making: Striking the right balance. https://www.thesunflowerlab.com/ai-vs-human-decision-making/

Sun, K., Roy, A., & Tobin, J. M. (2024). Artificial intelligence and machine learning: Definition of terms and current concepts in critical care research. *Journal of Critical Care*, *82*, 154792. DOI: 10.1016/j.jcrc.2024.154792 PMID: 38554543

Szabó, P., Ara, J., Halmosi, B., Sik-Lanyi, C., & Guzsvinecz, T. (2023). Technologies Designed to Assist Individuals with Cognitive Impairments. *Sustainability (Basel)*, *15*(18), 13490. DOI: 10.3390/su151813490

Tavory, T. (2024). Regulating AI in mental health: Ethics of care perspective. *JMIR Mental Health*, *11*(1), e58493. DOI: 10.2196/58493 PMID: 39298759

Teachman, B. A., Smith-Janik, S. B., & Saporito, J. (2007). Information processing biases and panic disorder: Relationships among cognitive and symptom measures. *Behaviour Research and Therapy*, *45*(4), 882–896. DOI: 10.1016/j.brat.2007.01.009 PMID: 17341415

Team, R. (2023a, December 6). *Pepper*. ROBOTS: Your Guide to the World of Robotics. https://robotsguide.com/robots/pepper

Team, R. (2023b, December 19). *Bandit*. ROBOTS: Your Guide to the World of Robotics. https://robotsguide.com/robots/bandit/

Technologies, T. (n.d.). AI and privacy risks: Challenges and solutions. https://www.trigyn.com/insights/ai-and-privacy-risks-challenges-and-solutions

The Ministry of Health and Family Welfare. (MOHFW, February 7th, 2025). *Advancing Mental Healthcare in India*. Retrieved from https://mohfw.gov.in/?q=pressrelease-206

The MIT Press. (2023). Prediction and judgment: Why artificial intelligence needs human oversight. *International Security*, *46*(3), 7. https://direct.mit.edu/isec/article/46/3/7/109668/Prediction-and-Judgment-Why-Artificial

The use of Technology-Assisted methods in therapy and psychological treatments. (2023, January 11). https://www.healthtechzone.com/topics/healthcare/articles/2023/01/11/454583-use-technology-assisted-methods-therapy-psychological-treatments.htm

Thieme, A., Wallace, J., Meyer, T. D., & Olivier, P. (2020). Designing for mental health: Understanding the interplay between technology and therapeutic practices. *Digital Health*, *6*, 1–15. DOI: 10.1177/2055207620936447\

Topol, A. (2019). High-performance medicine: The convergence of human and artificial intelligence. *Nature Medicine*, *25*(1), 44–56. DOI: 10.1038/s41591-018-0300-7 PMID: 30617339

Torous, J., Nicholas, J., Larsen, M., Firth, J., & Christensen, H. (2018). Clinical review of user engagement with mental health smartphone apps: Evidence, theory and improvements. *Evidence-Based Mental Health*, *21*(3), 116–119. DOI: 10.1136/eb-2018-102891 PMID: 29871870

T-Rex Group. (2023). *AI and mental health: Trends and examples.* https://itrexgroup.com/blog/ai-mental-health-examples-trends/

Tsai, C. H., Chen, P. C., Liu, D. S., Kuo, Y. Y., Hsieh, T. T., Chiang, D. L., Lai, F., & Wu, C. T. (2022). Panic attack prediction using wearable devices and machine learning: Development and cohort study. *JMIR Medical Informatics*, *10*(2), e33063. DOI: 10.2196/33063 PMID: 35166679

U.S. Food and Drug Administration. (2022). *Software as a medical device (SaMD): Clinical evaluation.* https://www.fda.gov/regulatory-information/search-fda-guidance-documents/software-medical-device-samd-clinical-evaluation

Ueda, D., Kakinuma, T., Fujita, S., Kamagata, K., Fushimi, Y., Ito, R., Matsui, Y., Nozaki, T., Nakaura, T., Fujima, N., Tatsugami, F., Yanagawa, M., Hirata, K., Yamada, A., Tsuboyama, T., Kawamura, M., Fujioka, T., & Naganawa, S. (2024). Fairness of artificial intelligence in healthcare: Review and recommendations. *Japanese Journal of Radiology*, *42*(1), 3–15. DOI: 10.1007/s11604-023-01474-3 PMID: 37540463

Ujike, S., Yasuhara, Y., Osaka, K., Sato, M., Catangui, E., Edo, S., ... & Mifune, K. (2019). Encounter of Pepper-CPGE for the elderly and patients with schizophrenia: an innovative strategy to improve patient's recreation, rehabilitation, and communication. *The Journal of Medical Investigation, 66*(1.2), 50-53. https://doi.org/DOI: 10.2152/jmi.66.50

United VARs. (n.d.). Mental health: The new frontier of technology. https://www.united-vars.com/en/blog/mental-health-the-new-frontier-of-technology

Usmani, S. S., Sharath, M., & Mehendale, M. (2022). Future of mental health in the metaverse. *General Psychiatry, 35*(4), e100825. DOI: 10.1136/gpsych-2022-100825 PMID: 36189180

Uzir, M. U. H., Al Halbusi, H., Lim, R., Jerin, I., Abdul Hamid, A. B., Ramayah, T., & Haque, A. (2021). Applied Artificial Intelligence and user satisfaction: Smartwatch usage for healthcare in Bangladesh during COVID-19. *Technology in Society, 67*, 101780. DOI: 10.1016/j.techsoc.2021.101780 PMID: 34697510

Vacaru, S., Fekete, K., Luyten, L. P., Saldien, J., & Bosse, T. (2024). Selecting optimal wearables for measuring physiological arousal in robot-delivered mindfulness-based exercises. *Advanced Robotics, 38*(1), 1–15. DOI: 10.1080/01691864.2024.2369797 PMID: 39507502

Valentine, L., D'Alfonso, S., & Lederman, R. (2023). Recommender systems for mental health apps: Advantages and ethical challenges. *AI & Society, 38*(4), 1627–1638. DOI: 10.1007/s00146-021-01322-w PMID: 35068708

Varghese, M. A., Sharma, P., & Patwardhan, M. (2024). Public Perception on Artificial Intelligence-Driven Mental Health Interventions: Survey Research. *JMIR Formative Research, 8*, e64380. DOI: 10.2196/64380 PMID: 39607994

Vayena, E., & Blasimme, A. (2018). Health research with big data: Time for systemic oversight. *The Journal of Law, Medicine & Ethics, 46*(1), 119–129. DOI: 10.1177/1073110518766026 PMID: 30034208

Vilaza, G. N., & McCashin, D. (2021). Is the automation of digital mental health ethical? Applying an ethical framework to chatbots for cognitive behaviour therapy. *Frontiers in Digital Health, 3*, 689736. DOI: 10.3389/fdgth.2021.689736 PMID: 34713163

VinBrain. (2023). *AI in mental health: A new revolution in diagnosis and treatment.* https://vinbrain.net/ai-in-mental-health-a-new-revolution-in-diagnosis-and-treatment

Vygotsky, L. S. (1978). *Mind in society: The development of higher psychological processes*. Harvard University Press.

Wanderley Espinola, C., Gomes, J. C., M^onica Silva Pereira, J., & dos Santos, W. P. (2022). Detection of major depressive disorder, bipolar disorder, schizophrenia and generalized anxiety disorder using vocal acoustic analysis and machine learning: An exploratory study. *Research on Biomedical Engineering*, *38*(3), 813–829. DOI: 10.1007/s42600-022-00222-2

Wang, R., Wang, W., DaSilva, A., Huckins, J. F., Kelley, W. M., Heatherton, T. F., & Campbell, A. T. (2019). Tracking depression dynamics in college students using mobile phone and wearable sensing. *Proceedings of the ACM on Interactive, Mobile, Wearable and Ubiquitous Technologies*, *3*(1), 1–26. DOI: 10.1145/3314398 PMID: 40144218

Wang, W., Chen, J., Hu, Y., Liu, H., Chen, J., Gadekallu, T. R., Garg, L., Guizani, M., & Hu, X. (2024). Integration of Artificial Intelligence and Wearable Internet of Things for Mental Health Detection. *International Journal of Cognitive Computing in Engineering*, *5*, 307–315. DOI: 10.1016/j.ijcce.2024.07.002

Wartel, S. G. (2023). *Facilitated Self-Help Guide: A Professional Guide for Technology-Assisted Behavioral Health Care* (Revised Edition). Stephen Wartel.

Watson, J. B., & Rayner, R. (1920). Conditioned emotional reactions. *Journal of Experimental Psychology*, *3*(1), 1–14. DOI: 10.1037/h0069608

Wiederhold, B. K., Miller, I., & Wiederhold, M. D. (2018). Augmenting behavioral healthcare: mobilizing services with virtual reality and augmented reality. *Digital Health: Scaling Healthcare to the World*, 123-137.

Wies, B., Landers, C., & Ienca, M. (2021). Digital mental health for young people: A scoping review of ethical promises and challenges. *Frontiers in Digital Health*, *3*, 697072. DOI: 10.3389/fdgth.2021.697072 PMID: 34713173

Wikipedia. (2023). *Wearable technology*. https://en.wikipedia.org/wiki/Wearable_technology

Williams, M. D. (2020, August 27). *10 trends for technology in counseling and psychotherapy*. https://www.wolterskluwer.com/en/expert-insights/10-trends-for-technology-in-counseling-and-psychotherapy

Williams, R. (2023). AI in clinical decision-making: Opportunities and risks. *PubMed Central*. https://pmc.ncbi.nlm.nih.gov/articles/PMC10482032/

Winfield, A. F., & Jirotka, M. (2017). The case for an ethical black box. In *Towards Autonomous Robotic Systems: 18th Annual Conference, TAROS 2017, Guildford, UK, July 19–21, 2017* [Springer International Publishing.]. *Proceedings*, *18*, 262–273.

Woebot Health. (2025, February 11). *Scalable enterprise solution for mental health | Woebot Health*. https://woebothealth.com/

Wolpe, J. (1961). *The systematic desensitization treatment of neuroses*. Stanford University Press. DOI: 10.1097/00005053-196103000-00001

Wong, H. (2021). Patients' perceptions of AI in clinical decision-making: An empirical study. *Journal of Medical Internet Research*, *23*(10), e25856. https://www.jmir.org/2021/11/e25856/

World Health Organization (WHO). (2020). *Guidelines on digital interventions for health system strengthening*. WHO. https://www.who.int/publications/i/item/9789241550505

World Health Organization (WHO). (2023, February 21). *Stress*. https://www.who.int/news-room/questions-and-answers/item/stress

World Health Organization. (2023a). Depressive Disorder (Depression). Retrieved from https://www.who.int/news-room/fact-sheets/detail/depression

World Health Organization. (2023b). Anxiety disorders. Retrieved from https://www.who.int/news-room/fact-sheets/detail/anxiety-disorders

World Health Organization. (2024). Leading the Future of Global Health with Responsible Artificial Intelligence. Retrieved from https://cdn.who.int/media/docs/default-source/digital-health-documents/who_brochure_ai-vision_web.pdf?

Wrede, S. J., Esch, T., & Michaelsen, M. M. (2023). Mindfulness in the Digital Workplace: an explorative study of the compatibility of mindfulness and technology. https://doi.org/DOI: 10.21203/rs.3.rs-2459776/v1

Wubineh, B. Z., Deriba, F. G., & Woldeyohannis, M. M. (2023). Exploring the opportunities and challenges of implementing artificial intelligence in healthcare: A systematic literature review. *Urologic Oncology*, *42*(3), 48–56. DOI: 10.1016/j.urolonc.2023.11.019 PMID: 38101991

Yalom, I. D. (2019). *When Nietzsche wept: A novel of obsession*. Basic Books.

Yang, X., Li, Z., & Sun, J. (2020). Effects of cognitive behavioral therapy–based intervention on improving glycaemic, psychological, and physiological outcomes in adult patients with diabetes mellitus: A meta-analysis of randomized controlled trials. *Frontiers in Psychiatry*, *11*, 711. DOI: 10.3389/fpsyt.2020.00711 PMID: 32848906

Yao, Y.. (2022). Digital Twin-Based Control for the Electric Vehicle Powertrain: An Industrial Application. *Electronics (Basel), 12*(3), 672. DOI: 10.3390/electronics12030672

Ybarra, C. E., & Turk, T. A. (2009). The evolution of trust in information technology alliances. *The Journal of High Technology Management Research, 20*(1), 62–74. DOI: 10.1016/j.hitech.2009.02.003

Ye, B. Y., Jiang, Z. Y., Li, X., Cao, B., Cao, L. P., Lin, Y., Xu, G.-Y., & Miao, G. D. (2016). Effectiveness of cognitive behavioral therapy in treating bipolar disorder: A n updated meta-analysis with randomized controlled trials. *Psychiatry and Clinical Neurosciences, 70*(8), 351–361. DOI: 10.1111/pcn.12399 PMID: 27177717

Yeh, P. L., Kuo, W. C., Tseng, B. L., & Sung, Y. H. (2025). Does the AI-driven Chatbot Work? Effectiveness of the Woebot app in reducing anxiety and depression in group counseling courses and student acceptance of technological aids. *Current Psychology (New Brunswick, N.J.), 44*(9), 1–13. DOI: 10.1007/s12144-025-07359-0

Young, Z., Moghaddam, N., & Tickle, A. (2020). The efficacy of cognitive behavioral therapy for adults with ADHD: A systematic review and meta-analysis of randomized controlled trials. *Journal of Attention Disorders, 24*(6), 875–888. DOI: 10.1177/1087054716664413 PMID: 27554190

Yu, L. (2023). Impact of AI on the doctor–patient relationship: A review. *BMC Medical Informatics and Decision Making, 23*(1), 62. https://bmcmedinformdecismak.biomedcentral.com/articles/10.1186/s12911-023-02162-y PMID: 37024897

Zafar, F., Fakhare Alam, L., Vivas, R. R., Wang, J., Whei, S. J., Mehmood, S., Sadeghzadegan, A., Lakkimsetti, M., & Nazir, Z. (2024). The Role of Artificial Intelligence in Identifying Depression and Anxiety: A Comprehensive Literature Review. *Cureus, 16*(3), e56472. DOI: 10.7759/cureus.56472 PMID: 38638735

Zamboni, L., Centoni, F., Fusina, F., Mantovani, E., Rubino, F., Lugoboni, F., & Federico, A. (2021). The effectiveness of cognitive behavioral therapy techniques for the treatment of substance use disorders: A narrative review of evidence. *The Journal of Nervous and Mental Disease, 209*(11), 835–845. DOI: 10.1097/NMD.0000000000001381 PMID: 34698698

Zhang, H. (2021). AI ethics principles in practice: Perspectives of designers in AI deployment. *arXiv Preprint.* https://arxiv.org/abs/2112.07467

Zhang, J. (2024). AI-assisted peer support in mental health: Empathy and ethical implications. *Journal of Mental Health (Abingdon, England), 45*(1), e58493. https://mental.jmir.org/2024/1/e58493

Zhang, Q. (2021). IAC framework for patient agency in AI-enabled healthcare: Ethical considerations. *arXiv Preprint*. https://arxiv.org/abs/2111.04456

Zhang, T., Schoene, A. M., Ji, S., & Ananiadou, S. (2022). Natural language processing applied to mental illness detection: A narrative review. *npj. Digital Medicine*, *5*(1), 46. DOI: 10.1038/s41746-022-00589-7 PMID: 35396451

Zhang, Y. (2019). AI applications in mental health: A survey of risks and challenges. *arXiv Preprint*. https://arxiv.org/abs/1907.07493

Zhang, Z., & Wang, J. (2024). Can AI replace psychotherapists? Exploring the future of mental health care. *Frontiers in Psychiatry*, *15*, 1444382. DOI: 10.3389/fpsyt.2024.1444382 PMID: 39544371

Zhao, M.. (2025). Understanding the Role of Artificial Intelligence in Medical Imaging: A Review. *Health Informatics Journal*, *29*(1), 22–39. DOI: 10.1186/s40359-025-02378-9

Zipfel, S., Lutz, W., Schneider, S., Schramm, E., Delgadillo, J., & Giel, K. E. (2024). The future of enhanced psychotherapy: Towards precision psychotherapy. *Psychotherapy and Psychosomatics*, *93*(4), 230–236. DOI: 10.1159/000539022 PMID: 38934154

About The Contributors

Asegul Hulus serves as an assistant professor while also being recognised as a fellow of the Higher Education Academy (FHEA). She is a distinguished researcher and author specialising in the field of computing. She holds memberships in esteemed organisations, including the Association for Computing Machinery (ACM) and the Institute of Electrical and Electronics Engineers (IEEE), among others. Additionally, she possesses skills in graphic design, video editing, audio production, and video game proficiency. Dr Hulus serves as a chair, chief editor, and peer reviewer for numerous journals, conferences, and publication houses. Furthermore, she holds a prominent position as one of the founding leaders in her field. Finally, she serves as a mentor for the Women4Cyber Foundation and holds various leadership positions in professional computing committees, such as the Association for Computing Machinery's Council on Women in Computing (ACM-W).

Hemachandran K is a distinguished academic and researcher in the fields of Artificial Intelligence, Machine Learning, and Business Analytics. He serves as the Director of the AI Research Centre and Associate Dean of the Graduate Program at the School of Business, Woxsen University. Dr. Hemachandran holds multiple prestigious positions, including Area Chair of the Analytics Department and named professorships: the Zita Zoltay Paprika Professor of Decision Sciences & Business Economics and the Course5i Chair Professor of Business Analytics and Machine Learning. His expertise spans Artificial Intelligence and Machine Learning, contributing significantly to the School of Business at Woxsen University.

Raul Villamarin Rodriguez is a professor and academic researcher specializing in Artificial Intelligence, Machine Learning, and Robotics at Woxsen University. His expertise in these cutting-edge fields is evidenced by his verified institutional email, highlighting his active engagement in both teaching and research. As a faculty member, Dr. Rodriguez contributes significantly to Woxsen University's academic programs and research initiatives in advanced technologies.

Manjeet Rege is a distinguished academic leader in the field of Software Engineering and Data Science. He serves as Professor and Chair of the Department of Software Engineering and Data Science, demonstrating his expertise and leadership in these crucial technological domains. Additionally, Dr. Rege holds the position of Director of the Center for Applied Artificial Intelligence, highlighting his significant role in advancing AI research and applications. His dual roles underscore Dr. Rege's comprehensive involvement in both the academic and practical aspects of software engineering, data science, and artificial intelligence. This combination of departmental leadership and research center directorship positions him at the forefront of technological education and innovation.

Mir Aadil is an Assistant Professor at Woxsen University, India. His research interests span across Natural Language Processing, Data Analytics, and Machine Learning. Dr. Aadil's academic expertise is further demonstrated through the diverse range of courses he teaches, including Data Sciences, Data Analytics, Machine Learning, Data Structures and Algorithms, and Theory of Automata and Formal Languages. He also provides instruction in programming languages such as C, C++, and Python. This combination of research focus and teaching portfolio positions Dr. Aadil at the intersection of theoretical computer science and practical applications in data-driven fields. His work contributes to both the academic advancement and the practical implementation of cutting-edge technologies in data science and artificial intelligence.

***

K M Anjalee is currently pursuing her Ph.D. at Gurukul Kangri University's Psychology Department. She holds a Master's degree in Psychology from Mahatma Gandhi Kashi Vidyapith and a PG Diploma in Counseling and Psychotherapy from BHU. Anjalee's expertise is highlighted by her qualification in the National Eligibility Test (NET). Her research expertise area is health psychology.

Basudha Bandyopadhyay is a licensed Rehabilitation Psychologist, (Mphil-RCI) with over 10 years of clinical experience. She currently works at the Child Development Centre in Hyderabad, where she serves as a Clinical Head. A seasoned professional in her field, Basudha has made significant contributions to various book chapters and papers, showcasing her expertise in rehabilitation psychology. Her research interests include Emotional Intelligence (EI) driven psychotherapy, assessment, and behavioural therapy.

Shouvik Basak is a Solution Architect at a leading global IT organisation. He specialises in architecting Cloud based solutions and Data Centre infrastructure technologies. By education, he has done M.Tech in Data Science and Engineering from BITS Pilani, India. He is a Data Science enthusiast with keen interest in developing AI based solutions and evolving data engineering architectures.

Debangshu Chanda is a Counseling Psychologist, interested in sex/sexuality/ gender based research, counseling and teaching Psychology innovatively.

Eshita Chatterjee is a postgraduate in Psychology from West Bengal State University and currently serves as a Project Assistant and Executive Assistant to the Vice-Chancellor at Woxsen University, Hyderabad. Specializing in Rehabilitation Psychology, she is passionate about contributing to the fields of child psychology, rehabilitation, and mental health disorders. She has published a research paper on dementia and authored a book chapter on the geriatric population. Additionally, she is actively engaged in various research endeavors exploring diverse psychological topics.

Sanchita Ghosh, Academically engaged with Psychology since the last 16 years. Had doctoral research on the junior participants of reality television shows. Worked in different research projects at reputed research institutions including Indian Statistical Institute and University of Calcutta. Started teaching in 2011 and taught in several reputed institutions, both undergraduate and postgraduate level, including Adamas University and West Bengal State University, the latter one being the current workplace. Working as a practicing psychologist for the last 19 years, the main driving force being passion for social psychology, media psychology and mental health. Other areas of research interest are positive psychology, Indian psychology, cognitive psychology etc. Has several publications and has presented papers in national and international conferences. Also organized several National seminars, conferences and first International seminar on peace psychology.

Susmita Halder is an accomplished academician, researcher, and clinician, currently serving as Professor of Psychology and Dean, Faculty of Arts and Social Studies at St. Xavier's University, Kolkata. She is also the Secretary of the Asian Cognitive Behavioral Therapy (CBT) Association. With over two decades of experience in teaching, clinical practice, and research, Dr. Halder has made notable contributions to clinical psychology—especially in CBT, cognitive training, aging, and suicide prevention. She formerly led the Department of Clinical Psychology at the Institute of Psychiatry, Kolkata, and has collaborated with institutions like Manipal College of Medical Science, Fortis Hospital, Medica Super Specialty

Hospital, and Amity University. She has authored multiple books on mental health and published over 130 research papers in national and international journals.

Nistha Jain is a Research Assistant at Woxsen University. Her research focuses on AI-based mental health care, psychometric assessments and clinical disorders. Nistha has contributed to multiple journal articles and has an aim to develop sustainable AI-driven mental healthcare solutions.

Akash Mahato, PhD is Associate Professor in Clinical Psychology and Head of the Amity Institute of Behavioral Health and Allied Sciences, Amity University Kolkata. Dr Akash has an experience of over 15 years in the industry and academia. He has contributed as faculty and Clinical Psychologist in several Postgraduate teaching departments in medical colleges and Universities the past. His interest areas lie in Evidence based Psychotherapy, therapeutic indication of neuro-cognitive functions, and community mental health. Dr Akash has over 50 publications in peer reviewed and indexed journals.

Moumita Mukherjee is an Assistant Professor of Psychology at the Department of Psychology, Faculty of Arts & Social Studies, St. Xavier's University, Kolkata, India. Dr. Mukherjee obtained her Bachelor's Degree with Psychology Honours, Master's Degree with Specialization in Experimental Psychology, and her Doctorate degree in Psychology from University of Calcutta, India. She has a second Master's Degree in Psychology with concentration in Cognition, Neuroscience and Social Psychology from University of South Florida, U.S.A. Dr. Mukherjee has over 15 years' experience in teaching in Higher Educational Institutions and research in both national and international settings. Her key interest areas of research are Cognitive and Experimental Psychology, and their applications and implications for professional workspaces.

Kavi Priya N is an Assistant Professor, Department of Faculty of Management Sciences.

Vijendra Nath Pathak, an Associate Professor, Department of Psychology, Parul Institute of Liberal Arts, Parul University, Vadadora, previously served as a Research Associate Fellow at DRDO's Defense Institute of Psychological Research in Delhi. He holds a Post-Doctoral Fellowship from the Indian Council of Social Science Research and completed his M.Phil. and Ph.D. in Psychology from Mahatma Gandhi Kashi Vidyapith. He specializes in health psychology, crafting interventions for healthier lifestyles, psychometrics, and military psychology. With 2 patents, 10 copyrights, 20+ publications, and 11 book chapters.

Shinjini Samajdar is an RCI-registered Clinical Psychologist. She is currently pursuing her PH.D. from St. Xavier's University Kolkata and working as visiting faculty member in the Department of Psychology at Amity University Kolkata. She has previously held positions as an assistant professor at the Sweekar Institute of Rehabilitation Sciences in Hyderabad. Ms. Samajdar's clinical expertise includes working with chronic medical conditions such as Chronic Obstructive Pulmonary Disease (COPD) and diabetes mellitus. With over 25 publications in peer-reviewed journals, she is a prolific contributor to psychological research. Her specializations include Cognitive Behavioral Therapy (CBT), elderly care, and cognitive retraining.

Bhuvishashri V is Pursuing Bachelor of Business Administration in Data Science and Artificial Intelligence at Woxsen University.

Index

A

B

C

D

R

S

T

U

W

www.ingramcontent.com/pod-product-compliance
Lightning Source LLC
LaVergne TN
LVHW080559300926
849162LV00005B/992